Writing Testbenches:
Functional Verification of HDL Models
Second Edition

Writing Testbenches:
Functional Verification of HDL Models
Second Edition

by
Janick Bergeron
Synopsys, Inc.

Kluwer Academic Publishers
Boston/Dordrecht/London

Distributors for North, Central and South America:
Kluwer Academic Publishers
101 Philip Drive
Assinippi Park
Norwell, Massachusetts 02061 USA
Telephone (781) 871-6600
Fax (781) 871-6528
E-Mail <kluwer@wkap.com>

Distributors for all other countries:
Kluwer Academic Publishers Group
Post Office Box 322
3300 AH Dordrecht, THE NETHERLANDS
Telephone 31 78 6576 000
Fax 31 78 6576 474
E-Mail <orderdept@wkap.nl>

Electronic Services <http://www.wkap.nl>

Library of Congress Cataloging-in-Publication

Bergeron, Janick.
Writing testbenches: functional verification of HDL models/
by Janick Bergeron.--2nd ed.
p.cm.
ISBN: 1-4020-7401-8
1.Computer hardware description languages. 2.Integrated
circuits--Verification. I. Title.

TK7885.7.B472003
621.3815—dc21   2003041995

TABLE OF CONTENTS

CHAPTER 3   *The Verification Plan*   *85*

CHAPTER 4 *High-Level Modeling* *121*

Writing Testbenches: Functional Verification of HDL Models

Table of Contents

Writing Testbenches: Functional Verification of HDL Models

ABOUT THE COVER

The cover of the first edition featured a photograph of the collapse of the Quebec bridge (the cantilever steel bridge in the background[1]) in 1907. The ultimate cause of the collapse was a major change in the design specification that was not verified. To save on construction cost, the engineer in charge of the project increased the span of the bridge from 1600 to 1800 feet, turning the project into the longest bridge in the world, without recalculating weights and stresses.

In those days, engineers felt they could span any distances, as ever longer bridges were being successfully built. But each technology eventually reaches its limits. Almost 100 years after its completion in 1918 (after a complete re-design and a second collapse!), the Quebec bridge is *still* the longest cantilever bridge in the world[2]. Even with all of the advances in civil engineering and composite material, cantilever bridging technology had reached its limits.

You cannot realistically hope to keep applying the same solution to ever increasing problems. Even an evolving technology has its

---

1. Photo: Phototèque, Transports Québec. Cover designed by Elizabeth Nephew (www.nephco.com).

2. The next longest cantilever bridge, completed in 1890 and composed of two spans of 1700 feet, is the Firth of Forth Rail Bridge in Scotland.

---

limit. Eventually, you will have to face and survive a revolution that will provide a solution that is faster and cheaper.

Replacing the Quebec bridge with another cantilever structure is estimated to cost over $600 million today. When it was decided to span the St-Lawrence river once more in 1970, the high cost of a cantilever structure caused a different technology to be used: a suspension bridge. The Pierre Laporte Bridge, visible in the foreground, has a span of 2,200 feet and was built at a cost of $45 million. It provides more lanes of traffic over a longer span at a lower cost and weight. It is better, faster and cheaper. The suspension bridge technology has replaced cantilever structures in all but the shortest spans.

The way most design teams are verifying their design today is reaching its limits. With multi-million gate designs requiring over 1,000 different testcases to verify, the number of testcases that will be required to verify the next generation of designs will become unmanageable and impossible to complete in time. A revolution in methodology is required.

Directed testcases, as described in the first edition, were the cantilever bridges of verification. Coverage-driven constrained-random transaction-level self-checking testbenches are the suspension bridges. This methodology revolution, enabled by the introduction of hardware verification languages such as *e* and OpenVera, make verifying a design better, faster and cheaper.

I'm hoping, with this second edition, to facilitate your transition from ad-hoc, directed testcase verification to a state-of-the-art verification methodology.

FOREWORD

The first edition of Janick Bergeron's *Writing Testbenches* is inarguably the most popular (and successful) contemporary verification textbook. The timing of Bergeron's desperately needed first edition was perfect. The verification challenge was (and continues to be) on everyone's mind with recent industry studies estimating that half of all chips designed today require one or more re-spins. More importantly, these studies indicate that 74% of all re-spins are due to functional errors. In addition to the challenges of achieving good silicon, in many cases verification now consumes significantly more resources in terms of time and labor than all other combined processes within the design flow. Clearly verification is the bottleneck in a project's time-to-profit goal.

There are many factors contributing to today's verification challenge, including the consumer's insatiable demand for new product features (resulting in more complicated design architectures), rising silicon capacity, as well as breakthroughs in design productivity. While silicon capacity continues to increase along the Moore's Law curve (enabling us to create very complex systems), the effort required to verify these designs has increased at an even greater, and thus alarming, rate—doubling roughly every six to nine months. In addition to rising silicon capacity, our ability to utilize this larger silicon capacity has also increased approximately tenfold within the past decade due to the breakthroughs in synthesis technology. Yet the ability to verify larger systems has not kept pace. Rather, verification productivity has experienced only incremental improvements during the same period. What is clearly needed in *verification* techniques and technology is the equivalent of a *synthesis* productivity breakthrough.

In the second edition of *Writing Testbenches*, Bergeron raises the verification level of abstraction by introducing coverage-driven constrained-random transaction-level self-checking testbenches— all made possible through the introduction of hardware verification languages (HVLs), such as *e* from Verisity and OpenVera from Synopsys. The state-of-art methodologies described in *Writing Testbenches* will contribute greatly to the much-needed equivalent of a *synthesis* breakthrough in verification productivity. I not only highly recommend this book, but also I think it should be required reading by anyone involved in design and verification of today's ASIC, SoCs and systems.

Harry Foster
Chief Architect
Verplex Systems, Inc.

PREFACE

If you survey hardware design groups, you will learn that between 60% and 80% of their effort is now dedicated to verification. Unlike synthesizeable coding, there is no particular coding style nor language required for verification. The freedom of using any language that can be interfaced to a simulator and of using any features of that language has produced a wide array of techniques and approaches to verification. The absence of constraints and historical lack of available expertise and references in verification has resulted in ad hoc approaches. The consequences of an informal verification process can range from a non-functional design requiring several re-spins, through a design with only a subset of the intended functionality, to a delayed product shipment.

WHY THIS BOOK IS IMPORTANT

Take a survey of the books about Verilog or VHDL currently available. You will notice that the majority of the pages are devoted to explaining the details of the languages. In addition, several chapters are focused on the synthesizeable—or RTL—coding style replete with examples. Some books are even devoted entirely to the subject of RTL coding.

When verification is addressed, only one or two chapters are dedicated to the topic. And often, the primary focus is to introduce more language constructs. Verification is usually presented in a very rudimentary fashion, using simple, non-scalable techniques that become tedious in large-scale, real-life designs.

The first edition of this book was the first book specifically devoted to functional verification techniques for hardware models. Since then, several other verification-only books have appeared. Major conferences now include verification tracks. Universities, in collaboration with industry, are now offering verification courses in their engineering curriculum. Pure verification EDA companies are now offering new tools to improve productivity and the overall design quality. All of these contribute to create a formal body of knowledge in design verification. Such a body of knowledge is an essential foundation to creating a science of verification and fueling progress in methodology and productivity.

In this second edition, I will present the latest verification techniques that are successfully being used to produce fully functional first-silicon ASICs, systems-on-a-chip (SoC), boards and entire systems. It builds on the content of the first edition—transaction-level self-checking testbenches—to introduce a revolution in functional verification: coverage-driven constrainable random testbenches.

WHAT THIS BOOK IS ABOUT

I will first introduce the necessary concepts and tools of verification, then I'll describe a process for planning and carrying out an effective functional verification of a design. I will also introduce the concept of coverage models that can be used in a coverage-driven verification process.

It will be necessary to cover some VHDL and Verilog language semantics that are often overlooked or oversimplified in textbooks intent on describing the synthesizeable subset. These unfamiliar semantics become important in understanding what makes a well-implemented and robust testbench and in providing the necessary control and monitor features. Once these new semantics are understood in a familiar language, the same semantics are presented in new verification-oriented languages.

I will also present techniques for applying stimulus and monitoring the response of a design, by abstracting the physical-level transactions into high-level procedures using bus-functional models. The architecture of testbenches built around these bus-functional models is important to create a layer of abstraction relevant to the function being verified and to minimize development and maintenance effort. I also show some strategies for making testbenches self-checking.

Creating random testbenches involves more than calling the *random()* function in whatever language is used to implement them. I will show how random stimulus generators, built on top of bus-functional models, can be architected and designed to be able to produce the desired stimulus patterns. Random generators must be easily externally constrained to increase the likelihood that a set of interesting patterns will be generated.

Behavioral modeling is another important concept presented in this book. It is used to parallelize the implementation and verification of a design and to perform more efficient simulations. For many, behavioral modeling is synonymous with synthesizeable or RTL modeling. In this book, the term "behavioral" is used to describe any model that adequately emulates the functionality of a design, usually using non-synthesizeable constructs and coding style.

WHAT PRIOR KNOWLEDGE YOU SHOULD HAVE

This book focuses on the functional verification of hardware designs using VHDL, Verilog, *e* or OpenVera. I expect the reader to have at least a basic knowledge of VHDL, Verilog, OpenVera or *e*. Ideally, you should have experience in writing models and be familiar with running a simulation using any of the available VHDL or Verilog simulators. There will be no detailed description of language syntax or grammar. It may be a good idea to have a copy of a language-focused textbook as a reference along with this book[1]. I do not describe a synthesizeable subset, nor limit the implementation of the verification techniques to using that subset. Verification is a complex task: The power of a language will be used to its fullest.

I also expect that you have a basic understanding of digital hardware design. This book uses several hypothetical designs from various application domains (video, datacom, computing, etc.). How these designs are actually specified, architected and then implemented is beyond the scope of this book. The content focuses on the specification, architecture, then implementation of the *verification* of these same designs.

READING PATHS

You should really read this book from cover to cover. However, if you are pressed for time, here are a few suggested paths.

If you are using this book as a university or college textbook, you should focus on Chapter 4 through 6 and Appendix A. If you are a junior engineer who has only recently joined a hardware design group, you may skip Chapters 3 and 7. But do not forget to read them once you have gained some experience.

Chapters 3 and 6, as well as Appendix A, will be of interest to a senior engineer in charge of defining the verification strategy for a project. If you are an experienced designer, you may wish to skip ahead to Chapter 3. If you are an experienced Verilog or VHDL user, you may wish to skip Chapter 4—but read it anyway, just to make sure your definition of "experienced" matches mine.

If you have a software background, Chapter 4 and Appendix A may seem somewhat obvious. If you have a hardware design and RTL coding mindset, Chapters 4 and 7 are probably your best friends.

---

1. For Verilog, I recommend *The Verilog Hardware Description Language* by Thomas & Moorby, 3rd edition or later (Kluwer Academic Publisher).

   For VHDL, I recommend *VHDL Coding Styles and Methodologies* by Ben Cohen (Kluwer Academic Publisher).

   For OpenVera, the *OpenVera Language Reference Manual* is available at http://Open-Vera.com. Vera users will find the *Vera Users Manual* available under $VERA_HOME/doc/vum.

   For *e*, Specman Elite users will find the *e Language Reference Manual* under the HELP menu. It can also be found at https://verificationvault.com.

If your responsibilities are limited to managing a hardware verification project, you probably want to concentrate on Chapter 3, Chapter 6 and Chapter 7.

CHOOSING A LANGUAGE

The first decision a design group is often faced with is deciding which language to use. As the author of this book, I faced the same dilemma. In many cases, the choice does not exist as the company has selected a language over others and has invested heavily into supporting that language in terms of licenses, training and intellectual property. But for small companies, companies in transition or companies without a central CAD group, the answer is usually dictated by the decision maker's own knowledge or personal preference.

VHDL vs. Verilog

In my opinion, VHDL and Verilog are inadequate by themselves, especially for verification. They are both equally poor for synthesizeable description. Some things are easier to accomplish in one language than in the other. For a *specific* model, one language is better than the other: One language has features that map better to the functionality to be modeled. However, as a general rule, neither is better than the other.

Some sections are Verilog only. In my experience, Verilog is a much abused language. It has the reputation for being easier to learn than VHDL, and to the extent that the learning curve is not as steep, it is true. However, all languages provide similar concepts: sequential statements, parallel constructs, structural constructs and the illusion of parallelism.

For all languages, these concepts *must* be learned. Because of its lax requirements, Verilog lulls the user into a false sense of security. The user believes that he or she knows the language because there are no syntax errors or because the simulation results appear to be correct. Over time, and as a design grows, race conditions and fragile code structures become apparent, forcing the user to learn these important concepts. Both languages have the same *area under the learning curve*. VHDL's is steeper but Verilog's goes on for much longer. Some sections in this book take the reader farther down the Verilog learning curve.

Hardware Verification Languages

Hardware verification languages (HVLs) are languages that were specifically designed to implement testbenches efficiently and productively. As I write this book, there are several to choose from. Commercial solutions include *e* from Verisity, OpenVera from Synopsys and RAVE from Forte Design. Open-source solutions include the SystemC Verification Library (SCV) from Cadence and Jeda from Juniper Networks. There are also a plethora of home-grown solutions based on Perl, SystemC, C++ or TCL. Verification extensions to the Verilog language are also being added in SystemVerilog. Not all support a coverage-driven constrainable random verification strategy (see "Coverage-Driven Random-Based Approach" on page 109) equally well. Many are still better suited to a directed test strategy (see "Directed Testbenches Approach" on page 104).

Switching from Verilog or VHDL to an HVL involves more than simply learning a new syntax. Although one can continue to use an HDL-like directed methodology with an HVL, using an HVL requires a shift in the way verification is approached and testbenches are implemented. The directed verification strategy used with Verilog and VHDL is the schematic capture of verification. Using an HVL with a constraint-driven random verification strategy is the synthesis of verification. When used properly, HVLs are an incredible productivity boost (see Figure 2-17 on page 63).

If this book had been written from scratch, I would not have bothered including Verilog or VHDL examples. Because they were already there and they can be useful as a foundation for understanding the new concepts provided by HVLs, I've decided to keep these examples. I also took advantage of this second edition to update the Verilog content to reflect the new Verilog-2001 standard.

And the Winner Is...

I know VHDL, Verilog, C++, *e* and OpenVera equally well. I work using all of them. I teach all of them. When asked which one I prefer, I usually answer that I was asked the wrong question. The right question should be, "Which one do I hate the least?" And the answer to that question is, "The one I'm not currently working with." When working in one language, you do not notice the things that are simple to describe or achieve in that language. Instead, you notice the frustrations and how it would be easy to do it if only you were using the *other* language.

Verification techniques transcend the language used. VHDL, Verilog, *e* and OpenVera are only implementation vehicles. All are used throughout the book, but examples are typically shown in only one language. I trust that a monolingual reader will be able to understand the example in the other languages, even though the syntax is slightly different. In areas where each language requires different approaches or methodologies, I will present each individually. However, I will make no attempt to cover all of the features of each language. This is a methodology book, not a language book.

I have selected *e* and OpenVera because they are the languages I know best and, at the time of writing, are the HVLs that best support a coverage-driven constrainable random verification process. The book was not written as a medium for comparing language features or the number of lines required to implement various functionality. The decision to use one language over another involves more than a mere side-by-side comparison of features and syntax.

The code syntax in all examples, any mentioned language or tool limitations and any example or discussion of tool output or feature is up-to-date and factually correct[2] at the time of writing. For VHDL, VHDL-93 and ModelSim™ 5.5.*e* were used. For Verilog, VCSi 6.1 and ModelSim 5.5.e were used. For OpenVera, Vera™ 6.0.0 was used. For *e,* Specman Elite™ 4.1 was used.

---

2. I welcome correction of any factual errors in the book via email at janick@bergeron.com. These corrections will be posted in the *errata* section of the book website.

A common complaint I received about the first edition was the lack of complete examples. You'll notice that in this edition, like the first, code samples are still provided only as excerpts. I fundamentally believe that this is a better way to focus the reader's attention on the important point I'm trying to convey. I do not want to bury you under pages and pages of complete but dry (and ultimately mostly irrelevant) source code. Instead, the entire source code for all examples can now be found in the *source* section at the following URL:

http://janick.bergeron.com/wtb

FOR MORE INFORMATION

If you want more information on topics mentioned in this book, you will find links to relevant resources in the book-companion Web site at the following URL:

http://janick.bergeron.com/wtb

In the *resources* area, you will find links to publicly available utilities, documents and tools that make the verification task easier. You will also find an errata section listing and correcting the errors that inadvertently made their way in this edition.[3]

ACKNOWLEDGEMENTS

My wife, Danielle, continued to give this book energy against its constant drain. Kyle Smith, my editor, once more gave it form from its initial chaos. Chris Macionski, Andrew Piziali, Chris Spear, Ben Cohen and Grant Martin, my technical reviewers, gave it light from its stubborn darkness. And FrameMaker, my word processing software, once more reliably went where no Word had gone before!

I also thank Mentor Graphics for supplying licenses for their ModelSim™ HDL simulator, Verisity Design Ltd. for supplying

---

3. If you know of a verification-related resource or an error in this book that is not mentioned in the Web site, please let me know via email at janick@bergeron.com. I thank you in advance.

licenses for their Specman Elite™ environment, and Synopsys Inc. for supplying licenses for their VCS™ and Vera™ simulators.

ModelSim is a trademark of Mentor Graphics Ltd. Specman Elite and *e*VC are trademarks of Verisity Design Ltd. The *e* language is the intellectual property of and copyrighted by Verisity Design Ltd. and used in this book under license. VCS and Vera are trademarks of Synopsys Inc. All other trademarks are the property of their respective owners.

Writing Testbenches: Functional Verification of HDL Models

"Everyone knows debugging is twice as hard as writing a program in the first place"

- Brian Kernighan
"Elements of Programming Style"
1974

CHAPTER 1   WHAT IS VERIFICATION?

Verification is not a testbench, nor is it a series of testbenches. Verification is a *process* used to demonstrate that the intent of a design is preserved in its implementation. We all perform verification processes throughout our daily lives: balancing a checkbook, tasting a simmering dish, associating landmarks with symbols on a map. These are all verification processes.

In this chapter, I introduce the basic concepts of verification, from its importance and cost, to making sure you are verifying that you are implementing what you want. We look at the differences between various verification approaches as well as the difference between testing and verification. I also show how verification is key to design reuse, and I show the challenges of verification reuse.

WHAT IS A TESTBENCH?

The term "testbench" usually refers to simulation code used to create a prdetermined input sequence to a design, then optionally to observe the response. A testbench is commonly implemented using VHDL, Verilog, *e* or OpenVera, but it may also include external data files or C routines.

Figure 1-1 shows how a testbench interacts with a *design under verification* (DUV). The testbench provides inputs to the design and watches any outputs. Notice how this is a completely closed system: no inputs or outputs go in or out. The testbench is effectively a model of the universe as far as the design is concerned. The verification challenge is to determine what input patterns to supply to the design and what is the expected output of a properly working design when submitted to those input patterns.

Figure 1-1.
Generic structure of a testbench and design under verification

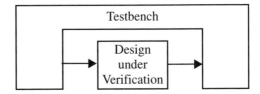

THE IMPORTANCE OF VERIFICATION

Most books focus on syntax, semantic and RTL subset.

If you look at a typical book on Verilog or VHDL, you will find that most of the chapters are devoted to describing the syntax and semantics of the language. You will also invariably find two or three chapters on synthesizeable coding style or Register Transfer Level (RTL) subset.

Most often, only a single chapter is dedicated to testbenches. Very little can be adequately explained in one chapter and these explanations are usually very simplistic. In nearly all cases, these books limit the techniques described to applying simple sequences of vectors in a synchronous fashion. The output is then verified using a waveform viewing tool. Most books also take advantage of the topic to introduce the file input mechanisms offered by the language, devoting yet more content to detailed syntax and semantics.

Given the significant proportion of literature devoted to writing synthesizeable VHDL or Verilog code compared to writing testbenches to verify their functional correctness, you could be tempted to conclude that the former is a more daunting task than the latter. The evidence found in all hardware design teams points to the contrary.

70% of design effort goes to verification.

Today, in the era of multi-million gate ASICs, reusable intellectual property (IP), and system-on-a-chip (SoC) designs, verification consumes about 70% of the design effort. Design teams, properly staffed to address the verification challenge, include engineers dedicated to verification. The number of verification engineers can be up to twice the number of RTL designers.

The Importance of Verification

Verification is on the critical path.

Given the amount of effort demanded by verification, the shortage of qualified hardware design and verification engineers, and the quantity of code that must be produced, it is no surprise that, in all projects, verification rests squarely in the critical path. The fact that verification is often considered after the design has been completed, when the schedule has already been ruined, compounds the problem. It is also the reason verification is currently the target of new tools and methodologies. These tools and methodologies attempt to reduce the overall verification time by enabling parallelism of effort, higher abstraction levels and automation.

Verification time can be reduced through parallelism.

If efforts can be parallelized, additional resources can be applied effectively to reduce the total verification time. For example, digging a hole in the ground can be parallelized by providing more workers armed with shovels. To parallelize the verification effort, it is necessary to be able to write—and debug—testbenches in parallel with each other as well as in parallel with the implementation of the design.

Verification time can be reduced through abstraction.

Providing higher abstraction levels enables you to work more efficiently without worrying about low-level details. Using a backhoe to dig the same hole mentioned above is an example of using a higher abstraction level.

Using abstraction reduces control over low-level details.

Higher abstraction levels are usually accompanied by a reduction in control and therefore must be chosen wisely. These higher abstraction levels often require additional training to understand the abstraction mechanism and how the desired effect can be produced.

Using a backhoe to dig a hole suffers from the same loss-of-control problem: The worker is no longer directly interacting with the dirt; instead the worker is manipulating levers and pedals. Digging happens much faster, but with lower precision and only by a trained operator. The verification process can use higher abstraction levels by working at the transaction- or bus-cycle levels (or even higher ones), instead of always dealing with low-level zeroes and ones.

Verification time can be reduced through automation.

Automation lets you do something else while a machine completes a task autonomously, faster and with predictable results. Automation requires standard processes with well-defined inputs and outputs. Not all processes can be automated. For example, holes must be dug in a variety of shapes, sizes, depths, locations and in varying soil conditions, which render general-purpose automation impossible.

Verification faces similar challenges. Because of the variety of functions, interfaces, protocols and transformations that must be verified, it is not possible to provide a general purpose automation solution for verification, given today's technology. It is possible to automate some portion of the verification process, especially when applied to a narrow application domain. For example, trenchers have automated digging holes used to lay down conduits or cables at shallow depths. Tools automating various portions of the verification process are being introduced. For example, there are tools that will automatically generate bus-functional models from a higher-level abstract specification.

Randomization can be used as an automation tool.

For specific domains, automation can be emulated using randomization. By constraining a random generator to produce valid inputs within the bounds of a particular domain, it is possible to automatically produce almost all of the interesting conditions. For example, the tedious process of vacuuming the bottom of a pool can be automated using a broom head that, constrained by the vertical walls, randomly moves along the bottom. After a few hours, only the corners and a few small spots remain to be cleaned manually. This type of automation process takes more computation time to achieve the same result, but it is completely autonomous, freeing valuable resources to work on other critical tasks. Furthermore, this process can be parallelized[1] easily by concurrently running several random generators. They can also operate overnight, increasing the total number of productive hours.

---

1. Optimizing these concurrent processes to reduce the amount of overlap is another question!

RECONVERGENCE MODEL

The *reconvergence model* is a conceptual representation of the verification process. It is used to illustrate what exactly is being verified.

Do you know what you are actually verifying?

One of the most important questions you must be able to answer is: What are you verifying? The purpose of verification is to ensure that the result of some transformation is as intended or as expected. For example, the purpose of balancing a checkbook is to ensure that all transactions have been recorded accurately and confirm that the balance in the register reflects the amount of available funds.

Figure 1-2.
Reconvergent
paths in
verification

Transformation

Verification

Verification is the reconciliation, through different means, of a specification and an output.

Figure 1-2 shows that verification of a transformation can be accomplished only through a second reconvergent path with a common source. The transformation can be any process that takes an input and produces an output. RTL coding from a specification, insertion of a scan chain, synthesizing RTL code into a gate-level netlist and layout of a gate-level netlist are some of the transformations performed in a hardware design project. The verification process reconciles the result with the starting point. If there is no starting point common to the transformation and the verification, *no* verification takes place.

The reconvergent model can be described using the checkbook example as illustrated in Figure 1-3. The common origin is the previous month's balance in the checking account. The transformation is the writing, recording and debiting of several checks during a one-month period. The verification reconciles the final balance in the checkbook register using this month's bank statement.

Figure 1-3.
Balancing a
checkbook is a
verification
process

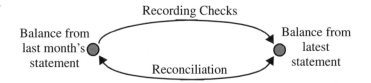

Recording Checks

Balance from
last month's
statement

Reconciliation

Balance from
latest
statement

THE HUMAN FACTOR

If the transformation process is not completely automated from end to end, it is necessary for an individual (or group of individuals) to interpret a specification of the desired outcome and then perform the transformation. RTL coding is an example of this situation. A design team interprets a written specification document and produces what they believe to be functionally correct synthesizeable HDL code. Usually, each engineer is left to verify that the code written is indeed functionally correct.

Figure 1-4.
Reconvergent
paths in
ambiguous
situation

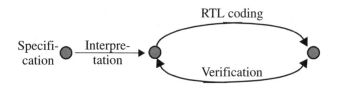

RTL coding

Specifi-
cation

Interpre-
tation

Verification

Verifying your own design verifies against your interpretation, not against the specification.

Figure 1-4 shows the reconvergent path model of the situation described above. If the same individual performs the verification of the RTL coding that initially required interpretation of a specification, then the common origin is that interpretation, *not* the specification.

In this situation, the verification effort verifies whether the design accurately represents the *implementer's interpretation* of that specification. If that interpretation is wrong in any way, then this verification activity will never highlight it.

Any human intervention in a process is a source of uncertainty and unrepeatability. The probability of human-introduced errors in a process can be reduced through several complementary mechanisms: automation, *poka-yoke* or redundancy.

Automation

Eliminate human intervention.

Automation is the obvious way to eliminate human-introduced errors in a process. Automation takes human intervention completely out of the process. However, automation is not always possible, especially in processes that are not well-defined and continue to require human ingenuity and creativity, such as hardware design.

Poka-Yoke

Make human intervention foolproof.

Another possibility is to mistake-proof the human intervention by reducing it to simple, and foolproof steps. Human intervention is needed only to decide on the particular sequence or steps required to obtain the desired results. This mechanism is also known as *poka-yoke* in Total Quality Management circles. It is usually the last step toward complete automation of a process. However, just like automation, it requires a well-defined process with standard transformation steps. The verification process remains an art that, to this day, does not yield itself to well-defined steps.

Redundancy

Have two individuals check each other's work.

The final alternative to removing human errors is redundancy. It is the simplest, but also the most costly mechanism. Redundancy requires every transformation resource to be duplicated. Every transformation accomplished by a human is either independently verified by another individual, or two complete and separate transformations are performed with each outcome compared to verify that both produced the same or equivalent output. This mechanism is used in high-reliability environments, such as airborne and space systems. It is also used in industries where later redesign and replacement of a defective product would be more costly than the redundancy itself, such as ASIC design.

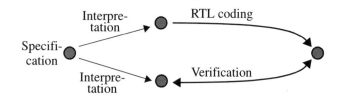

Figure 1-5. Redundancy in an ambiguous situation enables accurate verification

A different person should be in charge of verification.

Figure 1-5 shows the reconvergent paths model where redundancy is used to guard against misinterpretation of an ambiguous specification document. When used in the context of hardware design, where the transformation process is writing RTL code from a written specification document, this mechanism implies that a different individual must be in charge of the verification.

WHAT IS BEING VERIFIED?

Choosing the common origin and reconvergence points determines what is being verified. These origin and reconvergence points are often determined by the tool used to perform the verification. It is important to understand where these points lie to know which transformation is being verified. Formal verification, model checking, functional verification, and rule checkers verify different things because they have different origin and reconvergence points.

Formal Verification

Formal verification does not eliminate the need to write testbenches.

Formal verification is often misunderstood initially. Engineers unfamiliar with the formal verification process often imagine that it is a tool that mathematically determines whether their design is correct, without having to write testbenches. Once you understand what the end points of the formal verification reconvergent paths are, you know what exactly is being verified.

The application of formal verification falls under two broad categories: equivalence checking and model checking.

Equivalence Checking

Equivalence checking compares two models.

Figure 1-6 shows the reconvergent path model for equivalence checking. This formal verification process mathematically proves that the origin and output are logically equivalent and that the transformation preserved its functionality.

Figure 1-6.
Equivalence checking paths

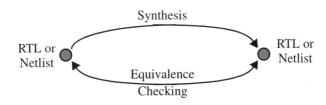

It can compare two netlists.

In its most common use, equivalence checking compares two netlists to ensure that some netlist post-processing, such as scan-chain insertion, clock-tree synthesis or manual modification[2], did not change the functionality of the circuit.

It can detect bugs in the synthesis software.

Another popular use of equivalence checking is to verify that the netlist correctly implements the original RTL code. If one trusted the synthesis tool completely, this verification would not be necessary. However, synthesis tools are large software systems that depend on the correctness of algorithms and library information. History has shown that such systems are prone to error. Equivalence checking is used to keep the synthesis tool honest. In some rare instances, this form of equivalence checking is used to verify that manually written RTL code faithfully represents a legacy gate-level design.

Less frequently, equivalence checking is used to verify that two RTL descriptions are logically identical, sometimes to avoid running lengthy regression simulations when only minor non-functional changes are made to the source code to obtain better synthesis results, or when a design is translated from an HDL to another.

2. Text editors remain the greatest design tools!

Equivalence checking found a bug in an arithmetic operator.

Equivalence checking is a true alternative path to the logic synthesis transformation being verified. It is only interested in comparing Boolean and sequential logic functions, not mapping these functions to a specific technology while meeting stringent design constraints. Engineers using equivalence checking found a design at Digital Equipment Corporation (now part of HP) to be synthesized incorrectly. The design used a synthetic operator that was functionally incorrect when handling more than 48 bits. To the synthesis tool's defense, the documentation of the operator clearly stated that correctness was not guaranteed above 48 bits. Since the synthesis tool had no knowledge of documentation, it could not know it was generating invalid logic. Equivalence checking quickly identified a problem that could have been very difficult to detect using gate-level simulation.

Model Checking

Model checking proves assertions about the behavior of the design.

Model checking is a more recent application of formal verification technology. In it, assertions or characteristics of a design are formally proven or disproved. For example, all state machines in a design could be checked for unreachable or isolated states. A more powerful model checker may be able to determine if deadlock conditions can occur.

Another type of assertion that can be formally verified relates to interfaces. Using a formal description language, assertions about the interfaces of a design are stated and the tool attempts to prove or disprove them. For example, an assertion might state that, given that signal ALE will be asserted, then either the DTACK or ABORT signal will be asserted eventually.

Figure 1-7.
Model
checking paths

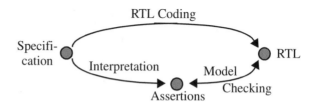

Writing Testbenches: Functional Verification of HDL Models

What Is Being Verified?

Knowing which
assertions to prove
and expressing
them correctly is
the most difficult
part.

The reconvergent path model for model checking is shown in
Figure 1-7 . The greatest obstacle for model checking technology is
identifying, through interpretation of the design specification,
which assertions to prove. Of those assertions, only a subset can be
proven feasibly. Current technology cannot prove high-level assertions about a design to ensure that complex functionality is correctly implemented. It would be nice to be able to assert that, given
specific register settings, a set of asynchronous transfer mode
(ATM) cells will end up at a set of outputs in some relative order.
Unfortunately, model checking technology is not at that level yet.

Functional Verification

Figure 1-8.
Functional
verification
paths

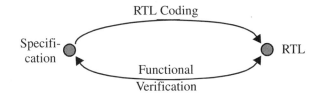

Functional verifi-
cation verifies
design intent.

The main purpose of functional verification is to ensure that a
design implements intended functionality. As shown by the reconvergent path model in Figure 1-8, functional verification reconciles
a design with its specification. Without functional verification, one
must trust that the transformation of a specification document into
RTL code was performed correctly, without misinterpretation of the
specification's intent.

You can prove the
presence of bugs,
but you cannot
prove their
absence.

It is important to note that, unless a specification is written in a formal language with precise semantics,[3] it is *impossible* to prove that
a design meets the intent of its specification. Specification documents are written using natural languages by individuals with varying degrees of ability in communicating their intentions. Any
document is open to interpretation. Functional verification, as a
process, can *show* that a design meets the intent of its specification.
But it cannot *prove* it. One can easily prove that the design does *not*
implement the intended function by identifying a single discrep-

---

3. Even if such a language existed, one would eventually have to show
   that this description is indeed an accurate description of the design
   intent, based on some higher-level ambiguous specification.

ancy. The converse, sadly, is not true: No one can prove that there are *no* discrepancies, just as no one can prove that flying reindeers or UFOs do not exist. (However, producing a single flying reindeer or UFO would be sufficient to prove the opposite!)

FUNCTIONAL VERIFICATION APPROACHES

Functional verification can be accomplished using three complementary approaches: black-box, white-box and grey-box.

Black-Box Verification

Black-box verification cannot look at or know about the inside of a design.

With a black-box approach, the functional verification is performed without any knowledge of the actual implementation of a design. All verification is accomplished through the available interfaces, without direct access to the internal state of the design, without knowledge of its structure and implementation. This method suffers from an obvious lack of visibility and controllability. It is often difficult to set up an interesting state combination or to isolate some functionality. It is equally difficult to observe the response from the input and locate the source of the problem. This difficulty arises from the frequent long delays between the occurrence of a problem and the appearance of its symptom on the design's outputs.

Testcase is independent of implementation.

The advantage of black-box verification is that it does not depend on any specific implementation. Whether the design is implemented in a single ASIC, RTL code, gates, multiple FPGAs, a circuit board or entirely in software, is irrelevant. A black-box functional verification approach forms a true conformance verification that can be used to show that a particular design implements the intent of a specification, regardless of its implementation.

My mother is a veteran of the black-box approach: To prevent us from guessing the contents of our Christmas gifts, she never puts any names on the wrapped boxes[4]. At Christmas, she has to correctly identify the content of each box, without opening it, so it can be given to the intended recipient. She has been known to fail on a few occasions, to the pleasure of the rest of the party!

---

4. To my wife's chagrin who likes shaking any box bearing her name.

In black-box verification, it is difficult to control and observe specific features.

The pure black-box approach is impractical in today's large designs. A multi-million gates ASIC possesses too many internal signals and states to effectively verify all of its functionality from its periphery. Critical functions, deep into the design, will be difficult to control and observe. Furthermore, a design fault may not readily present symptoms of a flaw at the outputs of the ASIC. For example, the black-box ASIC-level testbench in Figure 1-9 is used to verify a critical round-robin arbiter. If the arbiter is not completely fair in its implementation, what symptoms would be visible at the outputs? This type of fault could be only found through performance analysis of several long simulations to identify discrepancies between the actual throughput of a channel compared with its theoretical throughput.

Figure 1-9.
Black-box verification of a low-level feature

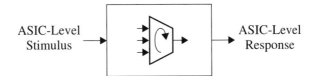

ASIC-Level Stimulus → → ASIC-Level Response

White-Box Verification

White box verification has intimate knowledge and control of the internals of a design.

As the name suggests, a white-box approach has full visibility and controllability of the internal structure and implementation of the design being verified. This method has the advantage of being able to set up an interesting combination of states and inputs quickly, or to isolate a particular function. It can then easily observe the results as the verification progresses and immediately report any discrepancies from the expected behavior.

White-box verification is tied to a specific implementation.

However, this approach is tightly integrated with a particular implementation. Changes in the design may require changes in the testbench. Furthermore, those testbenches cannot be used in gate-level simulations, on alternative implementations or future redesigns. It also requires detailed knowledge of the design implementation to know which significant conditions to create and which results to observe.

White-box techniques can augment black-box approaches.

White-box verification is a useful complement to black-box verification. This approach can ensure that low-level implementation-specific features behave properly, such as counters rolling over after reaching their end count value or datapaths being appropriately steered and sequenced. The white-box approach can be used only to verify the correctness of the functionality, while still relying on the black- or grey-box stimulus. For example, Figure 1-10 shows the black-box ASIC-level environment shown in Figure 1-9 augmented with white-box checks to verify the functional correctness of the round-robin arbiter. Should fairness not be implemented correctly, the white-box checks would immediately report a failure. The reported error would also make it easier to identify and confirm the cause of the problem. A 2% discrepancy in a channel throughput in three overnight simulations can be explained easily as a simple statistical error.

Figure 1-10.
White-box checks in black-box environment

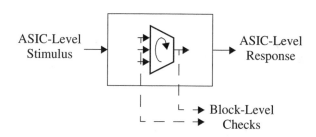

A pure white-box verification approach is often used on SoC design and system-level verification. A system is defined as a design composed of independently designed and verified components. The objective of system-level verification is to verify the system-level features, not re-verify the individual components. Because of the large number of possible states and the difficulty in setting up interesting conditions, system-level verification is often accomplished by treating it as a collection of black-boxes. The independently-designed components are treated as black-boxes, but the system itself is treated as a white-box, with full controllability and observability.

White-box is used in system-level verification.

Grey-Box Verification

Grey-box verification is a compromise between the aloofness of a black-box verification and the dependence on the implementation of white-box verification. The former may not fully exercise all parts of a design, while the latter is not portable.

Testcase may not be relevant on another implementation.

As in black-box verification, a grey-box approach controls and observes a design entirely through its top-level interfaces. However, the particular verification being accomplished is intended to exercise significant features specific to the implementation. The same verification of a different implementation would be successful, but the verification may not be particularly more interesting than any other black-box verification. A typical grey-box test case is one written to increase coverage metrics. The input stimulus is designed to execute specific lines of code or create a specific set of conditions in the design. Should the structure (but not the function) of the design change, this test case, while still correct, may no longer contribute toward better coverage.

Add functions to the design to increase controllability and observability

A typical grey-box strategy is to include some non-functional modifications to provide additional visibility and controllability. Examples include additional software-accessible registers to control or observe internal states, speed up a real-time counter, force the raising of exception or modify the size of the processed data to minimize verification time. These registers and features would not be used during normal operations, but they are often valuable during the integration phase of the first prototype systems.

White-box cannot be used in parallel with design.

The black-box and grey-box approaches are the only ones that can be used if the functional verification is to be implemented in parallel with the implementation using a behavioral model of the design (see "Behavioral Models" on page 375). Because there is no detailed implementation to know about beforehand, these two verification strategies are the only possible avenue.

TESTING VERSUS VERIFICATION

Testing verifies
manufacturing.

Testing is often confused with verification. The purpose of the former is to verify that the design was manufactured correctly. The purpose of the latter is to ensure that a design meets its functional intent.

Figure 1-11.
Testing vs.
Verification

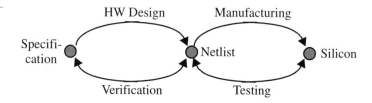

Figure 1-11 shows the reconvergent paths models for both verification and testing. During testing, the finished silicon is reconciled with the netlist that was submitted for manufacturing.

Testing verifies
that internal nodes
can be toggled.

Testing is accomplished through test vectors. The objective of these test vectors is not to exercise functions. It is to exercise physical locations in the design to ensure that they can go from 0 to 1 and from 1 to 0 and that the change can be observed. The ratio of physical locations tested to the total number of such locations is called *test coverage*. The test vectors are usually automatically generated to maximize coverage while minimizing vectors through a process called *automatic test pattern generation* (ATPG).

Thoroughness of
testing depends on
controllability and
observability of
internal nodes.

Testing and test coverage depends on the ability to set internal physical locations to either 1 or 0, and then observe that they were indeed appropriately set. Some designs have very few inputs and outputs, but these designs have a large number of possible states, requiring long sequences to observe and control all internal physical locations properly. A perfect example is an electronic wristwatch: It has three or four inputs (the buttons around the dial) and a handful of outputs (the digits and symbols on the display). However, if it includes chronometer and calendar functions, it has billions of possible state combinations (hundreds of years divided into milliseconds). At speed, it would take hundreds of years to take such a design through all of its possible states.

Scan-Based Testing

Linking all registers into a long shift register increases controllability and observability.

Fortunately, scan-based testability techniques help reduce this problem to something manageable. With scan-based tests, all registers inside a design are hooked-up in a long serial chain. In normal mode, the registers operate as if the scan chain was not there (see Figure 1-12(a)). In scan mode, the registers operate as a long shift register (see Figure 1-12(b)).

To test a scannable design, the unit under test is put into scan mode, then an input pattern is shifted through all of its internal registers. The design is then put into normal mode and a *single* clock cycle is applied, loading the result of the normal operation based on the scanned state into the registers. The design is then put into scan mode again. The result is shifted out of the registers (at the same time the next input pattern is shifted in), and the result is compared against the expected value.

Figure 1-12.
Scan-based
testing

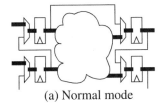 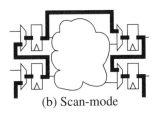

(a) Normal mode   (b) Scan-mode

Scan-based testing puts restrictions on design.

This increase in controllability and observability, and thus test coverage, comes at a cost. Certain restrictions are put onto the design to enable the insertion of a scan chain and the automatic generation of test patterns. Some of these restrictions include, but are not limited to: fully synchronous design, no derived or gated clocks and use of a single clock edge. The topic of design for testability is far greater and complex than this simple introduction implies. For more details, there are several excellent books[5] and papers[6] on the subject.

5. Abramovici, Breuer, and Friedman. *Digital System Testing and Testable Design.* IEEE. ISBN 0780310624

6. Cheung and Wang, *"The Seven Deadly Sins of Scan-Based Design,"* Integrated System Design, Aug. 1997, p50-56.

Hardware designers introduced to scan-based testing initially rebel against the restrictions imposed on them. They see only the immediate area penalty and their favorite design technique rendered illegal. However, the increased area and additional design effort are quickly outweighed when a design can be fitted with one or more scan chains, when test patterns are generated and high test coverage is achieved automatically, at the push of a button. The time saved and the greater confidence in putting a working product on the market far outweighs the added cost for scan-based design.

Design for Verification

Design practices need to be modified to accommodate testability requirements. Isn't it acceptable to modify those same design practices to accommodate verification requirements?

With functional verification requiring more effort than design, it is reasonable to require additional design effort to simplify verification. Just as scan chains are put in a design to improve testability without adding to the functionality, it should be standard practice to add non-functional structures and features to facilitate verification. This approach requires that verification be considered at the outset of a project, during its specification phase. Not only should the architect of the design answer the question, "What is this supposed to do?" but also, "How is this thing going to be verified?"

Typical design-for-verification techniques include well-defined interfaces, clear separation of functions in relatively independent units, providing additional software-accessible registers to control and observe internal locations and providing programmable multiplexers to isolate or bypass functional units.

DESIGN AND VERIFICATION REUSE

Today, design reuse is a fact of life. It is the best way to overcome the difference between the number of transistors that can be manufactured on a single chip and the number of transistors engineers can take advantage of in a reasonable amount of time. This difference is called the *productivity gap*. Design reuse was originally thought to be a simple concept that would be easy to put in practice. The reality proved—and continues to prove—to be more problematic.

Reuse Is About Trust

You won't use what you do not trust.

The major obstacle to design reuse is cultural. Engineers have little incentive and willingness to incorporate an unknown design into their own. They do not trust that the other design is as good or as reliable as one designed by themselves. The key to design reuse is gaining that trust.

Trust, like quality, is not something that can be added to a design after the fact. It must be built-in, through the best possible design practices. And it must be earned by standing behind a reusable design: Providing support services and building a relationship with the user. Once that trust is established, reuse will happen more often.

Proper functional verification demonstrates trustworthiness of a design.

Trustworthiness can be also demonstrated through a proper verification process. By showing the user that a design has been thoroughly and meticulously verified according to the design specification, trust can be built and communicated much faster. Functional verification is the only way to demonstrate that the design meets, or even exceeds, the quality of a similar design that an engineer could do himself or herself.

Verification for Reuse

Reusable designs must be verified to a greater degree of confidence.

If you create a design, you have a certain degree of confidence in your own abilities as a designer and implicitly trust its correctness. Functional verification is used only to confirm that opinion and to augment that opinion in areas known to be weak. If you try to reuse a design, you can rely only on the functional verification to build that same level of confidence and trust. Thus, reusable designs must be verified to a greater degree of confidence than custom designs.

All claims, possible configurations and uses must be verified.

Because reusable designs tend to be configurable and programmable to meet a variety of possible environment and applications, it is necessary to verify a reusable design under all possible configurations and for all possible uses. All claims made about the reusable design must be verified and demonstrated to users.

Verification Reuse

Testbench components can be reused also.

If portions of a design can be reused, portions of testbenches can be reused as well. For example, Figure 1-13 shows that the bus-functional model used to verify a design block (a) can be reused to verify the system that uses it (b).

Figure 1-13. Reusing BFMs from block-level testbench in system-level testbench

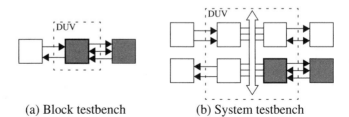

(a) Block testbench   (b) System testbench

Verification reuse has its challenges.

There are degrees of verification reuse, some easier to achieve, others facing difficulties similar to design reuse. Reusing BFMs across different testbenches and test cases for the same design is a simple process of properly architecting a verification environment. Reusing testbench components or test cases in a subsequent revision of the same design presents some difficulties in introducing the verification of the new features. Reusing a testbench component between two different projects or between two different levels of abstraction has many challenges that must be addressed when designing the component itself.

Salvaging is not reuse.

Salvaging is reusing a piece of an existing testbench that was not expressly designed to be reused. The suitability of the salvaged component will vary greatly depending on the similarities between the needs of the design to be verified and those of the original design. For example, a BFM that was designed to verify an interface block (as in Figure 1-13 (a)) may not be suitable for verifying a system using that interface block.

Block- and system-level testbenches put different requirements on a BFM.

Block-level verification must exercise the state machines and decoders used in implementing the interface protocol. This verification requires a transaction-level BFM with detailed controls of the protocol signals to vary timing or inject protocol errors. However, the system-level verification must exercise the high-level functionality that resides behind the interface block. This verification requires the ability to encapsulate high-level data onto the interface transactions. The desired level of controllability resides at a much higher level than the signal-level required to verify the interface block.

THE COST OF VERIFICATION

Verification is a necessary evil. It always takes too long and costs too much. Verification does not directly generate a profit or make money: After all, it is the design being verified that will be sold and ultimately make money, not the verification. Yet verification is indispensable. To be marketable and create revenues, a design must be functionally correct and provide the benefits that the customer requires.

As the number of errors left to be found decreases, the time—and cost—to identify them increases.

Verification is a process that is never truly complete. The objective of verification is to ensure that a design is error-free, yet one cannot prove that a design is error-free. Verification can show only the *presence* of errors, not their *absence*. Given enough time, an error *will* be found. The question thus becomes: Is the error likely to be severe enough to warrant the effort spent identifying it? As more and more time is spent on verification, fewer and fewer errors are found with a constant incremental effort expenditure. As verification progresses, it has diminishing returns. It costs more and more to find each remaining error.

Functional verification is similar to statistical hypothesis testing. The hypothesis under test is: "Is my design functionally correct?"

The answer can be either yes or no. But either answer could be wrong. These wrong answers are Type II and Type I mistakes, respectively.

Figure 1-14.
Type I & II
mistakes

	Fail	Pass
Bad Design		Type II (False Positive)
Good Design	Type I (False Negative)	

False positives must be avoided.

Figure 1-14 shows where each type of mistake occurs. Type I mistakes, or *false negatives*, are the easy ones to identify. The verification is finding an error where none exist. Once the misinterpretation is identified, the implementation of the verification is modified to change the answer from "no" to "yes," and the mistake no longer exists. Type II mistakes are the most serious ones: The verification failed to identify an error. In a Type II mistake, or *false positive* situation, a bad design is shipped unknowingly, with all the potential consequences that entails.

The United States Food and Drug Administration faces Type II mistakes on a regular basis with potentially devastating consequences: In spite of positive clinical test results, is a dangerous drug released on the market? Shipping a bad design may result in simple product recall or in the total failure of a space probe after it has landed on another planet.

With the future of the company potentially at stake, the 64-thousand dollar question in verification is: "How much is enough?" The functional verification process presented in this book, along with some of the tools described in the next chapter attempt to answer that question.

The 64-million dollar question is: "When will I be done?" Knowing where you are in the verification process, although impossible to establish with certainty, is much easier to estimate than how long it will take to complete the job. The verification planning process described in Chapter 3 creates a tool that enables a verification manager to better estimate the effort and time required to complete the task at hand, to the degree of certainty required.

SUMMARY

Verification is a process, not a set of testbenches.

Verification can be only accomplished through an independent path between a specification and an implementation. It is important to understand where that independence starts and to know what is being verified.

Verification can be performed at various levels of the design hierarchy, with varying degrees of visibility within those hierarchies. I prefer a black-box approach because it yields portable testbenches. Augment with grey and white-box testbenches to meet your goals.

Consider verification at the beginning of the design. If a function would be difficult to verify, modify the design to give the necessary observability and controllability over the function.

## CHAPTER 2

VERIFICATION TOOLS

As mentioned in the previous chapter, one of the mechanisms that can be used to improve the efficiency and reliability of a process is automation. This chapter covers tools used in a state-of-the-art functional verification environment. Some of these tools, such as simulators, are essential for the functional verification activity to take place. Others, such as linting or code coverage tools, automate some of the most tedious tasks of verification and help increase the confidence in the outcome of the functional verification.

Not all tools are mentioned in this chapter. It is not necessary to use all the tools mentioned.

It is not necessary to use all of the tools described here. Nor is this list exhaustive, as new application-specific and general purpose verification automation tools are regularly brought to market. As a verification engineer, your job is to use the necessary tools to ensure that the outcome of the verification process is not a Type II mistake, which is a false positive. As a project manager responsible for the delivery of a working product on schedule and within the allocated budget, your responsibility is to arm your engineers with the proper tools to do their job efficiently and with the necessary degree of confidence. Your job is also to decide when the cost of finding the next functional bug has increased above the value the additional functional correctness brings. This last responsibility is the heaviest of them all. Some of these tools provide information to help you decide when you've reached that point.

No endorsements of commercial tools.

I mention some commercial tools by name. They are used for illustrative purposes only and this does not constitute a personal endorsement. I apologize in advance to suppliers of competitive products I fail to mention. It is not an indication of inferiority, but rather an indication of my limited knowledge. All trademarks and service marks, registered or not, are the property of their respective owners.

LINTING TOOLS

Linting tools find common programmer mistakes.

The term "lint" comes from the name of a UNIX utility that parses a C program and reports questionable uses and potential problems. When the C programming language was created by Dennis Ritchie, it did not include many of the safeguards that have evolved in later versions of the language, like ANSI-C or C++, or other strongly-typed languages such as Pascal or ADA. lint evolved as a tool to identify common mistakes programmers made, letting them find the mistakes quickly and efficiently, instead of waiting to find them through a dreaded segmentation fault during execution of the program.

lint identifies real problems, such as mismatched types between arguments and function calls or mismatched number of arguments, as shown in Sample 2-1. The source code is syntactically correct and compiles without a single error or warning using gcc version 2.8.1.

Sample 2-1.
Syntactically
correct K&R
C source code

```
int my_func(addr_ptr, ratio)
    int   *addr_ptr;
    float ratio;
{
    return (*addr_ptr)++;
}

main()
{
    int my_addr;
    my_func(my_addr);
}
```

However, Sample 2-1 suffers from several pathologically severe problems:

1. The *my_func* function is called with only one argument instead of two.

2. The *my_func* function is called with an integer value as a first argument instead of a pointer to an integer.

Problems are found faster than at runtime.

As shown in Sample 2-2, the lint program identifies these problems, letting the programmer fix them before executing the program and observing a catastrophic failure. Diagnosing the problems at runtime would require a runtime debugger and would take several minutes. Compared to the few seconds it took using lint, it is easy to see that the latter method is more efficient.

Sample 2-2.
Lint output for Sample 2-1

```
src.c(3): warning: argument ratio unused in
function my_func
src.c(11): warning: addr may be used before set
src.c(12): warning: main() returns random value
to invocation environment
my_func: variable # of args.     src.c(4)    ::
src.c(11)
my_func, arg. 1 used inconsistently
src.c(4)    ::    src.c(11)
my_func returns value which is always ignored
```

Linting tools are static tools.

Linting tools have a tremendous advantage over other verification tools: They do not require stimulus, nor do they require a description of the expected output. They perform checks that are entirely static, with the expectations built into the linting tool itself.

The Limitations of Linting Tools

Linting tools can only identify a certain class of problems.

Other potential problems were also identified by lint. All were fixed in Sample 2-3, but lint continues to report a problem with the invocation of the *my_func* function: The return value is always ignored. Linting tools cannot identify all problems in source code. They can only find problems that can be statically deduced by looking at the code structure, not problems in the algorithm or data flow.

For example, in Sample 2-3, lint does not recognize that the uninitialized *my_addr* variable will be incremented in the *my_func* function, producing random results. Linting tools are similar to spell checkers; they identify misspelled words, but do not determine if the wrong word is used. For example, this book could have several instances of the word "with" being used instead of "width". It is a type of error the spell checker (or a linting tool) could not find.

Sample 2-3.
Functionally
correct K&R
C source code

```
int my_func(addr_ptr)
    int *addr_ptr;
{
    return (*addr_ptr)++;
}

main()
{
    int my_addr;
    my_func(&my_addr);
    return 0;
}
```

Many false negatives are reported.

Another limitation of linting tools is that they are often too paranoid in reporting problems they identify. To avoid making a Type II mistake—reporting a false positive, they err on the side of caution and report potential problems where none exist. This results in many Type I mistakes—or false negatives. Designers can become frustrated while looking for non-existent problems and may abandon using linting tools altogether.

Carefully filter error messages!

You should filter the output of linting tools to eliminate warnings or errors known to be false. Filtering error messages helps reduce the frustration of looking for non-existent problems. More importantly, it reduces the output clutter, reducing the probability that the report of a real problem goes unnoticed among dozens of false reports. Similarly, errors known to be true positive should be highlighted. *Extreme* caution must be exercised when writing such a filter: You must make sure that a true problem is not filtered out and never reported.

Naming conventions can help output filtering.

A properly defined naming convention is a useful tool to help determine if a warning is significant. For example, the report in Sample 2-4 about a latch being inferred on a signal whose name ends with "_lat" would be considered as expected and a false warning. All other instances would be flagged as true errors.

Sample 2-4.
Output from a
hypothetical
Verilog lint-
ing tool

```
Warning: file decoder.v, line 23: Latch
inferred on reg "address_lat".
Warning: file decoder.v, line 36: Latch
inferred on reg "next_state".
```

Do not turn off checks.	Filtering the output of a linting tool is preferable to turning off checks from within the source code itself or via the command line. A check may remain turned off for an unexpected duration, potentially hiding real problems. Checks that were thought to be irrelevant may become critical as new source files are added.
Lint code as it is being written.	Because it is better to fix problems when they are created, you should run lint on the source code while it is being written. If you wait until a large amount of code is written before linting it, the large number of reports—many of them false—will be daunting and create the impression of a setback. The best time to identify a report as true or false is when you are still intimately familiar with the code.
Enforce coding guidelines.	The linting process, through the use of user-defined rules, can also be used to enforce coding guidelines and naming conventions[1]. Therefore, it should be an integral part of the authoring process to make sure your code meets the standards of readability and maintainability demanded by your audience.

Linting Verilog Source Code

Linting Verilog source code catches common errors.	Linting Verilog source code ensures that all data is properly handled without accidentally dropping or adding to it. The code in Sample 2-5 shows a Verilog model that looks perfect, compiles without errors, but produces unintended results under some circumstances in Verilog-95.

| **Sample 2-5.** Potentially problematic Verilog code | ```
module tristate_buffer(in, out, enable);
parameter WIDTH = 8;
input   [WIDTH-1:0]  in;
output  [WIDTH-1:0]  out;
input                enable;

assign out = (enable) ? in : 'bz;

endmodule
``` |
| Problems may not be apparent under most conditions. | The problem is in the width mismatch in the continuous assignment between the output "*out*" and the constant "*bz*". The unsized constant is 32-bits wide (or a value of 32'hzzzzzzzz), while the output |

---

1. See Appendix A for a set of coding guidelines.
```

has a user-specified width. As long as the width of the output is less than or equal to 32, everything is fine: The value of the constant will be appropriately truncated to fit the width of the output.

However, in Verilog-95, the problem occurs when the width of the output is greater than 32 bits: Verilog-95 *zero-extends* the constant value to match the width of the output, producing the wrong result. The least significant 32-bits are set to high-impedance, while all the other more significant bits are set to zero. This "feature" has been fixed in Verilog-2001.

It is an error that could not be found in simulation, unless a configuration greater than 32 bits was used, *and* it produced wrong results at a time and place you were looking at. A linting tool finds the problem every time, in just a few seconds.

Linting VHDL Source Code

Because of its strong typing, VHDL does not need linting as much as Verilog. Much of the checks performed by a linting tool are required to be performed by the VHDL compiler. However, potential problems are still best identified using a linting tool.

Linting can find unintended multiple drivers.

For example, a common problem in VHDL is created by using the STD_LOGIC type. Since it is a resolved type, STD_LOGIC signals can have more than one driver. When modeling hardware, multiple driver signals are required in a single case: to model buses. In all other cases (which is over 99% of the time), a signal should have only one driver. The VHDL source shown in Sample 2-6 demonstrates how a simple typographical error can go undetected easily and satisfy the usually paranoid VHDL compiler.

Typographical errors can cause serious problems.

In Sample 2-6, both concurrent signal assignments labeled "*statement1*" and "*statement2*" assign to the signal "*s1*" (ess-one), while the signal "*sl*" (ess-ell) remains unassigned. Had I used the STD_ULOGIC type instead of the STD_LOGIC type, the VHDL toolset would have reported an error after finding multiple drivers on an unresolved signal. However, it is not possible to guarantee the STD_ULOGIC type is used for all signals with a single driver. A

Sample 2-6.
Erroneous
multiple
drivers

```
library ieee;
use ieee.std_logic_1164.all;
entity my_entity is
    port (my_input: in std_logic);
end my_entity;

architecture sample of my_entity is
    signal s1: std_logic;
    signal sl: std_logic;
begin
    statement1: s1 <= my_input;
    statement2: s1 <= not my_input;
end sample;
```

linting tool is still required to report multiple driver signals regard-less of the type, as shown in Sample 2-7.

Sample 2-7.
Output from a
hypothetical
VHDL linting
tool

```
Warning: file my_entity.vhd: Signal "s1" is
multiply driven.
Warning: file my_entity.vhd: Signal "sl" has no
drivers.
```

Use naming con-
vention to filter
output.

It would be up to the author to identify the signals that were intended to model buses and ignore the warnings about them. Using a naming convention for such signals facilitates recognizing warnings that can be safely ignored, and enhances the reliability of your code. An example of a naming convention, illustrated in Sample 2-8, would be to name any signals modeling buses with the "_bus" suffix[2].

Sample 2-8.
Naming con-
vention for
signals with
multiple
drivers

```
--
-- data_bus, addr_bus and sys_err_bus
-- are intended to be multiply driven
--
signal data_bus : std_logic_vector(15 downto 0);
signal addr_bus : std_logic_vector( 7 downto 0);
signal ltch_addr: std_logic_vector( 7 downto 0);
signal sys_err_bus: std_logic;
signal bus_grant  : std_logic;
```

2. See Appendix A for an example of naming guidelines.

The accidental multiple driver problem is not the only one that can
be caught using a linting tool. Others, such as unintended latch
inference in synthesizeable code, or the enforcement of coding
guidelines, can also be identified.

Linting OpenVera and *e* Source Code

Because of their strong typing, *e* and OpenVera do not need linting
as much as Verilog. But like Verilog, potential problems are still
best identified using a linting tool. For example, Sample 2-9 shows
a race condition between two concurrent execution branches that
will yield an unpredictable result (this race condition is explained in
details in the section titled "Write/Write Race Conditions" on
page 212). This type of error would be easily detectable by a linting
tool. Linting tools for OpenVera and *e* are starting to emerge.

Sample 2-9.
Race condition
on OpenVera
code

```
{
    integer i;
    fork
        i = 1;
        i = 0;
    join
}
```

Code Reviews

Although not technically linting tools, the objective of code reviews
is essentially the same: Identify functional and coding style errors
before functional verification and simulation. Linting tools can only
identify questionable language uses. They cannot check if the
intended behavior has been coded. In code reviews, the source code
produced by a designer is reviewed by one or more peers. The goal
is not to publicly ridicule the author, but to identify problems with
the original code that could not be found by an automated tool.
Reviews can identify discrepancies between the design intent and
the implementation. They also provide an opportunity for suggest-
ing coding improvements, such as better comments, better structure
or better organization.

A code review is an excellent venue for evaluating the maintainability of a source file, and the relevance of its comments. Other
qualitative coding style issues can also be identified. If the code is

well understood, it is often possible to identify functional errors or omissions.

Code reviews are not new ideas either. They have been used for many years in the software design industry. Detailed information on how to conduct effective code reviews can be found in the *resources* section at:

http://janick.bergeron.com/wtb

SIMULATORS

Simulate your design before implementing it.

Simulators are the most common and familiar verification tools. They are named simulators because their role is limited to approximating reality. A simulation is never the final goal of a project. The goal of all hardware design projects is to create real physical designs that can be sold and generate profits. Simulators attempt to create an artificial universe that mimics the future real design. This type of verification lets the designers interact with the design before it is manufactured, and correct flaws and problems earlier.

Simulators are only approximations of reality.

You must never forget that a simulator is an approximation of reality. Many physical characteristics are simplified—or even ignored—to ease the simulation task. For example, a four-state digital simulator assumes that the only possible values for a signal are 0, 1, unknown, and high-impedance. However, in the physical—and analog—world, the value of a signal is a continuous function of the voltage and current across a thin aluminium or copper wire track: an infinite number of possible values. In a discrete simulator, events that happen deterministically 5 ns apart may be asynchronous in the real world and may occur randomly.

Simulators are at the mercy of the descriptions being simulated.

Within that simplified universe, the only thing a simulator does is execute a description of the design. The description is limited to a well-defined language with precise semantics. If that description does not accurately reflect the reality it is trying to model, there is no way for you to know that you are simulating something that is different from the design that will be ultimately manufactured. Functional correctness and accuracy of models is a big problem as errors cannot be proven *not* to exist.

Stimulus and Response

Simulation requires stimulus.

Simulators are not static tools. A static verification tool performs its task on a design without any additional information or action required by the user. For example, linting tools are static tools. Simulators, on the other hand, require that you provide a facsimile of the environment in which the design will find itself. This facsimile is called a testbench. Writing this testbench is the main objective of this textbook. The testbench needs to provide a representation of the inputs observed by the design, so the simulator can emulate the design's responses based on its description.

The simulation outputs are validated externally, against design intents.

The other thing that you must not forget is that simulators have no knowledge of your intentions. They cannot determine if a design being simulated is correct. Correctness is a value judgment on the outcome of a simulation that must be made by you, the verification engineer. Once the design is subjected to an approximation of the inputs from its environment, your primary responsibility is to examine the outputs produced by the simulation of the design's description and determine if that response is appropriate.

Event-Driven Simulation

Simulators are never fast enough.

Simulators are continuously faced with one intractable problem: They are never fast enough. They are attempting to emulate a physical world where electricity travels at the speed of light and millions of transistors switch over one billion times in a second. Simulators are implemented using general purpose computers that can execute, under ideal conditions, up to one billion sequential instructions per second. The speed advantage is unfairly and forever tipped in favor of the physical world.

Outputs change only when an input changes.

One way to optimize the performance of a simulator is to avoid simulating something that does not need to be simulated. Figure 2-1 shows a 2-input XOR gate. In the physical world, if the inputs do not change (Figure 2-1(a)), even though voltage is constantly applied to the output, current is continuously flowing through the transistors (in some technologies), and the atomic particles in the semiconductor are constantly moving, the *interpretation* of the output electrical state as a binary value (either a logic 1 or a logic 0) does not change. Only if one of the inputs change (as in Figure 2-1(b)), does the output change.

Figure 2-1.
Behavior of an
XOR gate

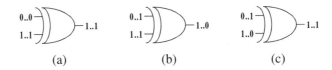

(a)   (b)   (c)

Change in values, called *events*, drive the simulation process.

Sample 2-10 shows a VHDL description (or model) of an XOR gate. The simulator could choose to execute this model continuously, producing the same output value if the input values did not change. An opportunity to improve upon that simulator's performance becomes obvious: do not execute the model while the inputs are constants. Phrased another way: Only execute a model when an input changes. The simulation is therefore driven by changes in inputs. If you define an input change as an *event*, you now have an *event-driven* simulator.

Sample 2-10.
VHDL model
for an XOR
gate

```
XOR_GATE: process (A, B)
begin
    if A = B then
        Z <= '0';
    else
        Z <= '1'
    end if;
end process XOR_GATE;
```

Sometimes, input changes do not cause the output to change.

But what if both inputs change, as in Figure 2-1(c)? In the logical world, the output does not change. What should an event-driven simulator do? For two reasons, the simulator should execute the description of the XOR gate. First, in the real world, the output of the XOR gate *does* change. The output might oscillate between 0 and 1 or remain in the "neither-0-nor-1" region for a few hundredths of picoseconds (see Figure 2-2). It just depends on how accurate you want your model to be. You could decide to model the XOR gate to include the small amount of time spent in the unknown (or x) state to more accurately reflect what happens when both inputs change at the same time.

Figure 2-2.
Behavior of an XOR gate when both inputs change

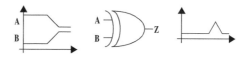

Descriptions between inputs and outputs are arbitrary.

The second reason is that the event-driven simulator does not know apriori that it is about to execute a model of an XOR gate. All the simulator knows is that it is about to execute a description of a 2-input, 1-output function. Figure 2-3 shows the view of the XOR gate from the simulator's perspective: a simple 2-input, 1-output black box. The black box could just as easily contain a 2-input AND gate (in which case the output might very well change if both inputs change), or a 1024-bit linear feedback shift register (LFSR).

Figure 2-3.
Event-driven simulator view of an XOR gate

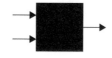

The mechanism of event-driven simulation introduces some limitations and interesting side effects that are discussed further in Chapter 4.

Acceleration options are often available in event-driven simulators

Simulation vendors are forever locked in a constant battle of beating the competition with an easier-to-use, faster simulator. It is possible to increase the performance of an event-driven simulator by simplifying some underlying assumptions in the design or in the simulation algorithm. For example, reducing delay values to identical unit delays or using two states (0 and 1) instead of four states (0, 1, x and z) are techniques used to speed-up simulation. You should refer to the documentation of your simulator to see what acceleration options are provided. It is also important to understand what are the consequences, in terms of reduced accuracy, of using these acceleration options.

Writing Testbenches: Functional Verification of HDL Models

Cycle-Based Simulation

Figure 2-4 shows the event-driven view of a synchronous circuit composed of a chain of three 2-input gates between two edge-triggered flip-flops. Assuming that Q1 holds a 0, Q2 holds a 1 and all other inputs remain constant, a rising edge on the clock input would cause an event-driven simulator to simulate the circuit as follows:

Figure 2-4.
Event-driven
simulator view
of a
synchronous
circuit

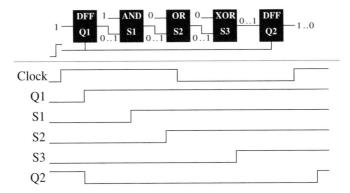

1. The event (rising edge) on the clock input causes the execution of the description of the flip-flop models, changing the output value of Q1 to 1 and of Q2 to 0, after a delay of 1 ns.

2. The event on Q1 causes the description of the AND gate to execute, changing the output S1 to 1, after a delay of 2 ns.

3. The event on S1 causes the description of the OR gate to execute, changing the output S2 to 1, after a delay of 1.5 ns.

4. The event on S2 causes the description of the XOR gate to execute, changing the output S3 to 1 after a delay of 3 ns.

5. The next rising edge on the clock causes the description of the flip-flops to execute, Q1 remains unchanged, and Q2 changes back to 1, after a delay of 1 ns.

Many intermediate events in synchronous circuits are not functionally relevant. To simulate the effect of a single clock cycle on this simple circuit required the generation of six events and the execution of seven models (some models were executed twice). If all we are interested in are the final states of Q1 and Q2, not of the intermediate combinatorial signals, then the simulation of this circuit could be optimized by acting only on the significant events for Q1 and Q2: the active edge of the clock. Phrased another way: Simulation is based on clock cycles. This is how cycle-based simulators operate.

The synchronous circuit in Figure 2-4 can be simulated in a cycle-based simulator using the following sequence:

Cycle-based simulators collapse combinatorial logic into equations.

1. When the circuit description is compiled, all combinatorial functions are collapsed into a single expression that can be used to determine all flip-flop input values based on the current state of the fan-in flip-flops.

For example, the combinatorial function between Q1 and Q2 would be compiled from the following initial description:

```
S1 = Q1 & '1'
S2 = S1 | '0'
S3 = S2 ^ '0'
```

into this final single expression:

```
S3 = Q1
```

The cycle-based simulation view of the compiled circuit is shown in Figure 2-5.

Figure 2-5.
Cycle-based
simulator view
of a
synchronous
circuit

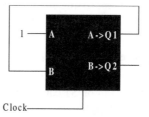

2. During simulation, whenever the clock input rises, the value of all flip-flops are updated using the input value returned by the pre-compiled combinatorial input functions.

The simulation of the same circuit, using a cycle-based simulator, required the generation of two events and the execution of a single model. The number of logic computations performed is the same in both cases. They would have been performed whether the "A" input changed or not. As long as the time required to perform logic computation is smaller than the time required to schedule intermediate events,[3] and there are many registers changing state at every clock cycle, cycle-based simulation will offer greater performance.

Writing Testbenches: Functional Verification of HDL Models

Cycle-based simulations have no timing information.	This great improvement in simulation performance comes at a cost: All timing and delay information is lost. Cycle-based simulators assume that the entire design meets the setup and hold requirements of all the flip-flops. When using a cycle-based simulator, timing is usually verified using a static timing analyzer.
Cycle-based simulators can only handle synchronous circuits.	Cycle-based simulators further assume that the active clock edge is the only significant event in changing the state of the design. All other inputs are assumed to be perfectly synchronous with the active clock edge. Therefore, cycle-based simulators can only simulate perfectly synchronous designs. Anything containing asynchronous inputs, latches or multiple-clock domains *cannot* be simulated accurately. Fortunately, the same restrictions apply to static timing analysis. Thus, circuits that are suitable for cycle-based simulation to verify the functionality are suitable for static timing verification to verify the timing.

Co-Simulators

	No real-world design and testbench is perfectly suited for a single simulator, simulation algorithm or modeling language. Different components in a design may be specified using different languages. A design could contain small sections that cannot be simulated using a cycle-based algorithm. Testbenches may (and should) be written using an HVL while the design is written in VHDL or Verilog.
Multiple simulators can handle separate portions of a simulation.	To handle the portions of a design that do not meet the requirements for cycle-based simulation, most cycle-based simulators are integrated with an event-driven simulator. As shown in Figure 2-6, the synchronous portion of the design is simulated using the cycle-based algorithm, while the remainder of the design is simulated using a conventional event-driven simulator. Both simulators

---

3. And they are. By a long shot.

(event-driven and cycle-based) are running together, cooperating to simulate the entire design.

Figure 2-6.
Event-driven
and cycle-
based co-
simulation

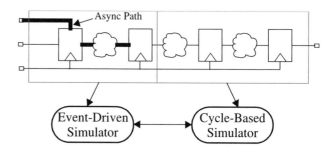

Other popular co-simulation environments provide VHDL and Verilog, HDL and HVL or digital and analog co-simulation. For example, Figure 2-7 shows the testbench (written in *e*) and a design co-simulated using Specman Elite and a HDL simulator.

Figure 2-7.
HVL and
HDL co-
simulation

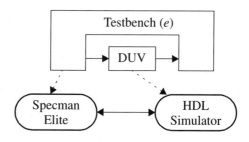

All simulators operate in locked-step.

During co-simulation, all simulators involved progress along the time axis in lock-step. All are at simulation time T_1 at the same time and reach the next time T_2 at the same time. This implies that the speed of a co-simulation environment is limited by the slowest simulator. Some experimental co-simulation environments implement *time warp* synchronization where some simulators are allowed to move ahead of the others.

Performance is decreased by the communication and synchronization overhead.

The biggest hurdle of co-simulation comes from the communication overhead between the simulators. Whenever a signal generated within a simulator is required as an input by another, the current value of that signal, as well as the timing information of any change in that value, must be communicated. This communication usually involves a translation of the event from one simulator into an (almost) equivalent event in another simulator. Ambiguities can arise during that translation when each simulation has different semantics. The difference in semantics is usually present: the semantic difference often being the requirement for co-simulation in the first place.

Translating values and events from one simulator to another can create ambiguities.

Examples of translation ambiguities abound. How do you map Verilog's 128 possible states (composed of orthogonal logic values and strengths) into VHDL's nine logic values (where logic values and strengths are combined)? How do you translate a voltage and current value in an analog simulator into a logic value and strength in a digital simulator? How do you translate an x or z value into a 2-state *e* value? How do you translate the timing of zero-delay events from Verilog (which has no strict concept of delta cycles)[4] to VHDL?

Co-simulation should not be confused with single-kernel simulation.

Co-simulation is when two (or more) simulators are cooperating to simulate a design, each simulating a portion of the design, as shown in Figure 2-8. It should not be confused with simulators able to read and compile models described in different languages. For example, Cadence's *NCSIM* simulator and Model Technology's *ModelSim* simulator can both simulate a design described using a mix of VHDL and Verilog. Synopsys's VCS simulator can simulate Verilog and a subset of OpenVera. As shown in Figure 2-9, all languages are compiled into a single internal representation or machine code and the simulation is performed using a single simulation engine.

Figure 2-8.
Co-simulator

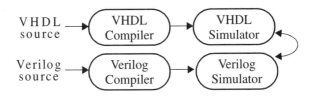

---

4. See "The Simulation Cycle" on page 194 for more details on delta cycles.

Figure 2-9.
Mixed-
language
simulator

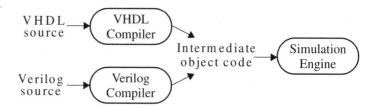

VERIFICATION INTELLECTUAL PROPERTY

You can buy IP for standard functions.

If you want to verify your design, it is necessary to have models for all the parts included in a simulation. The model of the RTL design is a natural by-product of the design exercise and the actual objective of the simulation. Models for embedded or external RAMs are also required, as well as models for standard interfaces and off-the-shelf parts. If you were able to procure the RAM, design IP, specification or standard part from a third party, you should be able to obtain a model for it as well. You may have to obtain the model from a different vendor than the one who supplies the physical part.

It is cheaper to buy models than write them yourself.

At first glance, buying a simulation model from a third-party provider may seem expensive. Many have decided to write their own models to save on licensing costs. However, you have to decide if this endeavor is truly economically fruitful: Are you in the modeling business or in the chip design business? If you have a shortage of qualified engineers, why spend critical resources on writing a model that does not embody any competitive advantage for your company? If it was not worth designing on your own in the first place, why is writing your own model suddenly justified?

Your model is not as reliable as the one you buy.

Secondly, the model you write has never been used before. Its quality is much lower than a model that has been used by several other companies before you. The value of a functionally correct and reliable model is far greater than an uncertain one. Writing and verifying a model to the *same degree of confidence* as the third-party model is always more expensive than licensing it. And be assured: No matter how simple the model is (such as a quad 2-input NAND gate, 74LS00), you'll get it wrong the first time. If not functionally, then at least with respect to timing or connectivity.

There are several providers of verification IP. Many are written using an HVL or C code; others are provided as non-synthesizeable

VHDL or Verilog source code. For intellectual property protection and licensing technicalities, most are provided as compiled binary models. Verification IP includes, but is not limited to functional models of external and embedded memories, bus-functional models for standard interfaces, protocol generators and analyzers, assertion sets for standard protocols and black-box models for off-the-shelf components and processors.

Hardware Modelers

What if you cannot find a model to buy?

You may be faced with procuring a model for a device that is so new or so complex, that no provider has had time to develop a reliable model for it. For example, at the time the first edition of this book was written, you could license full-functional models for the Pentium processor from at least two vendors. However, you could not find a model for the Pentium III. If you want to verify that your new PC board, which uses the latest Intel microprocessor, is functionally correct before you build it, you have to find some other way to include a simulation model of the processor.

You can "plug" a chip into a simulator.

Hardware modelers provide a solution for that situation. A hardware modeler is a small box that connects to your network. A *real, physical* chip that needs to be simulated is plugged into it. During simulation, the hardware modeler communicates with your simulator (through a special interface package) to supply inputs from the simulator to the device, then sends the sampled output values from the device back to the simulation. Figure 2-10 illustrates this communication process.

Figure 2-10.
Interfacing a hardware modeler and a simulator

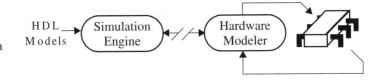

Timing of I/O signals still needs to be modeled.

Using a hardware modeler is not a trivial task. Often, an adaptor board must be built to fit the device onto the socket on the modeler itself. Also, the modeler cannot perform timing checks on the device's inputs nor accurately reflect the output delays. A timing shell performing those checks and delays must be written to more accurately model a device using a hardware modeler.

Hardware model-
ers offer better
simulation perfor-
mance.

Hardware modelers are also very useful when simulating a model
of the part at the required level of abstraction. A full-functional
model of a modern processor that can fetch, decode and execute
instructions could not realistically execute more than 10 to 50
instructions within an acceptable time period. The real physical
device can perform the same task in a few milliseconds. Using a
hardware modeler can greatly speed up board- and system-level
simulation.

WAVEFORM VIEWERS

Waveform view-
ers display the
changes in signal
values over time.

Waveform viewers are the most common verification tools used in
conjunction with simulators. They let you visualize the transitions
of multiple signals over time, and their relationship with other transitions. With such a tool, you can zoom in and out over particular
time sequences, measure time differences between two transitions,
or display a collection of bits as bit strings, hexadecimal or as symbolic values. Figure 2-11 shows a typical display of a waveform
viewer showing the inputs and outputs of a 4-bit synchronous
counter.

Figure 2-11.
Hypothetical
waveform
view of a 4-bit
synchronous
counter

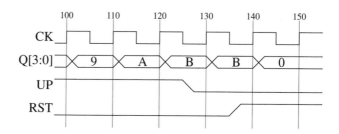

Waveform view-
ers are used to
debug simulations.

Waveform viewers are indispensable during the authoring phase of
a design or a testbench. With a viewer you can casually inspect that
the behavior of the code is as expected. They are needed to diagnose, in an efficient fashion, why and when problems occur in the
design or testbench. They can be used interactively during the simulation, but more importantly offline, after the simulation has completed. As shown in Figure 2-12, a waveform viewer can play back

the events that occurred during the simulation that were recorded in some trace file.

Figure 2-12.
Waveform viewing as post-processing

Recording waveform trace data decreases simulation performance.

Viewing waveforms as a post-processing step lets you quickly browse through a simulation that can take hours to run. However, keep in mind that recording trace information significantly reduces the performance of the simulator. The quantity and scope of the signals whose transitions are traced, as well as the duration of the trace, should be limited as much as possible. Of course, you have to trade-off the cost of tracing a greater quantity or scope of signals versus the cost of running the simulation over again to get a trace of additional signals that turn out to be required to completely diagnose the problem. If it is likely or known that bugs will be reported, such as the beginning of the project or during a debugging iteration, trace all the signals required to diagnose the problem. If no errors are expected, such as during regression runs, no signal should be traced.

Do not use a waveform viewer to determine if a design passes or fails.

In a functional verification environment, using a waveform viewer to determine the correctness of a design involves interpreting the dozens (if not hundreds) of wavy lines on a computer screen against some expectation. It can be an acceptable verification method used two or three times, for less than a dozen signals. As the number of signals and transitions increases, so does the number of relationships that must be checked for correctness. Multiply that by the duration of the simulation. Multiply again by the number of simulation runs. Very soon, the probability that a functional error is missed reaches one.

Some viewers can compare sets of waveforms.

Some waveform viewers can compare two sets of waveforms. One set is presumed to be a golden reference, while the other is verified for any discrepancy. The comparator visually flags or highlights any differences found. This approach has two significant problems.

How do you define a set of waveforms as "golden"?

First, how is the golden reference waveform set declared "golden"? If visual inspection is required, the probability of missing a significant functional error remains equal to one in most cases. The only time golden waveforms are truly available is in a redesign exercise, where cycle-accurate backward compatibility must be maintained. However, there are very few of these designs. Most redesign exercises take advantage of the process to introduce needed modifications or enhancements, thus tarnishing the status of the golden waveforms.

And are the differences really significant?

Second, waveforms are at the wrong level of abstraction to compare simulation results against design intent. Differences from the golden waveforms may not be significant. The value of all output signals is not significant all the time. Sometimes, what is significant is the relative relationships between the transitions, not their absolute position. The new waveforms may be simply shifted by a few clock cycles compared to the reference waveforms, but remain functionally correct. Yet, the comparator identifies this situation as a mismatch.

CODE COVERAGE

Did you forget to verify some function in your code?

Code coverage is a tool that can identify what code has been (and more importantly *not* been) executed in the design under verification. It is a methodology that has been in use in software engineering for quite some time. The problem with false positive answers (i.e., a bad design is thought to be good), is that they look identical to a true positive answer. It is impossible to know, with 100 percent certainty, that the design being verified is indeed functionally correct. All of your testbenches simulate successfully, but are there sections of the RTL code that you did not exercise and therefore not triggered a functional error? That is the question that code coverage can help answer.

Code must first be instrumented.

Figure 2-13 shows how a code coverage tool works. The source code is first *instrumented*. The instrumentation process simply adds checkpoints at strategic locations of the source code to record whether a particular construct has been exercised. The instrumentation method varies from tool to tool. Some may use file I/O features available in the language (i.e., use *$write* statements in Verilog or *textio.write* procedure calls in VHDL). Others may use special features built into the simulator.

Figure 2-13.
Code coverage
process

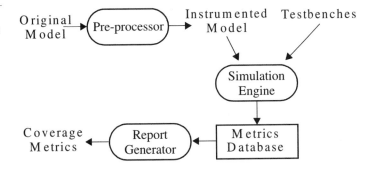

No need to instrument the testbenches.

Only the code for the design under verification is instrumented. The objective is to determine if you have forgotten to exercise some code in the design. The code for the testbenches need not be traced to confirm that it has executed. If a significant section of a testbench was not executed, it should be reflected in some portion of the design not being exercised. Furthermore, a significant portion of the testbench code is executed only if an error is detected. Code coverage metrics on testbench code are therefore of little interest.

Trace information is collected at runtime.

The instrumented code is then simulated normally using all available, uninstrumented, testbenches. The cumulative traces from all simulations are collected into a database. From that database, reports can be generated to measure various coverage metrics of the verification suite on the design.

Statement and block coverage are the same thing.

The most popular metrics are statement, path and expression coverage. Statement coverage can also be called block coverage, where a block is a sequence of statements that are executed if a single statement is executed. The code in Sample 2-11 shows an example of a statement block. The block named *acked* is executed entirely whenever the expression in the *if* statement evaluates to TRUE. So counting the execution of that block is equivalent to counting the execution of the four individual statements within that block.

Sample 2-11.
Block vs.
statement execution

```
if (dtack == 1'b1) begin: acked
    as    <= 1'b0;
    data  <= 16'hZZZZ;
    bus_rq <= 1'b0;
    state <= IDLE;
end
```

But block boundaries may not be that obvious.

Statement blocks may not be necessarily clearly delimited. In Sample 2-12, two statements blocks are found: one before (and including) the *wait* statement, and one after. The *wait* statement may have never completed and the process was waiting forever. The subsequent sequential statements may not have executed. Thus, they form a separate statement block.

Sample 2-12.
Blocks separated by a *wait* statement

```
address <= 16#FFED#;
ale    <= '1';
rw     <= '1';
wait until dtack = '1';
read_data := data;
ale    <= '0';
```

Statement Coverage

Did you execute all the statements?

Statement, line or block coverage measures how much of the total lines of code were executed by the verification suite. A graphical user interface usually lets the user browse the source code and quickly identify the statements that were not executed. Figure 2-14 shows, in a graphical fashion, a statement coverage report for a small portion of code from a model of a modem. The actual form of the report from any code coverage tool or source code browser will likely be different.

Figure 2-14.
Example of
statement
coverage

```
☑if (parity == ODD || parity == EVEN) begin
☐   tx <= compute_parity(data, parity);
☐   #(tx_time);
    end
☑tx <= 1'b0;
☑#(tx_time);
☑if (stop_bits == 2) begin
☑   tx <= 1'b0;
☑   #(tx_time);
    end
```

Why did you not execute all statements?

The example in Figure 2-14 shows that two out of the eight executable statements—or 25%—were not executed. To bring the statement coverage metric up to 100%, a desirable goal[5], it is necessary to understand what conditions are required to cause the execution of the uncovered statements. In this case, the parity must be set to either ODD or EVEN. Once the conditions have been determined, you must understand why they never occurred in the first place. Is it a condition that can never occur? Is it a condition that should have been verified by the existing verification suite? Or is it a condition that was forgotten?

It is normal for some statements not to be executed.

If it is a condition that can never occur, the code in question is effectively dead: It will never be executed. Removing that code is a definite option; it reduces clutter and increases the maintainability of the source code. However, a good defensive (and paranoid) coder often includes code that is not meant to be executed. This additional code simply monitors for conditions that should never occur and reports that an unexpected condition happened should the hypothesis prove false. This practice is very effective (see "Assertions" on page 64). Functional problems are positively identified near the source of the malfunction, without having to rely on the

---

5. But not necessarily achievable. For example, the *default* clause in a fully specified VHDL *case* statement should never be executed.

possibility that it produces an unexpected response at the right moment when you were looking for something else.

Sample 2-13.
Defensive pro-
gramming
technique

```
case (mode[1:0]) // synopsys full_case
2'b00: ...
2'b10: ...
2'b01: ...
// synopsys translate_off
// coverage off
default: $write("Case was not really full!\n");
// coverage on
// synopsys translate_on
endcase
```

Your model can tell you if things are not as assumed.

Sample 2-13 shows an example of defensive modeling in synthesizeable *case* statements. Even though there is a directive instructing the synthesis tool that the *case* statement describes all possible conditions, it is possible for an unexpected condition to occur during simulation. If that were the case, the simulation results would differ from the results produced by the hardware implementation, and that difference would go undetected until a gate-level simulation is performed, or the device failed in the system.

Do not measure coverage for code not meant to be executed.

It should be possible to identify code that was not meant to be executed and have it eliminated from the code coverage statistics. In Sample 2-13, significant comments are used to remove the defensive coding statements from being measured by our hypothetical code coverage tool. Some code coverage tools may be configured to ignore any statement found between synthesis translation on/off directives. It may be more interesting to configure a code coverage tool to ensure that code included between synthesis translate on/off directives is indeed *not* executed!

Add testcases to execute all statements.

If the conditions that would cause the uncovered statements to be executed should have been verified, it is an indication that one or more testbenches are either not functionally correct or incomplete. If the condition was entirely forgotten, it is necessary to add to an existing testbench, create an entirely new one or make additional runs with different seeds.

Path Coverage

There is more than one way to execute a sequence of statements.

Path coverage measures all possible ways you can execute a sequence of statements. The code in Sample 2-14 has four possible paths: the first *if* statement can be either true or false. So can the second. To verify all paths through this simple code section, it is necessary to execute it with all possible state combinations for both *if* statements: false-false, false-true, true-false, and true-true.

Sample 2-14.
Example of statement and path coverage

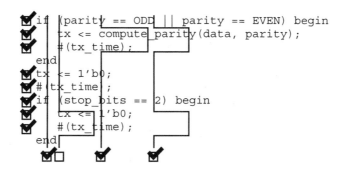

```
if (parity == ODD || parity == EVEN) begin
   tx <= compute_parity(data, parity);
   #(tx_time);
end
tx <= 1'b0;
#(tx_time);
if (stop_bits == 2) begin
   tx <= 1'b0;
   #(tx_time);
end
```

Why were some sequences not executed?

The current verification suite, although it offers 100% statement coverage, only offers 75% path coverage through this small code section. Again, it is necessary to determine the conditions that cause the uncovered path to be executed. In this case, a testcase must set the parity to neither ODD nor EVEN and the number of stop bits to two. Again, the important question one must ask is whether this is a condition that will ever happen, or if it is a condition that was overlooked.

Limit the length of statement sequences.

The number of paths in a sequence of statements grows exponentially with the number of control-flow statements. Code coverage tools give up measuring path coverage if their number is too large in a given code sequence. To avoid this situation, keep all sequential code constructs (in Verilog: *always* and *initial* blocks, *tasks* and *functions*; in VHDL: *processes*, *procedures* and *functions*) to under 100 lines.

Reaching 100% path coverage is very difficult.

Expression Coverage

There may be more than one cause for a control-flow change.

If you look closely at the code in Sample 2-15, you notice that there are two mutually independent conditions that can cause the first *if* statement to branch the execution into its *then* clause: parity being set to either ODD or EVEN. Expression coverage, as shown in Sample 2-15, measures the various ways paths through the code are executed. Even if the statement coverage is at 100%, the expression coverage is only at 50%.

Sample 2-15. Example of statement and expression coverage

```
if (parity == ODD || parity == EVEN) begin
    tx <= compute_parity(data, parity);
    #(tx_time);
end
tx <= 1'b0;
#(tx_time);
if (stop_bits == 2) begin
    tx <= 1'b0;
    #(tx_time);
end
```

Once more, it is necessary to understand why a controlling term of an expression has not been exercised. In this case, no testbench sets the parity to EVEN. Is it a condition that will never occur? Or was it another oversight?

Reaching 100% expression coverage is extremely difficult.

FSM Coverage

Statement coverage detects unvisited states.

Because each state in an FSM is usually explicitly coded using a choice in a *case* statement, any unvisited state will be clearly identifiable through uncovered statements. The state corresponding to an uncovered *case* statement choice was not visited during verification.

FSM coverage identifies state transitions.

Figure 2-15 shows a bubble diagram for an FSM. Although is has only five states, it has significantly more possible transitions: 14 possible transitions exist between adjoining states. State coverage of 100% can be easily reached through the sequence Reset, A, B, D, then C. However, this would yield only 36% transition coverage. To completely verify the implementation of this FSM, it is necessary to ensure the design operates according to expectation for all transitions.

Figure 2-15.
Example FSM
bubble
diagram

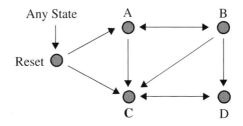

What about
unspecified states?

The FSM illustrated in Figure 2-15 only shows five specified states. Once synthesized into hardware, a 3-bit state register will be necessary (maybe more if a different state encoding scheme, such as one-hot, is used). This leaves three possible state values that were not specified. What if some cosmic rays zap the design into one of these unspecified states? Will the correctness of the design be affected? Logic optimization may yield decode logic that creates an island of transitions within those three unspecified states, never letting the design recover into specified behavior unless reset is applied. The issues of design safety and reliability and techniques for ensuring them are beyond the scope of this book. But it is the role of a verification engineer to ask those questions.

What Does 100% Code Coverage Mean?

Completeness
does not imply
correctness.

The short answer is: Everything you wrote was executed. Code coverage indicates how *thoroughly* your entire verification suite exercises the source code. But it does not provide an indication, in any way, about the *correctness* or *completeness* of the verification suite. Figure 2-16 shows the reconvergence model for automatically extracted code coverage metrics. It clearly shows that it does not help verify design intent, only that the RTL code, correct or not, was fully exercised.

Figure 2-16. Reconvergent paths in automated code coverage

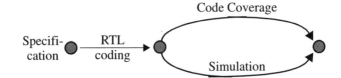

Results from code coverage tools should be interpreted with a grain of salt. They should be used to help identify corner cases that were not exercised by the verification suite or implementation-dependent features that were introduced during the implementation. You should also determine if the uncovered cases are relevant and deserve additional attention, or a consequence of the mindlessness of the coverage tool.

Code coverage lets you know if you are not done.

Code coverage indicates if the verification task is *not* complete through low coverage numbers. A high coverage number is by no means an indication that the job is over. Code coverage is an additional indicator for the completeness of the verification job. It can help increase your confidence that the verification job is complete, but it should not be your only indicator.

Code coverage tools can be used as profilers.

When developing models for simulation only, where performance is an important criteria, code coverage tools can be used for *profiling*. The aim of profiling is the opposite of code coverage. The aim of profiling is to identify the lines of codes that are executed most often. These lines of code become the primary candidates for performance optimization efforts.

FUNCTIONAL COVERAGE

Did you forget to verify some condition?

Functional coverage is another tool to help ensure that a bad design is not hiding behind passing testbenches. Although this methodology has been in use at some companies for quite some time, it is a recent addition to the arsenal of general-purpose verification tools. Functional coverage records relevant metrics (e.g., packet length, instruction opcode, buffer occupancy level) to ensure that the verification process has exercised the design through all of the interesting values. Whereas code coverage measures how much of the implementation has been exercised, functional coverage measures how much of the original design specification has been exercised.

It complements code coverage.

High functional coverage does not necessarily correlate with high code coverage. Whereas code coverage is concerned with recording the mechanics of code execution, functional coverage is concerned with the intent or purpose of the implemented function. For example, the decoding of a CPU instruction may involve separate *case* statements for each field in the opcode. Each *case* statement may be 100% code-covered due to combinations of field values from previously decoded opcodes. However, the particular combination involved in decoding a specific CPU instruction may not have been exercised.

It will detect errors of omission.

Sample 2-16 shows a *case* statement decoding a CPU instruction. Notice how the decoding of the RTS instruction is missing. If I relied solely on code coverage, I would be lulled in a false sense of completeness by having 100% coverage of this code. For code coverage to report a gap, the unexercised code must a priori exist. Functional coverage does not rely on actual code. It will report gaps in the recorded values whether the code to process them is there or not.

Sample 2-16. Example of coding error undetectable by code coverage

```
type OPCODE_TYP is [ADD, SUB, JMP, RTS, NOP];
...
case (OPCODE) is
when ADD => ...
when SUB => ...
when JMP => ...
when others => ...
end case;
```

It must be manually defined.

Code coverage tools were quickly adopted into verification processes because of their low adoption cost. They require very little additional action from the user: At most, execute one more command before compiling your code. Functional coverage, because it is a measure of values deemed to be interesting and relevant, must be manually specified. Since relevance and interest are qualities that are extracted from the intent of the design, functional coverage is not something that can be automatically extracted from the RTL source code. Your functional coverage metrics will be only as good as what you implement.

Metrics are collected at runtime and graded.

Like code coverage, functional coverage metrics are collected at runtime, during a simulation run. The values from individual runs are collected into a database or separate files. The functional coverage metrics from these separate runs are then merged for offline analysis. The marginal coverage of individual runs can then be graded to identify which runs contributed the most toward the overall functional coverage goal. These runs are then given preference in the regression suite, while pruning runs that did not significantly contribute to the objective.

Coverage data can be used at runtime.

Functional coverage tools usually provide a set of runtime procedures that let a testbench dynamically query a particular functional coverage metric. The testbench can then use the information to modify its current behavior. For example, it could increase the probability of generating values that have not been covered yet. It could decide to abort the simulation should the functional coverage not have significantly increased since the last query.

Although touted as a powerful mechanism by vendors, it is no silver bullet. Implementing the dynamic feedback mechanism is not easy: You have to correlate your stimulus generation process with the functional coverage metric, and ensure that one will cause the other to converge toward the goal. Dynamic feedback works best when there is a direct correlation between the input and the measured coverage, such as instruction types. It may be more efficient to achieve your goal with three or four runs of a simpler testbench without dynamic feedback than with a single run of a much more complex testbench.

Item Coverage

Did I generate all interesting and relevant values?

Item coverage is the recording of individual scalar values. It is a basic function of all functional coverage tools. The objective of the coverage metric is to ensure that all interesting and relevant values have been observed going to, coming out of, or in the design. Examples of item coverage include, but are not limited to, packet length, instruction opcode, interrupt level, bus transaction termination status, buffer occupancy level, bus request patterns and so on.

Define what to sample.

It is extremely easy to record functional coverage and be inundated with vast amounts of coverage data. But data is not the same thing as information. You must restrict coverage to only (but all!) values that will indicate how thoroughly your design has been verified. For example, measuring the value of the read and write pointers in a FIFO is fine if you are concerned about the full utilization of the buffer space and wrapping around of the pointer values. But if you are interested in the FIFO occupancy level (Was it ever empty? Was it ever full? Did it overflow?), you should measure and record the *difference* between the pointer values.

Define where to sample it.

Next, you must decide where in your testbench or design is the measured value accurate and relevant. For example, you can sample the opcode of an instruction at several places: at the output of the code generator, at the interface of the program memory, in the decoder register or in the execution pipeline. You have to ensure that a value, once recorded, is indeed processed or committed as implied by the coverage metric.

For example, if you are measuring opcodes that were executed, they should be sampled in the execution unit. Sampling them in the decode unit could result in false samples when the decode pipeline is flushed on branches or exceptions. Similarly, sampling the length of packets at the output of the generator may yield false samples: If a packet is corrupted by injecting an error during its transmission to the design in lower-level functions of the testbench, it may be dropped.

Define when to sample it.

Values are sampled at some point in time during the simulation. It could be at every clock cycle, whenever the address strobe signal is asserted, every time a request is made or after randomly generating a new value. You must carefully chose your sampling time. Over-sampling will decrease simulation performance and consume database resources without contributing additional information.

The sampled data must also be stable so race conditions must be avoided between the sampled data and the sampling event (see "Read/Write Race Conditions" on page 209). To reduce the probability that a transient value is being sampled, functional coverage tools may delay the sampling of values to the end of their simulation cycle, before time is about to advance (see "The Simulation Cycle" on page 194 and "The Co-Simulation Cycle" on page 196).

High-level testbench functions usually operate on different values in zero-time within the same simulation cycle. If the functional coverage tool delays its sampling to the end of the simulation cycle, it will not be possible to sample all of the intermediate values. Only the last value will be recorded, resulting in under-sampling.

Define *why* we cover it.

If functional coverage is supposed to measure interesting and relevant values, it is necessary to define what makes those values so interesting and relevant. For example, measuring the functional coverage of a 32-bit address will yield over 4 billion "interesting and relevant" values. Not all values are created equal—but most are. Values may be numerically different but functionally equivalent. By identifying those functionally equivalent values into a single set, you can reduce the number of interesting and relevant values to a more manageable size. For example, based on the decoder architecture, addresses 0x00000001 through 0x7FFFFFFF and addresses 0x80000000 through 0x8FFFFFFE are functionally equivalent, reducing the number of relevant and interesting values to 4 sets (min, 1 to mid, mid to max-1, max).

It can detect invalid values.

If you can define sets of equivalent values, it is possible to define sets of invalid or unexpected values. Functional coverage can be used as an error detecting tool, just like an *if* statement in your testbench code. However, you should not rely on functional coverage to detect invalid values. Functional coverage is an optional runtime tool that may not be turned on at all times. If functional coverage is not enabled to improve simulation performance and if a value is defined as invalid in only the functional coverage, then an invalid value may go undetected.

It can report holes.

The ultimate purpose of functional coverage is to identify what remains to be done. During analysis, the functional coverage tool can compare the number of value sets that contain at least one sample against the total number of value sets. Any value set that does not contain at least one sample is a hole in your functional coverage. By enumerating the empty value sets, you can focus on the holes in your test cases and complete your verification sooner rather than continue to exercise functionality that has already been verified.

For this enumeration to be possible, the total number of value sets must be relatively small. For example, it is practically impossible to fill the coverage for a 32-bit value without broad value sets. The number of holes will be likely in the millions, making enumeration impossible. You should strive to limit the number of possible value sets as much as possible. For example, Specman Elite has a default limit of 16 value sets for hole enumeration.

Cross Coverage

Did I generate all interesting combination of values?

Whereas item coverage is concerned with individual scalar values, cross coverage measures the presence or occurrence of combinations of values. It helps answer questions like, "Did I inject a corrupted packet on all ports?" "Did we execute all combinations of opcodes and operand modes?" and "Did this state machine visit each state while that buffer was full, not empty and empty?" Cross coverage can involve more than two scalar items. However, the number of possible value sets grows factorially with the number of crossed items.

Similar to item coverage.

Mechanically, cross coverage is identical to item coverage. Specific values are sampled at specific locations at specific points in time with specific value sets. The only difference is that two or more values are sampled instead of one.

Can be done in post-processing step.

It may be possible to perform offline cross coverage by post-processing the item coverage metrics. In some tools, it is the only cross coverage mechanism available. To enable offline cross-coverage analysis, it is necessary to sample the simulation time along with each individual value sample. The recording of additional information such as simulation time increases the size of the coverage data and reduces runtime performance. Most tools make gathering additional cross coverage information optional, and this feature is usually turned off by default. As shown in Sample 2-17, the cross coverage report is generated by identifying the values that were sampled at the same simulation time.

Sample 2-17. Example of offline cross-coverage analysis

Packet Length	Packet Valid	Length x Valid		
			good	bad
short @10	good@10	short	X	
long @20	bad @20	medium	X	
medium@30	good@30	long		X

Offline cross-coverage reports may yield false samples.

As long as each item value is sampled at different points in time, offline cross-coverage analysis works fine. When recording values located inside an RTL design or a bus-functional model sampled at clock edges, there can be only one value per simulation time. However, covering values in higher-level testbench functions, which operate in zero-time, may result in multiple values sampled at the same simulation time. Sample 2-18 shows the same item coverage as before, but sampled when all of the packets were generated at the same time. This approach yields an incorrect offline cross-coverage report. The same cross-coverage measure, if collected at runtime, would yield a correct report.

Sample 2-18. Example of invalid offline cross-coverage analysis

Packet Length	Packet Valid	Length x Valid		
			good	bad
short @10	good@10	short	X	X
long @10	bad @10	medium	X	X
medium@10	good@10	long	X	X

Transition Coverage

Did I generate all interesting *sequences* of values?

Whereas cross coverage is concerned with combination scalar values at the same point in time, transition coverage measures the presence or occurrence of sequences of values. Transition coverage helps answer questions like, "Did I perform all combinations of back-to-back read and write cycles?" "Did we execute all combinations of arithmetic opcodes followed by test opcodes?" and "Did this state machine traverse all significant paths?" Transition coverage can involve more than two consecutive values of the same scalar item. However, the number of possible value sets grows factorially with the number of transition states.

Similar to item coverage.
Mechanically, transition coverage is identical to item coverage. Specific values are sampled at specific locations at specific points in time with specific value sets. The only difference is that a sample is said to have occurred in a value set after two or more consecutive item samples instead of one. The other difference is that transition can overlap, hence two transition samples may be composed of the same item sample.

Similar to FSM path coverage.
Conceptually, transition coverage is identical to FSM path coverage (see "FSM Coverage" on page 52). Both record the consecutive values at a particular location of the design (for example, a state register), and both compare against the possible set of paths. But unlike FSM coverage tools, which are limited to state registers in RTL code, transition coverage can be applied to any sampled value in testbenches and behavioral models.

Transition coverage reflects intent.
Because transition coverage is (today) manually specified from the intent of the design or the implementation, it provides a true independent path to verifying the correctness of the design and the completeness of the verification. It can detect invalid transitions as well as specify transitions that may be missing from the implementation of the design.

What Does 100% Functional Coverage Mean?

It indicates completeness, not correctness.
Functional coverage indicates which interesting and relevant conditions were verified. It provides an indication of the *thoroughness* of the implementation of the verification plan. Unless some value sets are defined as invalid, it cannot provide an indication, in any way, about the *correctness* of those conditions or of the design's response to those conditions. Functional coverage metrics are only as good as the functional coverage model you have defined. Coverage of 100% means that you've covered all of the coverage points you included in the simulation. It makes no statement about the *completeness* of your functional coverage model.

Results from functional coverage tools should also be interpreted with a grain of salt. Since they are generated by additional testbench code, they have to be debugged and verified for correctness before being trusted. They will help identify additional interesting conditions that were not included in the verification plan.

If a metric is not interesting, don't measure it.

It is extremely easy to define functional coverage metrics and generate many reports. If coverage is not measured according to a specific purpose, you will soon drown under megabytes of functional coverage reports. And few of them will ever be close to 100%. It will also become impossible to determine which report is significant or what is the significance of the holes in others. The verification plan (see the next chapter) should serve as the functional specification for the coverage models, as well as for the rest of the verification environment. If a report is not interesting or meaningful to look at, if you are not eager to look at a report after a simulation run, then you should question its existence.

Functional coverage lets you know if you are done.

When used properly, functional coverage becomes a formal specification of the verification plan. Once you reach 100% functional coverage, it indicates that you have created and exercised all of the relevant and interesting conditions you originally identified. It confirms that you have implemented everything in the verification plan. However, it does not provide any indication of the completeness of the verification plan itself or the correctness of the design under such conditions.

VERIFICATION LANGUAGES

VHDL and Verilog are simulation languages, not verification languages.

Verilog was designed with a focus on describing low-level hardware structures. Verilog-2001 only recently introduced support for basic high-level data structures. VHDL was designed for very large design teams. It strongly encapsulates all information and communicates strictly through well-defined interfaces. Very often, these limitations get in the way of an efficient implementation of a verification strategy. VHDL and Verilog also lack features important in efficiently implementing a modern verification process.

Verification languages can raise the level of abstraction.

As mentioned in Chapter 1, one way to increase productivity is to raise the level of abstraction used to perform a task. High-level languages, such as C or Pascal, raised the level of abstraction from assembly-level, enabling software engineers to become more productive. Similarly, computer languages specifically designed for verification are able to raise the level of abstraction compared to general-purpose simulation languages. Hardware verification languages maintain important concepts necessary to interact with hardware: time, concurrency and instantiation. They also offer features that help raise the level of abstraction: complex data types, object-orientedness with inheritance and temporal assertions.

Verification language can automate verification.

If the main benefit of a hardware verification language was the higher level of abstraction and object-orientedness, then $C++$ would have long been identified as the best solution[6]: It is free and widely known. The main benefit of HVLs, as shown in Figure 2-17, is their capability of automating a portion of the verification by randomly generating stimulus, collecting functional coverage to identify holes then the ability to add constraints easily to create more stimulus targeted to fill those holes. To support this productivity cycle, some HVLs offer constrainable random generation, functional coverage measurement and a code extension mechanism.

Figure 2-17.
HVL productivity cycle

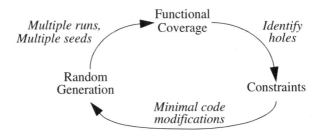

Several verification languages exist.

At the time of this writing, commercial verification language solutions include *e* from Verisity, *OpenVera* from Synopsys and *RAVE* from Forte. Open-source or public-domain solutions are available: the *SystemC Verification Library*[7] from Cadence and *Jeda* from Juniper Networks. Accelera is working on introducing HVL functionality in *SystemVerilog*. There is also a plethora of home-grown proprietary solutions based on $C++$, Perl or TCL.

The definition of what makes a language an HVL is still nebulous. Of the languages mentioned, most do not include all of the features identified in the productivity cycle. The large number of verification language solutions confirms that the industry recognizes the

6. $C++$ still lacks a native concept of time, concurrency and instantiation.

7. Formerly known as *TestBuilder.*

limitations of Verilog and VHDL for verification. It is also a classic characteristic of a young market, before coalescing around one or two market leaders and de-facto standards.

You must learn the basics of verification before learning verification languages.

In this book, I use VHDL and Verilog as the first implementation medium for the basic components of the verification infrastructure and to introduce the concept of self-checking transaction-level directed testbenches. Even though HVLs make implementing such testbenches easier (especially the self-checking part), you still need to plan your verification, define your verification objectives, design its strategy and architecture, design the stimulus, determine the expected response and compare the actual output. These are concepts that can be learned and applied using VHDL or Verilog.

Coverage-driven constrained random approach requires HVLs.

All HVLs can be used as if they were souped-up Verilog or VHDL languages. In fact, most HVL solutions are just that. But if your HVL contains all of the features required to support the HVL productivity cycle, the verification process must be approached—and implemented—in a different fashion.

This change is just like taking advantage of the productivity offered by logic synthesis tools: It requires an approach different from schematic capture. To successfully implement a coverage-driven constrained random verification approach, you need to modify the way you plan your verification, design its strategy and implement the testcases. Because Verilog and VHDL lack all of the required features, *e* and OpenVera will be used to illustrate these concepts.

ASSERTIONS

Assertions detect conditions that should always be true.

An assertion boils down to an *if* statement and an error message should the expression in the *if* statement become false. Assertions have been used in software design for many years: the *assert()* function has been part of the ANSI C standard from the beginning. In software for example, assertions are used to detect conditions such as NULL pointers or empty lists. VHDL has had an *assert* statement from day one too, but it was never a popular construct—except to terminate a simulation (see Sample 5-44 on page 264 for an example).

Assertions

Hardware assertions require a form of temporal language.

A software assertion simply checks that, at the time the *assert* statement is executed, the condition evaluates to TRUE. This simple zero-time test is not sufficient for supporting assertions in hardware designs. In hardware, functional correctness usually involves behavior over a period of time. Some hardware assertions such as, "This state register is one-hot encoded." or "This FIFO never overflows." can be expressed as immediate, zero-time expressions. But checking simple hardware assertions such as, "This signal must be asserted for a single clock period." or "A request must always be followed by a grant or abort within 10 clock cycles." require that the assertion condition be evaluated over time. Thus, assertions require the use of a temporal language to be able to describe relationships over time.

There are two classes of assertions.

Assertions fall in two broad classes: those specified by the designer and those specified by the verification engineer.

- Implementation assertions are specified by the designers.

- Specification assertions are specified by the verification engineers.

Implementation assertions verify assumptions.

Implementation assertions are used to formally encode the designer's assumptions about the interface of the design or conditions that are indications of misuse or design faults. For example, the designer of a FIFO would add assertions to detect if it ever overflows or underflows or that, because of a design limitation, the *write* and *read* pulses are ever asserted at the same time. Because implementation assertions are specified by the designer, they will not detect discrepancies between the functional intent and the design. But implementation assertions will detect discrepancies between the design assumptions and the implementation.

Specification assertions verify intent.

Specification assertions formally encode expectations of the design based on the functional intent. These assertions are used as a functional error detection mechanism and supplement the error detections performed in the self-checking section of testbenches. Specification assertions are typically *white-box* strategies because the relationships between the primary inputs and outputs of a modern design are too complex to be described in today's temporal languages. For example, rather than relying on the scoreboard to detect that an arbiter is not fair, it is much simpler to perform this check using a block-level assertion.

Assertion specifi-
cation is a com-
plex topic.

This simple introduction to assertions does not do justice to the richness and power—and ensuing complexity—of assertion. Entire books ought to be (and probably are being) written about the subject.

Simulation Assertions

The OVL started
the storm.

Assertions took the hardware design community by storm when Foster and Bening's book[8] introduced the concept using a library of predefined Verilog modules that implemented all of the common design assertions. The library, available in source form as the *Open Verification Library,*[9] was a clever way of using Verilog to specify temporal expressions. Foster, then at Hewlett-Packard, had a hidden agenda: Get designers to specify design assertions he could then try to prove using formal methods. Using Verilog modules was a convenient solution to ease the adoption of these assertions by the designers. The reality of what happened next proved to be even more fruitful.

They detect errors
close in space and
time to the fault.

If a design assumption is violated during simulation, the design will not operate correctly. The cause of the violation is not important: It could be a misunderstanding by the designer of the block or the designer of the upstream block or an incorrect testbench. The relevant fact is that the design is failing to operate according to the original intent. The symptoms of that low-level failure are usually not visible (if at all) until the affected data item makes its way to the outputs of the design and is flagged by the self-checking structure.

An assertion formally encoding the design assumption immediately fires and reports a problem at the time it occurs, in the area of the design where it occurs. Debugging and fixing the assertion failure (whatever the cause) will be a lot more efficient than tracing back the cause of a corrupted packet. In one of Foster's projects, 85% of the design errors where caught and quickly fixed using simulated assertions.

---

8. Harry Foster and Lionel Bening, "*Principles of Verifiable RTL Design,*" second edition, Kluwer Academic Publisher, ISBN 0-7923-7368-5.

9. See http://verificationlib.org.

Formal Assertion Proving

Is it possible for an assertion to fire?

Simulation can show only the presence of bugs, never prove their absence. The fact that an assertion has never reported a violation throughout a series of simulation does not mean that it can never be violated. Tools like code and functional coverage can satisfy us that a portion of a design was thoroughly verified—but there will (and should) always be a nagging doubt.

Model checking can mathematically prove or disprove an assertion.

Formal tools called *model checker* or *assertion provers* can mathematically prove that, given an RTL design and some assumptions about the relationships of the input signals, an assertion will always hold true. If a counter example is found, the formal tool will provide details on the sequence of events that leads to the assertion violation. It is then up to you to decide if this sequence of events is possible, given additional knowledge about the environment of the design.

Some assertions are used as assumptions.

Given total freedom over the inputs of a design, you can probably violate any and all assertions about its implementation. Fortunately, the usage of the inputs of a design are subject to limitation and rules to ensure proper operation of the design. Furthermore, these input signals usually come from other designs that do not behave (one hopes!) erratically. When proving some assertions on a design, it is thus necessary to supply assertions on the inputs or state of the design. The latter assertions are not proven. Rather, they are assumed to be true and used to constrain the solution space for the proof.

Assumptions need to be proven too.

The correctness of a proof depends on the correctness of the assumptions[10] made on the design inputs. Should any assumption be wrong, the proof no longer stands. An assumption on a design's inputs thus becomes an assertion to be proven on the upstream design supplying those inputs.

Semi-formal tools combine model checking with simulation.

Semi-formal tools are hybrid tools that combine formal methods with simulation. Semi-formal tools are an attempt to bridge the gap between a familiar technology (simulation) and the fundamentally

---

10. The formal verification community calls these input assertions "constraints." I used the term "assumptions" to differentiate them from random-generation constraints, which are randomization concepts.

different formal tools. They use intermediate simulation information—such as the current state of a design—as a starting point for proving or disproving assertions.

Use formal methods to prove cases uncovered in simulation.

Formal verification does not replace simulation or make it obsolete. Simulation (including simulated assertions) is the lawnmower of the verification garden: It is still the best tool for covering broad swaths of functionality and for weeding out the easy-to-find and some not-so-easy-to-find bugs. Formal verification puts the finishing touch on those hard-to-reach corners in critical and important design sections and ensures that the job is well done. Using functional coverage metrics collected from simulation (for example, request patterns on an arbiter), identifies conditions that remain to be verified. If those conditions would be difficult to create within the simulation environment, using these conditions as assumptions, proves the correctness of the design for the remaining uncovered cases.

REVISION CONTROL

Are we all looking at the same thing?

One of the major difficulties in verification is to ensure that what is being verified is actually what will be implemented. When you compile a Verilog source file, what is the guarantee that the design engineer will use that *exact same file* when synthesizing the design?

When the same person verifies and then synthesizes the design, this problem is reduced to that person using proper file management discipline. However, as I hope to have demonstrated in Chapter 1, having the same person perform both tasks is not a reliable functional verification process. It is more likely that separate individuals perform the verification and synthesis tasks.

Files must be centrally managed.

In very small and closely knit groups, it may be possible to have everyone work from a single directory, or to have the design files distributed across a small number of individual directories. Everyone agrees where each other's files are, then each is left to his or her own device. This situation is very common and very dangerous: How can you tell if the designer has changed a source file and maybe introduced a functional bug since you last verified it?

It must be easy to get at all the files, from a single location. This methodology is not scalable either. It quickly breaks down once the team grows to more than two or three individuals. And it does not work at all when the team is distributed across different physical or geographical areas. The verification engineer is often the first person to face the non-scalability challenge of this environment. Each designer is content working independently in his or her own directories. Individual designs, when properly partitioned, rarely need to refer to some other design in another designer's working directory. As the verification engineer, your first task is to integrate all the pieces into a functional entity. That's where the difficulties of pulling bits and pieces from heterogeneous working environments scattered across multiple file servers become apparent.

The Software Engineering Experience

HDL models are software projects! For about 25 years, software engineering has been dealing with the issues of managing a large number of source files, authored by many different individuals, verified by others and compiled into a final product. Make no mistake: Managing an HDL-based hardware design project is no different than managing a software project.

Free and commercial tools are available. To help manage files, software engineers use source control management systems. Some are available, free of charge, either bundled with the UNIX operating systems (RCS, CVS, SCCS), or distributed by the GNU project (RCS, CVS) and available in source form at:

```
ftp://prep.ai.mit.edu/pub/gnu
```

Commercial systems, some very sophisticated, are also available.

All source files are centrally managed. Figure 2-18 shows how source files are managed using a source control management system. All accesses and changes to source files are mediated by the management system. Individual authors and users interact solely through the management system, not by directly accessing files in working directories.

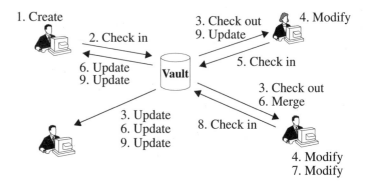

Figure 2-18. Data flow in a source control system

The history of a file is maintained.

Source code management systems maintain not only the latest version of a file, but also keep a complete history of each file as separate *versions*. Thus, it is possible to recover older versions of files, or to determine what changed from one version to another. It is a good idea to frequently *check in* file versions. You do not have to rely on a backup system if you ever accidentally delete a file. Sometimes, a series of modifications you have been working on for the last couple of hours is making things worse, not better. You can easily roll back the state of a file to a previous version known to work.

The team owns all the files.

When using a source management system, files are no longer owned by individuals. Designers may be nominally responsible for various sections of a design, but anyone—with the proper permissions—can make any change to any file. This lets a verification engineer fix bugs found in RTL code without having to rely on the designer, busy trying to get timing closure on another portion of the design. The source management system mediates changes to files either through exclusive locks, or by merging concurrent modifications.

Configuration Management

Each user works from a *view* of the file system.

Each engineer working on a project managed with a source control system has a private *view* of all the source files (or a subset thereof) used in the project. Figure 2-19 shows how two users may have two different views of the source files in the management system. Views need not be always composed of the latest versions of all the files. In fact, for a verification engineer, that would be a hindrance. Files checked in on a regular basis by their authors may include syntax errors, be simple placeholders for future work, or be totally broken. It would be very frustrating if the model you were trying to verify kept changing faster than you could identify problems with it.

Figure 2-19. User views of managed source files

design.v 1.53
cpuif.v   1.28
tb.v   1.38

Vault

design.v 1.1..1.56
cpuif.v   1.1..1.32
tb.v   1.1..1.49

design.v 1.41
cpuif.v   1.17
tb.v   1.38

Configurations are created by tagging a set of versions.

All source management systems use the concept of symbolic tags that can be attached to specific versions of files. You may then refer to particular versions of files, or set of files, using the symbolic name, without knowing the exact version number they refer to. In Figure 2-19, the user on the left could be working with the versions that were tagged as "ready to simulate" by the author. The user on the right, the system verification engineer, could be working with the versions that were tagged as "golden" by the ASIC verification engineer.

Configuration management translates to tag management.

Managing releases becomes a problem of managing tags, which can be a complex task. Table 2-1 shows a list of tags that could be used in a project to identify the various versions of a file as it progresses through the design process. Some tags, such as the "Version_M.N" tag, never move once applied to a specific version. Others, such as the "Submit" tag, move to newer versions as the development of the design progresses. Before moving a tag, it may be a good idea to leave a trace of the previous position of a tag. One possible mechanism for doing so is to append the date to the tag name. For example, the old "Submit" version gets tagged with the new tag

"Submit_000302" on March 2nd, 2000 and the "Submit" tag is moved to the latest version.

Table 2-1.
Example tags for release management

Tag Name	Description
Submit	Ready to submit to functional verification. Author has verified syntax correctness and basic level of functionality.
Bronze	Passes a basic set of functional testcases. Release is sufficiently functional for integration.
Silver	Passes all functional testcases.
Gold	Passes all functional testcases and meets coding coverage guidelines (requires additional corner-case testcases).
To_Synthesis	Ready to submit to synthesis. Usually matches "Silver" or "Gold".
To_Layout	Ready to submit to layout. Usually matches "Gold".
Version_M.N	Version that was manufactured. Matches corresponding "To_Layout" release. Future versions of the same chip will move tags beyond this point.
ON_YYMMDD	Some meaningful release on the specified date.

Working with Releases

Views can become out-of-date as new versions of files are checked into the source management system database and tags are moved forward.

Releases are specific configurations.

The author of the RTL for a portion of the design would likely always work with the latest version of the files he or she is actively working on, checking in and updating them frequently (typically at relevant points of code development throughout the day and at the end of each day). Once the source code is syntactically correct and its functionality satisfies the designer (by using a few ad hoc testbenches), the corresponding version of the files are tagged as ready for verification.

Users must update their view to the appropriate release.

You, as the verification engineer, must be constantly on the lookout for updates to your view. When working on a particularly difficult testbench, you may spend several days without updating your view to the latest version ready to be verified. That way, you maintain a consistent view of the design under test and limit changes to the testbenches, which you make. Once the actual verification and debugging of the design starts, you probably want to refresh your view to the latest "ready-to-verify" release of the design before running a testbench.

Update often.

When using a concurrent development model where multiple engineers are working in parallel on the same files, it is important to check in modifications often, and update your view to merge concurrent modifications even more often. If you wait too long, there is a greater probability of collisions that will require manual resolution. The concept of concurrently modifying files then merging the differences sounds impossibly risky at first. However, experience has shown that different functions or bug fixes rarely involve modification to the same lines of source code. As long as the modifications are separated by two or three lines of unmodified code, merging will proceed without any problems. Trust me, concurrent development is the way to go!

You can be notified of new releases.

An interesting feature of some source management systems is the ability to issue email notification whenever a significant event occurs. For example, such a system could send e-mail to all verification engineers whenever the tag identifying the release that is ready for verification is moved. Optionally, the e-mail could contain a copy of the descriptions of the changes that were made to the source files. Upon receiving such an e-mail, you could make an informed decision about whether to update your view immediately.

ISSUE TRACKING

All your bug are
belong to us!

The job of any verification engineer is to find bugs. Under normal conditions, you should expect to find functional irregularities. You should be *really* worried if no problems are being found. Their occurrence is normal and do not reflect the abilities of the hardware designers. Even the most experienced software designers write code that includes bugs, even in the simplest and shortest routines. Now that we've established that bugs *will* be found, how will you deal with them?

Bugs must be
fixed.

Once a problem has been identified, it *must* be resolved. All design teams have informal systems to track issues and ensure their resolutions. However, the quality and scalability of these informal systems leaves a lot to be desired.

What Is an Issue?

Is it worth worry-
ing about?

Before we discuss the various ways issues can be tracked, we must first consider what is an issue worth tracking. The answer depends highly on the tracking system used. The cost of tracking the issue should not be greater than the cost of the issue itself. However, do you want the tracking system to dictate what kind of issues are tracked? Or, do you want to decide on what constitutes a trackable issue, then implement a suitable tracking system? The latter position is the one that serves the ultimate goal better: Making sure that the design is functionally correct.

An issue is *anything* that can affect the functionality of the design:

1. Bugs found during the execution of a testbench are clearly issues worth tracking.

2. Ambiguities or incompleteness in the specification document should also be tracked issues. However, typographical errors definitely do not fit in this category.

3. Architectural decisions and trade-offs are also issues.

4. Errors found at all stages of the design, in the design itself or in the verification environment should be tracked as well.

5. If someone thinks about a new relevant testcase, it should be filed as an issue.

When in doubt, track it.	It is not possible to come up with an exhaustive list of issues worth tracking. Whenever an issue comes up, the only criterion that determines whether it should be tracked, should be its effect on the correctness of the final design. If a bad design can be manufactured when that issue goes unresolved, it *must* be tracked. Of course, all issues are not created equal. Some have a direct impact on the functionality of the design, others have minor secondary effects. Issues should be assigned a priority and be addressed in order of that priority.
You may choose not to fix an issue.	Some issues, often of lower importance, may be consciously left unresolved. The design or project team may decide that a particular problem or shortcoming is an acceptable limitation for this particular project and can be left to be resolved in the next incarnation of the product. The principal difficulty is to make sure that the decision was a conscious and rational one!

The Grapevine System

Issues can be verbally reported.	The simplest, and most pervasive issue tracking system is the *grapevine*. After identifying a problem, you walk over to the hardware designer's cubicle (assuming you are not the hardware designer as well!) and discuss the issue. Others may be pulled into the conversation or accidentally drop in as they overhear something interesting being debated. Simple issues are usually resolved on the spot. For bigger issues, everyone may agree that further discussions are warranted, pending the input of other individuals. The priority of issues is implicitly communicated by the insistence and frequency of your reminders to the hardware designer.
It works only under specific conditions.	The grapevine system works well with small, closely knit design groups, working in close proximity. If temporary contractors or part-time engineers are on the team, or members are distributed geographically, this system breaks down as instant verbal communications are not readily available. Once issues are verbally resolved, no one has a clear responsibility for making sure that the solution will be implemented.
You are condemned to repeat past mistakes.	Also, this system does not maintain any history. Once an issue is resolved, there is no way to review the process that led to the decision. The same issue may be revisited many times if the implementation of the solution is significantly delayed. If the proposed resolution turns out to be inappropriate, the team may end up going

in circles, repeatedly trying previous solutions. Without history, you are condemned to repeat it. There is no opportunity for the team to learn from its mistakes. Learning is limited to individuals, and to the extent that they keep encountering similar problems.

The Post-It System

Issues can be tracked on little pieces of paper.
When teams become larger, or when communications are no longer regular and casual, the next issue tracking system that is used is the 3M Post-It™ note system. It is easy to recognize at a glance: Every team member has a number of telltale yellow pieces of paper stuck around the periphery of their computer monitor.

If the paper disappears, so does the issue.
This evolutionary system only addresses the lack of ownership of the grapevine system: Whoever has the yellow piece of paper is responsible for its resolution. This ownership is tenuous at best. Many issues are "resolved" when the sticky note accidentally falls on the floor and is swept away by the janitorial staff.

Issues cannot be prioritized.
With the Post-It system, issues are not prioritized. One bug may be critical to another team member, but the owner of the bug may choose to resolve other issues first simply because they are simpler and because resolving them instead reduces the clutter around his computer screen faster. All notes look alike and none indicate a sense of urgency more than the others.

History will repeat itself.
And again, the Post-It system suffers from the same learning disabilities as the grapevine system. Because of the lack of history, issues are revisited many times, and problems are recreated repeatedly.

The Procedural System

Issues can be tracked at group meetings.
The next step in the normal evolution of issue tracking is the procedural system. In this system, issues are formally reported, usually through free-form documents such as e-mail messages. The outstanding issues are reviewed and resolved during team meetings.

Only the biggest issues are tracked.
Because the entire team is involved and the minutes of meetings are usually kept, this system provides an opportunity for team-wide learning. But the procedural system consumes an inordinate amount of precious meeting time. Because of the time and effort involved in tracking and resolving these issues, it is usually reserved for the

most important or controversial ones. The smaller, less important—but much more numerous—issues default back to the grapevine or Post-It note systems.

Computerized System

Issues can be tracked using databases.

A revolution in issue tracking comes from using a computer-based system. In such a system, issues must be seen through to resolution: Outstanding issues are repeatedly reported loud and clear. Issues can be formally assigned to individuals or list of individuals. Their resolution need only involve the required team members. The computer-based system can automatically send daily or weekly status reports to interested parties.

A history of the decision making process is maintained and archived. By recording various attempted solutions and their effectiveness, solutions are only tried once without going in circles. The resolution process of similar issues can be quickly looked-up by anyone, preventing similar mistakes from being committed repeatedly.

But it should not be easier to track them verbally or on paper.

Even with its clear advantages, computer-based systems are often unsuccessful. The main obstacle is their lack of comparative ease-of-use. Remember: The grapevine and Post-It systems are readily available at all times. Given the schedule pressure engineers work under and the amount of work that needs to be done, if you had the choice to report a relatively simple problem, which process would you use:

1. Walk over to the person who has to solve the problem and verbally report it.

2. Describe the problem on a Post-It note, then give it to that same person (and if that person is not there, stick it in the middle of his or her computer screen).

3. Enter a description of the problem in the issue tracking database and never leave your workstation?

It should not take longer to submit an issue than to fix it.

You would probably use the one that requires the least amount of time and effort. If you want your team to use a computer-based issue tracking system successfully, then select one that causes the smallest disruption in their normal work flow. Choose one that is a simple or transparent extension of their normal behavior and tools they already use.

I was involved in a project where the issue tracking system used a proprietary X-based graphical interface. It took about 15 seconds to bring up the entire interface on your screen. You were then faced with a series of required menu selections to identify the precise division, project, system, sub-system, device and functional aspect of the problem, followed by several other dialog boxes to describe the actual issue. Entering the simplest issue took *at least* three to four minutes. And the system could not be accessed when working from home on dial-up lines. You can guess how successful that system was...

Email-based systems have the greatest acceptance.

The systems that have the most success invariably use an e-mail-based interface, usually coupled with a Web-based interface for administrative tasks and reporting. Everyone on your team uses e-mail. It is probably already the preferred mechanism for discussing issues when members are distributed geographically or work in different time zones. Having a system that simply captures these e-mail messages, categorizes them and keeps track of the status and resolution of individual issues (usually through a minimum set of required fields in the e-mail body or header), is an effective way of implementing a computer-based issue tracking system.

METRICS

Metrics are essential management tools.

Managers love metrics and measurements. They have little time to personally assess the progress and status of a project. They must rely on numbers that (more or less) reflect the current situation.

Metrics are best observed over time to see trends.

Metrics are most often used in a static fashion: "What are the values today?" "How close are they to the values that indicate that the project is complete?" The odometer reports a static value: How far have you travelled. However, metrics provide the most valuable information when observed over time. Not only do you know where you are, but also you can know how fast you are going, and what direction you are heading. (Is it getting better or worse?)

Historical data should be used to create a baseline.

When compared with historical data, metrics can paint a picture of your learning abilities. Unless you know how well (or how poorly) you did last time, how can you tell if you are becoming better at your job? It is important to create a baseline from historical data to determine your productivity level. In an industry where the manufacturing capability doubles every 18 months, you cannot afford to maintain a constant level of productivity.

Metrics can help assess the verification effort.

There are several metrics that can help assess the status, progress and productivity of functional verification. One has already been introduced: code coverage.

Code-Related Metrics

Code coverage may not be relevant.

Code coverage measures how thoroughly the verification suite exercises the source code being verified. That metric should climb steadily toward 100% over time. From project to project, it should climb faster, and get closer to 100%.

However, code coverage is not a suitable metric for all verification projects. It is an effective metric for the smallest design unit that is individually specified (such as an FPGA, a reusable component or an ASIC). But it is ineffective when verifying designs composed of sub-designs that have been independently verified. The objective of that verification is to confirm that the sub-designs are interfaced and cooperate properly, not to verify their individual features. It is unlikely (and unnecessary) to execute all the statements.

The number of lines of code can measure implementation efficiency.

The total *number of lines of code* that is necessary to implement a verification suite can be an effective measure of the effort required in implementing it. This metric can be used to compare the productivity offered by new verification languages or methods. If they can reduce the number of lines of code that need to be written, then they should reduce the effort required to implement the verification.

Lines-of-code ratio can measure complexity.

The *ratio of lines of code* between the design being verified and the verification suite may measure the complexity of the design. Historical data on that ratio could help predict the verification effort for a new design by predicting its estimated complexity.

Code change rate should trend toward zero.

If you are using a source control system, you can measure the *source code changes* over time. At the beginning of a project, code changes at a very fast rate as new functionality is added and initial versions are augmented. At the beginning of the verification phase, many changes in the code are required by bug fixes. As the verification progresses, the rate of changes should decrease as there are fewer and fewer bugs to be found and fixed. Figure 2-20 shows a plot of the expected code change rate over the life of a project. From this metric, you are able to determine if the code is becoming stable, or identify the most unstable sections of a design.

Figure 2-20.
Ideal code
change rate
metric over
time

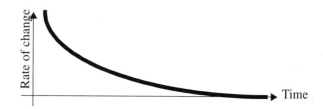

Quality-Related Metrics

Quality is subjective, but it can be measured indirectly.

Quality-related metrics are probably more directly related with the functional verification than other productivity metrics. Quality is a subjective value, yet, it is possible to find metrics that correlate with the level of quality in a design. This is much like the number of customer complaints or the number of repeat customers can be used to judge the quality of retail services.

Functional coverage can measure testcase completeness.

Functional coverage measures the range and combination of input and output values that were submitted to and observed from the design, and of selected internal values. By assigning a weight to each functional coverage metric, it can be reduced to a single functional coverage grade measuring how thoroughly the functionality of the design was exercised. By weighing the more important functional coverage measures more than the less important ones, it gives a good indicator of the progress of the functional verification. This metric should evolve rapidly toward 100% at the beginning of the project then significantly slow down as only hard-to-reach functional coverage points remain.

A simple metric is the number of known issues.

The easiest metric to collect is the *number of known outstanding issues*. The number could be weighed to count issues differently according to their severity. When using a computer-based issue tracking system, this metric, as well as trends and rates, can be easily generated. Are issues accumulating (indicating a growing quality problem)? Or, are they decreasing and nearing zero?

Code will be worn out eventually.

If you are dealing with a reusable or long-lived design, it is useful to measure the *number of bugs found during its service life*. These are bugs that were not originally found by the verification suite. If the number of bugs starts to increase dramatically compared to historical findings, it is an indication that the design has outlived its useful life. It has been modified and adapted too many times and needs to be re-designed from scratch. Throughout the normal life cycle of a reusable design, the number of outstanding issues exhibits a behavior as shown in Figure 2-21.

Figure 2-21.
Number of
outstanding
issues
throughout the
life cycle of a
design

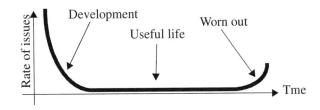

Interpreting Metrics

Whatever gets measured gets done.

Because managers rely heavily on metrics to measure performance (and ultimately assign reward and blame), there is a tendency for any organization to align its behavior with the metrics. That is why you must be extremely careful to select metrics that faithfully represent the situation and are correlated with the effect you are trying to measure or improve. If you measure the number of bugs found and fixed, you quickly see an increase in the number of bugs found and fixed. But do you see an increase in the quality of the code being verified? Were bugs simply not previously reported? Are designers more sloppy when writing their code since they'll be rewarded only when and if a bug is found and fixed?

Make sure metrics are correlated with the effect you want to measure.

Figure 2-22 shows a list of file names and current version numbers maintained by two different designers. Which designer is more productive? Do the large version numbers from the designer on the left indicate someone who writes code with many bugs that had to be fixed? Or, are they from a cautious designer who checkpoints changes often?

Figure 2-22. Using version numbers as a metric	alu_e.vhd	1.15	cpuif_e.vhd	1.2

<table>
<tr><td rowspan="6">**Figure 2-22.**
Using version
numbers as a
metric</td><td>alu_e.vhd</td><td>1.15</td><td>cpuif_e.vhd</td><td>1.2</td></tr>
<tr><td>alu_rtl.vhd</td><td>1.234</td><td>cpuif_rtl.vhd</td><td>1.4</td></tr>
<tr><td>decoder_e.vhd</td><td>1.12</td><td>regfile_e.vhd</td><td>1.1</td></tr>
<tr><td>decoder_rtl.vhf</td><td>1.155</td><td>regfile_rtl.vhf</td><td>1.7</td></tr>
<tr><td>dpath_e.vhd</td><td>1.7</td><td>addr_dec_e.vhd</td><td>1.3</td></tr>
<tr><td>dpath_rtl.vhd</td><td>1.176</td><td>addr_dec_rtl.vhd</td><td>1.6</td></tr>
</table>

On the other hand, Figure 2-23 shows a plot of the code change rate for each designer. What is your assessment of the code quality from designer on the left? It seems to me that the designer on the right is not making proper use the revision control system.

Figure 2-23. Using code change rate as a metric

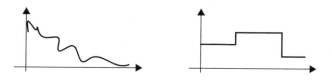

Writing Testbenches: Functional Verification of HDL Models

SUMMARY

Despite their reporting many false errors, linting and other static code checking tools are still the most efficient mechanism for finding certain classes to problems.

Simulators are only as good as the model they are simulating. Simulators offer many performance enhancing options and the possibility to co-simulate with other languages or simulators.

Assertion-based verification is a powerful addition to any verification methodology. This approach allows the quick identification of problems, where and when they occur.

Hardware verification languages offer an increase in productivity because of their specialization to the verification task and their support for coverage-driven random-based verification.

Use code and functional coverage metrics to provide a quantitative assessment of your progress. Do not focus on reaching 100% at all cost, nor should you consider the job done when you've reached your coverage goals.

Use a source control system and an issue tracking system to manage your code and bug reports.

CHAPTER 3   THE VERIFICATION PLAN

In this chapter, I describe the verification plan as a specification of the functional verification testcases and of the testbench infrastructure that will be necessary to support them. It is used to define what is first-time success, how a design is verified and which testbenches are written.

The design project that sits before you will propel your company to new levels of market share and profitability. A few system architects have designed and specified a system that should meet performance and cost goals. Several design leaders, using the system specification, have been working on writing detailed functional specification documents for each of the ASICs and FPGAs that are required to build this new product. Teams of hot-shot hardware designers are being assembled to implement each ASIC or FPGA. Using the detailed specification documents for each device, they are coming up with a detailed implementation schedule. So far, it appears that the project will meet its production deadline.

You are in charge of the verification for this design. Not only must this product be on time, but also it must be functionally correct. Your company's reputation depends on it. You have been asked by the project manager to produce a detailed schedule for the verification and define your staffing requirements. How can you determine either?

THE ROLE OF THE VERIFICATION PLAN

Traditionally, verification is an ad-hoc process.

In a traditional verification process, your decision would be simple. In fact, your own position would not exist. Verification would be left to each hardware designer to do as they wish. It would be performed as time allows. And everybody's fingers would be crossed hoping that system integration would be smooth and that the board designs would not need too many patch wires. Many devices would be implemented in FPGAs, trading additional per-unit costs for flexibility in fixing problems later found during system integration.

Tools exist to help determine when you are done.

The tools described in the previous chapter will help during your verification effort. Code coverage, functional coverage, bug discovery rate and code change rates are metrics that indicate how much progress you have made toward your goal. But they are like stock market indices or batting averages: They provide a snapshot of the current situation and, if recorded over time, show trends and progression. However, they cannot be used to predict the future.[1]

Specifying the Verification

You need a tool to determine when you will be done.

Today's question is about producing a schedule. You must determine, as reliably as possible, when the verification will be completed to the required degree of confidence. Unless you have a detailed specification of the work that needs to be accomplished, you cannot determine how many people you need, nor how long it is going to take or even if you are doing work that needs to be done. That's what the verification plan is about.

Start from the design specification.

Before you can decide on a plan of attack for the verification, a specification document for the design to be verified must exist. And it must exist in written form. "Folklore" specifications that describe the design as, "The same thing as we did before, but at twice the clock rate and with these additional features." are insufficient. Often, the design specification is implemented using two separate documents written at different abstraction levels.

---

1. However, many financial and sports analysts make a good living predicting an essentially random process or explaining, after the fact, why everybody was wrong.

- The first is the architectural specification, which details the functional requirements of the device.

- The second is the design specification, which describes the particular implementation of the architecture down to the block level.

The verification plan can start to be written once the architectural specification document is complete. It can be augmented with implementation-specific testcases once the implementation document is complete.

The specification document is the golden reference. The specification document is the common source for the verification and implementation efforts. It is the golden reference and the rule of law. Later, when discrepancies are found between the response expected by the testbench and the one produced by the design under verification, the specification document arbitrates and decides which one has the correct answer. If necessary, the specification document should be elaborated to remove any ambiguity. The specification document must exist before the implementation. The implementation must follow the specification. If the specification depends on or is a consequence of the implementation, it will be impossible to verify because the specification will change every time the implementation changes.

The verification plan is the specification document for the verification effort. Today's million-gate ASIC designs cannot proceed without a detailed specification document being written first. With the verification effort being 100% to 200% of the RTL design effort, why should it proceed without a specification document of its own? The verification plan is the specification document for the verification effort.

Defining First-Time Success

If, and only if, it is in the plan, will it be verified. The verification plan provides a forum for the entire design team to define what *first-time success* is. It is a mechanism that ensures all essential features are appropriately verified. If you want first-time success, you must identify which features must be exercised under which conditions and what the expected response should be. The verification plan documents which features are a priority and which ones are optional. In the face of schedule pressure, the decision to drop features from the first-time success requirements becomes a conscious one. The alternative is to live with whatever happens to work when the decision to ship the design cuts off the verification effort like a guillotine. Some of the features, essential for market acceptance, might fall in the basket.

From the verification plan, a detailed schedule can be created.

The verification plan creates a line in the sand that cannot be crossed without endangering the success of the project in the market place. Once the plan is written, you know how many testcases must be created, how complex they need to be and how they depend on each other. You can define a detailed verification schedule, and allocate tasks to resources, parallelizing verification as much as possible. Once the RTL passes all of the testcases, and you are satisfied with the coverage and bug-rate metrics, the design can be shipped. Not before.

The team owns the verification plan.

It is important for everyone involved with the design project to realize that they have a stake in the verification plan. The responsibility of an RTL designer is not to design RTL. That's only a means to an end. His or her responsibility is to produce a working design. The entire design team must contribute to the verification plan, to make sure that it is complete and correct.

This process is not revolutionary.

The process used to write a verification plan is not new. It has been used for decades by NASA, the FAA and aerospace companies to ensure that the ultra-reliable systems they were implementing met their specifications. This process has been used for software as well as for hardware designs.

LEVELS OF VERIFICATION

Verification can be performed at various levels of granularity.

The first question, when planning the verification, is to determine the level of granularity for the verification effort. A design is potentially composed of several levels. Some have a physical partition, such as printed circuit boards, FPGAs and ASICs. Others have a logical partition, such as synthesized units, reusable components or sub-systems.

As illustrated in Figure 3-1, each level of verification is best suited for a particular application and objective. The nature of design with reusable components shifts where stand-alone unit-level verification ends and system-level verification starts within the physical hierarchy, compared with a more traditional design process. Design with reuse does not diminish the need for verification. The unit-to-system boundary becomes a logical one instead of a physical one.

Figure 3-1.
Application of different levels of verification

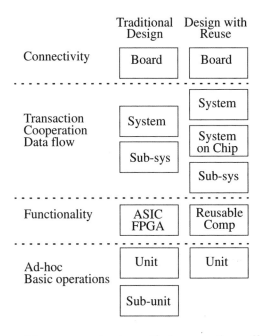

	Traditional Design	Design with Reuse
Connectivity	Board	Board
Transaction Cooperation Data flow	System / Sub-sys	System / System on Chip / Sub-sys
Functionality	ASIC FPGA	Reusable Comp
Ad-hoc Basic operations	Unit / Sub-unit	Unit

Deciding between levels of granularity involves trade-offs.

Smaller partitions are easier to verify because they offer greater controllability and observability. It is easier to set up interesting conditions and state combinations and to observe if the response is as expected. With larger partitions, the integration of the smaller partitions it contains is implicitly verified at the cost of lower controllability and observability.

Verifying at a given level of granularity requires stable interfaces.

Because the verification requires a significant implementation effort, any partition being verified must have relatively stable interfaces and intended functionality. If the interfaces keep changing, or functionality keeps being moved from one partition to another, the testbenches will constantly need to be changed with little progress being made. Once you've decided on specific partitions to be verified, their interface and overall functionality must be specified early on and remain as stable as possible. Ideally, each verified partition should have its own specification document or, at a minimum, its own section in the specification document.

Unit-Level Verification

Implementation determines the content of this partition.

Design units are logical partitions. They are created to facilitate the implementation or the synthesis process. They vary from the relatively small (e.g., FIFOs and state machines) to the complex (e.g., PCI slave interface and DSP datapaths). Their interfaces and functionality tend to vary a lot over time, as implementation details highlight shortcomings in the initial design. They usually do not have an independent specification document to verify against either.

Use ad-hoc verification for design units.

Because these design units are usually in a constant state of flux, they are better left to an ad-hoc verification process. The designer himself verifies the basic operation of the unit. The objective of this verification is to ensure that there are no syntax errors in the RTL code, and that basic functionality is operational. It is not to create a regressionable test suite and obtain high code coverage.

They are too numerous to verify formally.

The high number of design units in any project makes a verification process implemented at that level too time consuming. Each would require a custom verification environment, as described in Chapters 5 and 6. The precious verification resources would spend an inordinate amount of time creating stimulus generators and response monitors for a myriad of ever-changing interfaces. Writing a lot of simple testbenches is just as much work, if not more, as writing a few complex ones. And verification at the ASIC- or FPGA-level would still be required to verify the integration of these design units.

Unit-level verification may be required in large devices.

In today's very large and complex ASICs and FPGAs, it may not be possible to obtain the necessary functional coverage when verifying from the ASIC or FPGA partition. And not all units are created equal. For the highly sensitive and complex functional units, it may be more efficient to perform unit-level verification to have sufficient levels of controllability and observability and reach the desired level of confidence. Ideally, each functional unit verified at the unit level should have its own specification document.

Architect the design to facilitate unit-level verification.

If your design is so complex that you have to perform some unit-level verification, it should be designed to make that unit-level verification as relevant and complete as possible. Partition the design so the features to be verified are completely contained within a unit and can be verified on a stand-alone basis. Once verified, these features can be assumed to work during the verification of the higher levels. If the features to be verified at the unit level require interaction with other units, they have to be re-verified at a higher-level where the features are fully contained, to ensure that the integration correctly implements them.

Reusable Components Verification

Reusable designs are independent of any particular use.

Reusable components are designed to an independent specification. They are intended to be used as-is and unchanged in many different designs. Their reusability can be limited to a single product, the entire product family, or they could be applicable to any product requiring their functionality. They must be designed—and thus verified—independent of any one usage. It is a good idea to use assertions (see "Assertions" on page 64) to specify restrictions and requirements on the inputs of reusable components. They help ensure that the reused components is always used as intended.

Verification components can be reused as well.

Reusable components are usually designed using standardized interfaces. These interfaces can be designed to standard on-chip buses, or industry-standard external physical interfaces. The verification components used to stimulate and monitor these interfaces can be themselves reused across the various verification environments used to verify different reusable components. The verification effort can be leveraged across multiple components, thus minimizing the overall investment in verification. Chapter 6 will detail how to architect a testbench to promote the creation and use of reusable verification components.

Reusable components need a regression test suite.

Reusable components are used in many designs. When they are modified, either to fix problems that were found, or to enhance their functionality, you must make sure that they remain backward compatible. This is accomplished by implementing a *regression suite* that verifies the compliance of the component after any modification. Checking the equivalence of the new version with the previous version using formal verification would not really work unless the modifications were not functional. Adding functionality or fixing problems, by definition, makes the new version of the design not equivalent to the previous one.

They need thorough functional coverage.

Components will not be reused if potential users do not have confidence that they perform better and more reliably than one they could design themselves. This confidence can be obtained only by demonstrating the correctness and robustness of the components through a thorough, well documented verification process.

ASIC and FPGA Verification

The physical partition is an ideal verification level.

ASICs and FPGAs are physical partitions. They form a natural partition for verification because their interfaces and functionality do not change very much after the initial specification of the system and the completion of their specification documents.

They may have to be treated as systems.

The ever increasing densities offered by semiconductor technology enables ever increasing integration of complex functionality into a single device. To manage this complexity from a design and verification standpoint, devices are often designed as a collection of independently designed and verified components, usually reusable but not necessarily so. In that case, the ASIC is called a System-on-a-Chip (SoC) and its verification resembles a system-level verification process, as described in the next section. The bulk of the functional verification is performed using unit-level verification.

FPGAs now require an ASIC-like verification process.

Traditionally, FPGAs were able to survive an ad-hoc, or even a completely missing, verification process. Their ease of programmability, often without additional component costs, allowed their functionality to be modified up to the last minute. But today's million-gate FPGAs, even with only 50% effective usage, can implement functions that are too complex to verify and debug during integration. Their functionality must be verified from the RTL code, before synthesis and implementation.

System-Level Verification

A system need not follow physical boundaries.

Everybody's definition of a *system* is different. In this book, a system is a *logical* partition composed of independently verified components. A system could thus be composed of a few reusable components and cover a subset of an SoC ASIC. A system could also be composed of several ASICs physically located on separate printed circuit boards, as illustrated in Figure 3-2.

Figure 3-2.
Logical
system
partition

System

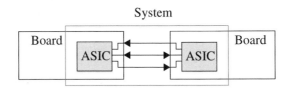

The verification focuses on interaction.

Individual components are specified and designed by separate individuals or teams with assumptions about how they will interact with other components. These assumptions made by different people are a prime source of bugs. System-level verification thus focuses on the interactions among the individual components instead of the functionality implemented in each one. The latter is better verified at the component-level verification. The system verification engineer has to rely on the individual components being functionally correct.

The testcase defines the system.

Since systems are logical partitions, they can be composed of any number of components, regardless of their physical location. Which system to use and verify depends on the testcases that are determined to be interesting and significant. To minimize the simulation overhead, it is preferable to use the smallest possible system necessary to execute the specified testcase. However, the number of possible systems being very large, a set of "standard" systems should be defined. The same system is used for many testcases even if, in some cases, some of the included components are not required.

Board-Level Verification

Board-level models are generated from the board design tool.

The primary objective of board-level verification is to confirm that the "system" captured by the board design tool is correct. Unlike a logical system model, the model for the board design must be automatically generated by the board capture tool. When verifying the board design, or any other physical partition, you must make sure that what is being verified is what will be manufactured. There must be a direct link between the captured design and what is simulated. Automatic generation of the board-level model from the capture tool provides that link. A logical system model has no such restriction: It can be manually generated for the system of interest.

Many components on a board do not fit in a digital simulation environment.

The main difficulty with board-level models is obtaining suitable models for all the components. That is where third-party sources and hardware modelers are useful (see "Verification Intellectual Property" on page 42). Also, generating a model out of a board design tool involves introducing approximations. For example, how do you represent capacitors in a digital simulation environment? Analog devices, connectors, opto-couplers and other components used in board-level designs do not translate easily in a digital simulation environment either.

Board-level parasitics can be modeled.

The generated model may include models for board-level parasitics that may affect the functional correctness of the board. As the speed of signals in a board increases, transmission line effects are becoming important. ASICs can no longer be designed without consideration of their eventual use on a circuit board.

VERIFICATION STRATEGIES

Decide on a black- or white-box approach for various levels of granularity.

Given the functionality that needs to be verified, you must decide on a strategy for carrying out the verification. You must decide on the level of granularity where verification will be accomplished. You must also decide on the invasiveness of the verification approach that will be used for each level of granularity. Testcases can be either white-box or black-box, depending on the visibility and knowledge you have of the internal implementation of each unit under verification (see "Black-Box Verification" on page 12 and "White-Box Verification" on page 13).

Decide on the level of abstraction where the tescases will be specified.

You also need to decide the level of abstraction where the bulk of the verification will be performed. The higher levels of abstraction usually apply to coarser granularity of design under verification. With higher levels of abstraction, you have less detailed control over the timing and coordination of the stimulus and response, but it is much easier to generate a lot of stimulus and verify long responses. If detailed controls are required to perform certain testcases, it may be necessary to work at a lower level of abstraction.

A processor inter-face could be veri-fied at the cycle or device driver level.	For example, verifying a processor interface can be accomplished at the individual read and write cycle levels. But that requires each testcase to have an intimate knowledge of the memory-mapped registers and how to program them. That same interface could be driven at the device driver level. The testcase would have access to a set of high-level procedural calls to perform complete operations. Each operation is composed of many individual read and write cycles to specific memory-mapped registers, but the testcase is removed from these implementation details.

Verifying the Response

Plan how you will check the response.	Deciding how to apply the stimulus is relatively easy. You are under complete control of its timing and content. It is verifying the response that is difficult. You must plan how you will determine the expected response, then how to verify that the design provided the response you expected. The section titled, "Self-Checking Testbenches" on page 341 suggests several techniques for implementing output verification.
Some responses are difficult to ver-ify in the simula-tion.	Throughout this book, implementing self-checking testbenches is recommended (see "Simple Output" on page 252). But, it can sometimes be difficult for a testbench to verify a response that can be recognized immediately as right or wrong by a human. For example, verifying a graphic engine involves checking the output picture for expected content. A self-checking simulation would be very good at verifying individual pixels in the picture. But a human would be more efficient in recognizing a filled red circle. The verification strategy must find a way to automate these types of testcases.
Detect errors as early as possible.	It may be more efficient to have the simulation produce a set of outputs that can be later compared against a set of reference outputs. The result of a simulation can be further processed outside of the simulator to determine success or failure. However, it is more efficient to detect problems as early as possible. When the response is checked within the simulation, the error is identified while the model is near the state that produced the error. It is easier to identify and fix the error.

FROM SPECIFICATION TO FEATURES

Identify features.

The first step in writing a verification plan is to identify the features that will be verified. From the specification document, you enumerate all the features that are described and thus must be verified. Other team members, especially the system architects and RTL designers, contribute additional features to be verified. These additional features may not have been obvious in the specification to someone unfamiliar with the purpose or characteristics of the design. Other features may become significant once a particular implementation is chosen. In *The Art of Verification*[2], Haque, Michelson and Khan propose using a methodical approach for extracting significant and relevant features to verify by first looking at the interfaces, then the functions, then finally the corner cases implied by the chosen architecture.

Enumerate interface-based features.

For every interface on the design to be verified, enumerate every feature it suggests that must be verified. The interfaced-based features can be obtained by asking questions such as:

- What transactions must be applied?
- What range of values?
- What sequences of transaction?
- What are the relevant transaction densities?
- What protocol violations should the design be able to sustain?
- What are the relevant interactions between this interface and other interfaces or internal design structures?
- Do transactions on an interface need to be synchronized with those of another interface?

A subset of the interface-based feature list for a Universal Asynchronous Receiver Transmitter (UART) is shown in Sample 3-1.

---

2. Faisal Haq,ue, Jon Michelson and Khizar Khan, "*The Art of Verification with VERA,*" http://www.verificationcentral.com

Sample 3-1.
Some of the interface-based features of a UART design

1. The Clear-To-Send (CTS) pin must be asserted when the UART can accept a new word to be transmitted via the CPU interface.

2. The Data Terminal Ready (DTR) pin must be asserted when there is a received word ready to be read by the CPU interface.

3. Read and write cycles to addresses other than 0 through 4 are ignored.

4. Back-to-back read/read, read/write, write/write and write/read cycles within the address space are supported.

5. All bits in the configuration registers are readable, writable and non-volatile.

Identify function-based features.

Following the major data paths through the design[3], enumerate every transformation and decision that must be verified. The function-based features can be obtained by asking questions such as:

- What are all the relevant configurations?

- What are the possible transformations that can be performed on the data?

- What are the possible sequences of transformation?

- What are the sensitive data values for triggering transformations?

- What are the sensitive values that affect each transformation?

- Where should the transformed data end up?

- How is the data ordering affected?

- What error detection mechanisms exist and how are they triggered?

- How do error mechanisms report errors?

- What happens to erroneous data?

A subset of the function-based feature list for a UART is shown in Sample 3-2.

---

3. As specified in the architecture specification document, *not* in the implementation.

Sample 3-2.
Some of the
function-based
features of a
UART design

1. Data bits are sent and received serially with the least significant bit first.

2. Data bytes are sent in the same order in which they were written.

3. Data bytes are read in the same order in which they were received.

4. Parity is generated according to configured mode.

5. Parity is checked according to configured mode.

List architecture-based features.

Finally, based on detailed knowledge of the architecture of the design, identify the conditions that will stress the design and push it toward its limit. The architecture-based features can be obtained by asking questions such as:

- Can I overflow or underflow a buffer? If so, what should happen?

- Where are the resource bottlenecks?

- Can multiple requests for these resources occur at the same time?

- Can a transformation path affect, prevent or block another?

A subset of the architecture-based feature list for a UART is shown in Sample 3-3.

Sample 3-3.
Some of the
architecture-
based features
of a UART
design

1. Receiving one more byte while the receive buffer is full will cause that byte to be dropped.

2. The Clear-to-Send (CTS) signal reflects the status of the transmit buffer (asserted when not full).

3. Data is received and transmitted in full-duplex.

Label each feature.

Features should be labeled and have a short description. The feature should be described in terms of what conditions need to be verified and the expected result, not how it is to be implemented. Each feature should be cross-referenced to the section or paragraph in the specification document that describes it in detail. Ideally, the specification document should also contain a cross-reference to the feature list in the verification plan. Specify features for the proper level of verification. The feature label should be used in error messages when it is found to be violated. Including feature labels in error messages will help in identifying what was assumed to have gone wrong and in assessing if the behavior is indeed incorrect.

Assign features to a suitable verification level. When enumerating features, be careful to include them in the verification plan for the proper verification level. Some features are better verified at the component (unit, reusable or ASIC) level, while others must be verified at the system level. Very often, there will be a large number of features concerned with verifying a critical function or block in your design. If the unit implementing that function or containing that block is not being verified independently, now is the time to reconsider your verification approach. It may be an indication that the unit needs to be verified independently to achieve the necessary level of confidence.

Component-Level Features

They are fully contained within the unit being verified. A component can be a unit, a reusable component, or an entire ASIC. Component-level features are fully contained within the component being verified. They do not involve system-level interaction with other components. Their correctness can be determined without depending on a subsequent verification of the integration of the component into a high-level system.

The bulk of the features will be component-level features. These features are assumed to be functional when the component is used in a system-level verification.

System-Level Features

Minimize system-level features. A system can be a subset of an ASIC, a few ASICs from different boards, an entire board design or the complete product. Because of the large size and long runtime of system-level simulations, it is necessary to minimize the features verified at this level. Whenever something is identified as a system-level feature, question whether it can be verified as a component-level feature instead. For example, in the design illustrated in Figure 3-3, the *MX* ASIC can select between the data from ASICs *ID0* or *ID1* under software control. Is the switching feature a system-level feature? The answer is *no*. The switching feature is entirely contained within the *MX* ASIC and is thus a component-level feature.

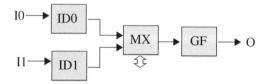

System-level features are usually limited to connectivity, flow-control and inter-operability. For example, the connectivity from the input ports to the output port would be a system-level feature. In verifying the connectivity, it is necessary to switch the input from the *ID0* stream to the *ID1* stream. But the switching is not the primary objective of the verification and would be assumed to work.

System-level features include connectivity, flow control and inter-operability.

Another system testcase would be verifying that full input FIFOs in the *MX* ASIC creates back-pressure through the *ID0* and *ID1* ASICs and stops the flow of data until the congestion clears.

Error Types to Look For

Assume design tools do not introduce functional errors.

When listing features to be verified, there is an implicit assumption about the errors that are likely to occur and should be found. Functional verification must focus on finding functional errors in the design. It is not the responsibility of functional verification to make sure that the design tools are bug-free. Functional simulation ensures that the design was implemented as specified without interpretation errors or problems introduced by the designers. For example, running all functional testbenches on the gate-level netlist only verifies that the synthesis tool works properly. Formal verification and static timing analysis are better tools to accomplish this task.

Likely errors are different based on the capture tool used.

The types of errors that can be made are different when using different capture tools. When schematic capture tools are used, connectivity errors, such as reversed bit orders in a bus, or mis-connected individual bits within a bus, are very common. In an RTL coding and logic synthesis environment, this type of error is not likely to occur: If a bus is properly connected, either all the bits work, or none do. Linting tools can detect some connectivity problems such as multiple drivers on a wire or an output that goes nowhere and would be a better strategy for identifying these types of problems.

Look for functional errors.

Common errors in a synthesis-based design flow include wrong polarities, protocol violations or incorrect computations. The type of stimulus that proved useful in the days of schematic capture, such as walking ones and zeroes may not be as useful in an RTL design verification. A pair of patterns of alternating ones and zeroes, for example "0xAAAA" followed by "0x5555", is usually sufficient.

Using *signatures* in the data stream is another efficient technique to detect functional errors. A signature can be as simple as a sequential number to help detect missing or repeating data items. A signature can also encode either the source or the expected destination of a data item. For example, the data associated with an address in a write cycle could contain a portion of the address and an identification of the bus master issuing the cycle. The section titled, "Data Tagging" on page 343 details how to use signatures to verify a class of designs.

Prioritize

Prioritize the features.

Not all features are created equal. Once they are enumerated, they must be prioritized. Some features are *must-have* for the design to properly function or to meet the demands of the market. This is the stage that defines first-time success. These features must operate properly for the design to be shipped. The completion of the verification of these features gates the successful completion of the project and the testbenches verifying these features are often on the critical path. The *must-have* features need to be thoroughly verified for all possible configuration and usage options.

Less important features receive less attention.

The *should-have* features are secondary for the commercial success of the design. They may simply offer expansion capabilities or differentiation from the competition. The main objective is to verify their basic functionality for correct operation. If time and resources allow, more detailed verification of these features may be accomplished. The verification of these features may be cancelled if schedule pressure forces the reallocation of resources to the verification of more important features.

The Verification Plan

Some features are
verified only as
time allows.

The *nice-to-have* features are purely optional. They are verified
only as time allows, usually in a primitive fashion. The reality of
today's design schedule almost guarantees that they'll never be verified!

Make an informed
decision when cut-
ting back on the
verification effort.

The prioritization of the features to be verified lets a project manager make informed decisions when schedule pressures make it
necessary to eliminate some planned activities. The verification
effort can be trimmed starting with features that were predetermined to be less important. If a greater impact of the project completion date is required and *must-have* features are dropped from
the verification, the decision will be a conscious one as these prior-
ities were clearly identified as critical to the initial marketing objectives. Cutting the verification effort of *must-have* features requires a
conscious re-evaluation of the marketing objectives for the project.

Design for Verification

Hard-to-verify fea-
tures will be iden-
tified.

At this stage of the verification planning, hard-to-verify features
will be identified. They can be difficult to verify because the chosen
partition lacks suitable controllability or observability of the features. An example would be the verification that an embedded 64-
bit counter properly rolls over and that the processing algorithm
works properly across the roll-over point. The difficulty may be
because of a poor choice in verification granularity. In that case, a
smaller partition containing the hard-to-verify features should be
used. The difficulty may also be due to the choice of implementation architecture or an artifact of the design itself. If a smaller partition cannot be used, or would not ease the verification of these
features, a grey- or white-box approach must be taken.

Modify the design
to aid verification.

The advantage of planning the verification up front is that you can
still influence the implementation of the design. If some features
prove to be to difficult to verify given the current architecture and
feature set of the design, have the design modified to include additional features to aid in their verification. Hardware design engineers will no doubt complain about adding functionality that is not
really needed by the design. However, if the alternative is to create
a design you cannot verify, what choice do they have? These features have always proven to be useful during lab integration of sample parts.

Writing Testbenches: Functional Verification of HDL Models

Provide state pre-load functions.

If the design contains long counters or other state conditions which require hundreds or thousands of cycles to reach from reset, make sure they can be pre-loaded to an arbitrary value via a memory-mapped register. Ideally, their current value should be available for read back through the same register set. In the previous example, a series of 8 bytes in the address space of the design could be allocated to pre-loading and reading back the value of the 64-bit counter.

Provide datapath by-pass paths.

The correct implementation of long data paths can also be difficult to verify if you do not have detailed control over all the operands. For example, speech synthesizers are simple digital signal processing designs with a datapath that shapes random noise[4]. You have complete control over the coefficients applied to the data samples to form specific sounds. However, you do not have control over one critical element: the primary input data value. That's a random number. To properly verify the operation of this datapath, you need control over its initial input value.

Figure 3-4. Verifiable datapath for a speech synthesizer

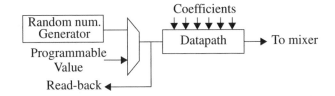

As shown in Figure 3-4, the design should include a mechanism to use a programmable constant input value instead of a random number as input to the datapath. Conversely, you should also be able to read back the output of the random number generator to ensure that it is indeed producing random numbers.

Pop quiz: Why is the read-back point located after the multiplexer that selects between the normal operation using the random number generator and the programmable static value, and not at the output of the random number generator?[5]

---

4. It is used to produce consonant sounds, such as the *sh* sound. It is then mixed with a shaped-base frequency used to produce vowel sounds, such as the *a* sound, which hopefully creates intelligible speech.

Provide sample points.	If observability is the problem but not controllability, adding sample points readable through memory-mapped registers can help ease the verification of some features. If the address space allocated to the design is at a premium, these sample points could be multiplexed into a single address location, using a second address to select which point is currently being sampled.
Provide error injection mechanism.	If the design includes error and exception detection mechanisms, you may want to have provisions to force the detection of an error or exception. For example, verifying the maskability of interrupts is very time consuming if the design has to be coerced into every exception condition. The same task is rendered considerably easier if a simple register write can manually raise the same interrupts. Of course, the task of verifying that the exception condition raises the interrupt remains. The decision to include error injection should be carefully considered. If it is for hardware verification only, it may not be properly documented for the software engineers. This feature may be accidentally turned on when a device driver writes a value that was thought to be inoffensive.

DIRECTED TESTBENCHES APPROACH

	With directed testbenches, individual features are verified using individual testbenches. The stimulus is manually crafted to exercise that feature. The response is verified against the symptoms that would appear should the feature not be correctly implemented.
Use for small number of testcases.	Before you embark on the directed testbenches path, you need to consider its lack of scalability. This approach can be managed and completed if the total number of testcases is in the low hundreds. But as the number of testcases grows, so does the number of testbenches. A project with over a thousand identified testcases would require over a year to complete using a directed approach. For a larger number of testcases, some form of testbench automation is necessary to complete the task within an acceptable time frame. Currently, the best method of testbench automation is the coverage-driven random-based approach (see page 109).

5. You want to verify that, when the datapath is put into normal operation mode, the multiplexer is functionally correct and the input value is indeed coming from the random number generator.

Group into Testcases

Group features
with similar verifi-
cation require-
ments.

Features naturally fall into groups. Some features require similar configuration, granularity or verification strategy to perform their verification. To maximize productivity, these features should be grouped together and assigned to the same verification engineer. For example, all features related to the CPU interface should be grouped together. As another example, verifying the baud rate, number of data bits and parity generation of a UART falls within the same group. Each group of feature verification forms a *testcase*.

Cross-reference
into the feature
list.

Each testcase should be labeled and given a short description of its objective. Its description should contain a list of the features verified in this testcase. The feature list should also be annotated with cross-references to the testcases where a particular feature is being verified. If a feature does not have a cross-reference to a testcase, it is not being verified.

Define dependen-
cies.

The description of a testcase should also contain a list of the features assumed to be operational and functionally correct. From these dependencies, you can determine the order in which the testcases must be written, and identify any parallelism opportunities in the testbench development effort.

Specify the
testcase stimulus.

The sequence and characteristics of the stimulus for the testcase must also be described. For example, describe the various operations or bus cycles that must be performed. It is a good idea to fill all non-relevant or background data with random values or transactions.

Specify the accep-
tance criteria.

More than just the expected response, the testcase specification must state how the response will be determined as valid. This includes expected values, timing and protocol. For example, the output of a packet processor could be determined as correct solely on the basis of the destination address matching the output port where it appeared. Or, a more stringent requirement could be specified, such as packets from different sources showing up in the proper order and interleaved with a proper distribution.

Specify what
errors to look for.

One of the more explicit ways of describing acceptance criteria is to state exactly which errors to look for. For example, making sure that a packet comes out with a correct CRC value. Another example is to describe events that are mutually exclusive, such as the assertion of the *full* and *empty* flags in a FIFO. Being explicit about what errors to look for lets a verification engineer, who is not intimately familiar with the design, implement a highly reliable testbench.

Inject errors to make sure they are detected.

Never trust a testbench that does not produce error messages. Every testcase should include some error injection mechanism to make sure that errors are detected and reported by the testbench. The absence of an error message would be a failure condition for that testcase. For example, a testcase verifying the parity generation in a UART should purposefully misconfigure the parity in the UART to make sure that the testbench detects a wrong parity. Of course, the testbench must not abort the simulation as soon as the error message is issued.

From Testcases to Testbenches

Testcases naturally fall into groups.

Just like features, testcases naturally fall into groups. They require a similar configuration of the design, use the same abstraction level for the stimulus and response, generate similar stimulus, determine the validity of the response using a similar strategy, or verify closely-related features. For example, the testcase verifying that a UART properly transmits data can be grouped with the testcase that verifies its configuration controls. Both need similar stimulus (a variety of data words to transmit), and both verify the correctness of the output in a similar fashion (is the data value identical, with no parity error).

Group testcases into testbenches.

Each group of testcases is then divided into testbenches. A popular division, the one used in this book, is one testcase per testbench. The minimization of Verilog compilation time, or the time spent back-annotating a large gate-level netlist with a correspondingly large Standard Delay File (SDF) may dictate that a minimum number of testbenches be created by grouping several testcases into a single testbench.

Cross-reference testbenches with testcases.

Each testbench should be labeled and uniquely identified. This identifier should be used as the filename where the top-level code for the testbench is implemented. For each testbench, enumerate the list of testcases it implements. Then cross-reference each testbench into the testcase list. The description of a testcase should contain the name of the testbench where it is implemented. If a testbench is not identified, a testcase has not yet been implemented.

Allocate each group to an engineer.

Regardless of the division of testcases into testbenches, allocate each group of testcases to a verification engineer. Testcases in the same group have similar implementation requirements. They can build on the implementation of previous testcases in the group. The first testbench takes the longest to write. But as engineers responsible for each testcase group gain experience and debug their verification infrastructure, a lot can be reused, often through cut-and-paste, in subsequent testbenches. The name of the individual to whom a testbench has been assigned should be recorded in the verification plan. That person is responsible for implementing the testbench according to its specification.

Verifying Testbenches

How do you verify that testbenches implement the verification plan?

The purpose of the verification effort and writing testbenches is to verify that a design meets its specification. If the verification plan is the specification for the verification effort, how do you verify that the testbenches implement their specification? How can you prevent a significant portion of a testcase from being skipped because of human error? Testbenches often include temporary code structures to bypass large sections to speed up the debugging of a critical section. How can you make sure that they are taken out, returning the testbench to implementing the entire set of testcases it is supposed to contain?

Verify testbenches through peer reviews.

As described in "The Human Factor" on page 6, one way to verify a transformation performed by a human (in this case, writing a testbench from a specification), is to provide redundancy. Once completed, testbenches should be reviewed by other verification engineers to ensure that they implement the specification of the testcases they contain. For more details, refer to section "Code Reviews" on page 32. The simulation output log should also be reviewed to ensure that the execution of the testbench follows the specification as well. To that effect, the testbench should produce regular notice messages. It should state what stimulus is about to be generated, and what error or response is being checked. The output log should ultimately contain, in a bullet form, the specification of the testcases that have been executed.

Use functional coverage.

Another redundant path is functional coverage measurement. By specifying, through a functional coverage model, what you expect a directed testcase to accomplish, you can obtain a positive confirmation that the testcase was indeed executed. After a directed test-bench is run, the functional coverage metrics should meet 100% of the goal. Since the stimulus was manually coded, it is deterministic and should fill 100% of the relevant and interesting value sets. For example, a directed testcase that is supposed to iterate through all possible values for a particular configuration register can be covered by recording all values written to the address corresponding to that configuration register.

Measuring Progress

Testcase completion measures progress.

In a directed testbench approach, progress is measured using a simple table. On one dimension, all of the testcases are listed. On the other, the current status of each testcase is tracked throughout its lifetime: assigned, coded, running, passing, reviewed. Figure 3-5 shows the progress of a directed testbench approach. Initially, little progress is made because the verification infrastructure is being developed and the design is being debugged. Once the first testcase completes successfully, the progress will accelerate as less and less bugs remain to be found, and more and more verification infrastructure code is reused. This acceleration may not translate into an accelerated testcase completion rate as testcases become increasingly complex to implement.

Figure 3-5. Progress of a directed testbench approach

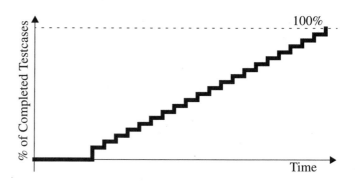

When are you done?

The completion of all testcases does not necessarily indicate that the verification task is over. Code coverage metrics can indicate that the original set of testcases is not as thorough as imagined and additional testcases must be created to increase the code coverage scores to more acceptable levels. In reality, "done" is usually defined when you have to ship the design and you are confident enough that the must-have features are working properly.

COVERAGE-DRIVEN RANDOM-BASED APPROACH

The HVL productivity cycle ("Verification Languages" on page 62) rests on constrained random verification. You can use HVLs to implement a directed testcase approach as described in the previous section. Their high-level programming language constructs would facilitate the implementation of testcases. In fact, the HVLs with lesser support for functional coverage and constrained random generation are designed for that purpose. However, it would only increase the slope of the testcase completion curve somewhat (Figure 3-5), not alter the nature of the curve. Changing the curve itself requires changing how verification is approached.

Random verification still provides valid stimulus.

Random verification does not mean that you randomly apply zeroes and ones to every input signal in the design. This would not represent an accurate usage of the design and would not accomplish anything. With random verification, the inputs are subjected to valid individual operations, such as a read cycle or an ethernet packet. It is the sequence and timing of these operations and the content of the data transferred that is random. Through the addition of constraints, a random testbench can be steered toward exercising specific features.

Measuring Progress

There are too many testcases.

Today's multi-million gate ASICs contain hundreds of features to be verified for hundreds of different combinations of data values. Assuming a bug-free design and a team of highly productive engineers who can code and debug a self-checking testbench in three days, a team of 10 verification engineers (a rarity by today's standards) would require over seven months to implement 500 testcases. The number of testcases cannot be reduced. Throwing more engineers at the problem quickly produces diminishing returns. The only way to reduce the verification time is to write more testcases in less time. In other words, exercise the same functionality with less code.

Testcases exercise more than the target feature.

Although each testbench, when verifying a testcase, considers the target feature in isolation, applying stimulus to the design exercises other features at the same time. Since progress, in a directed approach, is tracked by associating features with testbenches, how can you track progress against features that are not explicitly coded and verified in a testbench?

Measure functional coverage, not features.

The solution is to measure progress against functional coverage points that will identify whether a feature has been exercised. The objective becomes filling a functional coverage model of your design rather than writing a series of testcases. You could fill this coverage model using large directed testbenches. Or you could let a random testbench create the testcases and exercise the features for you. Figure 3-6 shows progress using a coverage-driven approach with a random testbench against the traditional directed approach. The former trades-off longer initial testbench development time for more productive feature coverage in the long run. The promised ultimate productivity gain should not be measured on this qualitative plot: It depends highly on your commitment to this approach, and your experience in writing random generators that can be constrained easily ("Random Stimulus" on page 354).

Figure 3-6. Progress of a coverage-driven testbench approach

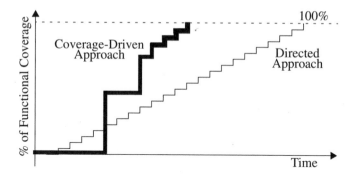

You will develop more confidence.

Directed testcases can only find bugs you were looking for. Random simulations will create conditions that you have not thought of when writing your verification plan. They create unexpected conditions or hit corner cases. They also reduce the bias introduced by the verification engineer when coding directed testbenches. Instead of creating input sequences that are easy or familiar to code, they create more realistic stimulus. Because your design will have been exercised under a larger number of conditions (compared to a directed approach during the same time period), the overall quality of the design will be higher.

This approach requires commitment.	Using a constraint-driven approach requires commitment. Under pressure, it is too easy to fall back to writing directed testcases. A critical component of this approach is that you need to simulate your testbench and your design to know how much functional coverage you have achieved. If the RTL model is not available on time (and it never is), how can you debug your self-checking random testbench? How can you show that the verification team is making progress towards its functional coverage goals? The easy answer is to start writing testcases as directed pseudo-random testbenches that implicitly fill functional coverage points. That puts you back on the staircase curve. A better approach is to stage the RTL delivery to enable simulations as early as possible and to use behavioral models. For more details on behavioral models, see "Behavioral Models" on page 375.

From Features to Functional Coverage

Start with functional coverage.	In a coverage-driven approach, functional coverage measurement is used to identify which testcases were executed instead of explicitly coding those testcases. Thus, it is important to implement functional coverage models and collect functional coverage measurement right from the start. Functional coverage is not like code coverage. The latter is often added to the verification process toward the end to measure how thoroughly the code is being exercised and to identify implementation code that was not exercised. Functional coverage is used from the beginning of the project to record which testcases and conditions were automatically created by the random generator. If you are not using functional coverage in tandem with your random environment, I'm afraid you are only doing directed testcases with random stimulus filling.
Measure symptoms of data indicative of feature.	Each feature presents a characteristic or symptom in the input data stream, the design configuration or the internal state of the design that must be exercised. Functional coverage must identify, then record, those characteristics and symptoms. Sample 3-4 shows a description of the functional coverage items used to identify that the interface-based features identified in Sample 3-1 have been exercised.
Define your goal.	The functional coverage tool can help you measure your progress only if your goals are explicitly defined. It will also make analysis of the functional coverage easier. The progress will be measured

Sample 3-4. Functional coverage for interfaced-based features of a UART design	1. Level of the Clear-to-Send (CTS) pin.
	2. Level of the Data-Ready (DTR) pin.
	3. CPU cycle kind crossed with address.
	4. CPU cycle kind transition.
	5. CPU cycle kind crossed with address crossed with data.

against a constant goal. If the goals are intellectually defined every time you analyze a functional coverage report, then these goals are subject to human error. There will also be a tendency to minimize the importance of holes toward the end of the project as you sub-consciously justify your progress against the looming deadline. Sample 3-5 describes the functional coverage goals for each of the functional coverage points identified in Sample 3-4. Different features may use the same coverage point but imply a broader goal.

Sample 3-5. Functional coverage goals for interfaced-based features of a UART design	1. A least one value of 0 and 1 observed.
	2. At least one value of 0 and 1 observed.
	3. At least one read and write cycle with address greater 4.
	4. All combination of read and write cycles.
	5. At least one read and write cycle for each address equal to configuration register and with individual bits equal to 0 and 1.

Understand the complexity of the goal.

Don't make your goal more accurate or precise that it needs to be. The more value sets that must be filled to meet your goal, the more work it is going to be. With cross-coverage, the number of values that must be filled grows exponentially. For example, it is not real-istic to attempt to cover all possible values for a 32-bit address for both read and write cycles: That is over 8 billion values. Define value sets for equivalent values or combination of values to mini-mize the number of samples required to meet your goal. Functional coverage tools have a limit on the total number of value sets that must be filled to provide a measure of the coverage. For example, Specman Elite will give you a percentage score of your functional coverage space, if and only if, the goal has a maximum of 16 value sets to be filled[6].

Question, reduce, inform.

It is very easy to collect a large number of functional coverage metrics. But the more functional coverage data you have, the harder it becomes to analyze the results. Always question the relevance of a functional coverage point. If you start to ignore some coverage point reports or are not looking forward to the next report, you should probably not collect it. There is a fundamental difference between data and information. Haphazard functional coverage points only provide data that must be analyzed. Well-chosen functional coverage points with well-defined goals provide information that is immediately meaningful. For example, if a FIFO must be exercised across its operating range, measuring the values of the read and write pointers would be data. Measuring the *difference* between the read and write pointers with goals stated as empty, full and neither would be much more meaningful. Crossed with some critical pointer regions (such as roll-over points), this latter functional coverage point will provide much more relevant information.

Functional coverage definition is an evolving art.

Developing a good functional coverage model of your verification plan is not easy. This section has described the necessary steps only in the broadest of terms. Functional coverage modeling is a topic that can and should be developed into a science with well-defined processes. It would require a book onto itself!

From Features to Testbench

Identify how correctness will be determined.

Note that the functional coverage points described above do not make any reference to the correctness of the results. Correctness is the responsibility of the self-checking portion of the testbench. Given the features that must be verified, you have to determine how its correctness is going to be confirmed. The process is similar to identifying the expected response in a directed testcase exercising that feature. The difference is that you do not know the timing or ordering of the stimulus that will trigger the feature. Errors can be detected by a failure of the random testbench to operate properly, by explicitly comparing output data against expected data in the self-ckecking structure, or white-box assertions on the design itself. The list of error detection mechanisms becomes a detailed specification of the self-checking random testbench. Sample 3-6 shows

---

6. That number is configurable. This is the default value.

the error detection mechanism that will confirm the correctness of the features identified in Sample 3-1.

Sample 3-6.
Error detection for interfaced-based features of a UART design

1. Data source will wait for Clear-to-Send (CTS) pin to be asserted before writing the next data to send. If it is not functional, no data will be transmitted.

2. Data sink will wait for the Data-Terminal-Ready (DTR) pin to be asserted before reading the next received data. If it is not asserted, no data will be received.

3. Covered by #4.

4. Verify that read cycles return expected values given the previous values written, writability of bits and reset value, size and presence of registers.

5. Covered by #4.

Identify termination mechanisms.

It is easy to terminate a directed testcase: Once you are done applying the stimulus necessary to exercise the target feature, exit from the simulator. But a random testbench is not about exercising a single feature. How do you know when to stop? You have to plan for several termination mechanisms that can be triggered or turned off through additional constraints on the random testbench or a simple procedural call at the beginning of a simulation run. Any one termination mechanism, once triggered will cause the orderly shutdown of the simulation.

There are several popular termination mechanisms. A watchdog timer is useful to prevent deadlocked simulations: It must be reset at regular interval, otherwise the simulation is terminated. A time bomb helps prevent run-away simulations: It will terminate the simulation after a predetermined amount of time. An idle detector will stop the simulation when all of the output interfaces have been idle for some time. A simple data item counter will terminate the simulation after a specified number of data items was sent to or received from the design. Functional coverage feedback can terminate the simulation if the metrics are not significantly increasing or if the coverage goal has been reached.

You can run for a long time, or you can run many times.

You can generate a lot of random data using two strategies: You can run a random source for a long time, or you can run a random source many times, for a short time, each time with a different seed. If the random source is truly random and the seeds are chosen as not to repeat a previous sequence, then the quality of the resulting random data should be the same. However, the effects on a simulation of each strategy are quite different.

Plan for many short runs.

A long simulation will be cumbersome to reproduce if the error is detected toward the end. Furthermore, since a single simulation run will typically use a single configuration of the device, you will have less opportunities to verify different configurations. Your device may also find itself in a particular corner of the state space and remain in that corner. Continuing to apply more stimulus under those conditions is unlikely to yield increased functional coverage.

Using the many-short-runs strategy, you can reproduce a problem more quickly, run many more configurations and quickly traverse the state space. Since HVLs are separate from Verilog and VHDL, it is not necessary to recompile the design to run a different simulation. If your device requires a lot of simulation cycles to reach certain states after reset, consider state-forcing mechanisms as described in "Design for Verification" on page 102.

From Features to Generators

Identify stimulus requirements.

To be able to fill your coverage goals, it will be necessary for the random generators to generate the necessary data, with the necessary characteristics and necessary timing. It is very simple to generate a single packet or instruction with random content. But this simple random generation approach is likely insufficient if you need the ability to generate packets of different lengths, lots of consecutive packets of the same lengths, straight-through instruction sequences, nested loop structures, invalid or corrupted data or synchronized data across multiple random streams.

A random generator that will be able to exercise the required features does not happen by accident. It has to be designed and architected to produce the required data sequences. Sample 3-7 shows the random generator requirements necessary to exercise the features identified in Sample 3-1.

Sample 3-7.
Generator
requirements
for interfaced-
based features
of a UART
design

1. Generate send data stream.

2. Generate receive data stream.

3. Generate read or write cycles.

4. Covered by #3.

5. Covered by #3.

Constraints
become preferable
to more seeds.

What if, after multiple random simulations, some of your functional coverage points remain unfilled? You could run more simulations with additional seeds, or you could add constraints to your testbench to increase the probability (hopefully to 100%) of filling at least one of the remaining functional coverage points. The latter, although requiring more work on your part, is likely to be the more productive avenue, especially if hundreds of previous runs failed to produce the necessary inputs to fill those coverage points. The gap in your functional coverage measurements could also be a symptom of a functional problem in your random generators or verification environment. It is possible that a lingering constraint is preventing the generation of input sequences that will cause the functional coverage points to be filled.

Identify constrain-
able dimensions.

When architecting your random generator, it is necessary to consider the available constraint mechanisms. Traditionally, different random streams were produced by physically altering the code of the random generator. As shown in Figure 3-7, altering the code of the random generator effectively created a different random generator for each testbench.

Figure 3-7.
Different
random
generators

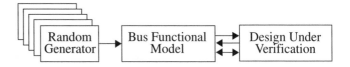

To minimize the amount of duplicated code and the amount of new code that must be written to fill additional functional coverage points (and thus to be more productive), it is better to design a random generator that can be constrained easily, from the outside, as illustrated in Figure 3-8. Writing a random generator that can be constrained easily from the outside does not happen by accident.

The section titled "Random Stimulus" on page 354 shows how to write such generators. Sample 3-8 shows the constraint mechanisms that must be available in the generators to exercise the features in Sample 3-1.

Figure 3-8.
Constraining a
single random
generator

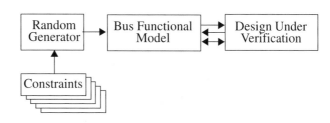

Sample 3-8.
Constraint
requirements
for interfaced-
based features
of a UART
design

1. No constraints.

2. No constraints.

3. Must be able to constrain address.

4. Must be able to constrain type of cycle in sequences of cycles.

5. Must be able to constrain address and data.

Randomly gener-
ate the device or
testbench configu-
ration.

It is easy to conceive of randomly generating data streams throughout a simulation. But you can just as easily randomly generate data that is used only once, at the beginning of the simulation. For example, configurable or programmable devices are often verified using only a few configurations hard-coded in the testbench or an external file. And most simulations are usually run using one of those configurations.

Why not randomly generate the device configuration then download it into the device? The device configuration descriptor is then used by the self-checking structure to predict the response according to the current configuration. Similarly, you could randomly generate the configuration of the testbench. For example, when verifying an ethernet switch, why not randomly generate the number of devices on each port, their speed and their station MAC addresses? Then use functional coverage measurement on the randomly generated configuration to know which configurations and combinations of configuration parameters were verified.

Constrain configu-
rations if neces-
sary.

Your self-checking structure does not yet support all possible device configurations? Or, you are unable to "compile" all possible configuration into register writes? Or, you are migrating from a Verilog testbench that can use only two configurations through *$readmemh* tasks? No problem. Simply constrain the configuration generator to generate only the supported configurations. Once you are able to support additional configurations, remove the constraints accordingly.

Directed Testcases

Identify low-prob-
ability testcases.

There are some features that will have a low probability of being exercised through random stimulus. For example, verifying that interrupt bits are maskable would require that the mask bits be randomly set to 1 and 0 while the associated interrupt bits were 1 and 0 (i.e., fill the cross-coverage of the interrupt bit value with the mask bit value, for all interrupt bits). Given that interrupts usually signal exceptional events in a design, it will likely take a very long time for random stimulus to completely verify this feature. These features a probably better verified using directed testcases.

May be imple-
mented using con-
straints.

Writing a directed testcase does not necessarily imply using a directed procedural implementation. If the random generators were designed to be highly constrainable, it is possible to constrain them so much that they will produce directed stimulus. For example, to verify the maskability of a particular interrupt, you would constrain the CPU cycle generator to write a 0 then a 1 in the appropriate position at the interrupt mask register address. You would then constrain the data generators to cause the interrupt condition. Once the condition is detected, constrain the CPU cycle generator to write a 0 and 1 in the mask bit again. An assertion would verify that the external interrupt would be asserted only when the interrupt condition is not masked. However, if the most productive approach is to write a directed procedural testcase, the random environment can be suspended to allow access to the transaction layer of the bus-functional model.

The first testcases
are the simplest
but also the tough-
est.

When a new version of the design first hits the verification team, it is subjected to a few simple testcases. The objective of these testcases is to verify that the basic functionality of the design operates correctly. Once it passes these initial debug tests, it will be subjected to high volumes of traffic to thoroughly verify the design.

These first debug testcases, although very simple, are the toughest ones to pass. You may spend weeks running the same simple tests. Because they are used on immature code, they catch the most bugs. These early debug testcases usually involve performing a write cycle followed by a read cycle, or transmitting a single packet,or executing a few straight-through instructions.

Debug testcases can be random.

Because of their simplicity, you could be tempted to write the first debug testcases as directed testcases. Given well-designed generators, they are usually much simpler to write as constrained tests. For example, constraining the test to two cycles, where the first one must be a write cycle, the second must be a read cycle and both addresses must be the same. Or, constrain the packet generator to generating only one packet for the entire simulation. Alternatively, constrain the instruction generator to generate only arithmetic opcodes without any branches. Once the initial debug tests pass successfully, you remove these constraints and let the random environment loose on the design.

SUMMARY

Write a verification plan. It is the specification for all testcases and supporting testbench functions. Implement and verify from a common specification. Do not verify an implementation.

Define the various levels of granularity used to verify the design: block, unit, component, FPGA, ASIC, subsystem, board. Trade off greater visibility and controllability for fewer testbenches and more integration tests.

Define the self-checking strategy that will be used to detect errors.

Identify features from the design specification, and enumerate which features must be verified.

Consider verification early in the design phase. Architect the design as needed to make it as easy to verify as possible.

You can use a directed testbench approach if the number of testcases is small. For each feature, specify a testcase. Implement each testcase in a separate testbench.

Define a functional coverage model from the enumerated features. From those same features, identify the degrees of freedom and constraint dimensions of the generators required to generate the stimulus that will exercise each feature.

CHAPTER 4   HIGH-LEVEL MODELING

A skilled verification engineer must break the "RTL mindset" that most hardware engineers, out of necessity, have grown into. To efficiently accomplish the verification task, you must be well versed in behavioral (i.e., non-synthesizeable and highly algorithmic) descriptions. To reliably and correctly use the behavioral constructs of any modeling language, it is necessary to understand the side effects of the simulation algorithm and the limitations of the language—and to understand ways to circumvent those side effects and limitations. This understanding was not required to write RTL models successfully.

BEHAVIORAL VERSUS RTL THINKING

In this section, I illustrate the differences between the approaches to writing an RTL model and to writing a behavioral model.

Many guide-lines help code RTL models.

All experienced hardware design engineers are very comfortable with writing synthesizeable models. The models conform to a well-defined subset of the VHDL or Verilog languages and follow one of a few coding styles. Numerous RTL coding guidelines have been published.[1] They help designers obtain efficient implementations: low area, high speed or low power. Guidelines, such as the ones shown in Sample 4-1, can help a novice designer avoid undesirable hardware components, such as latches, internal buses or tristate buffers. More importantly, guidelines can help maintain identical behavior between the synthesizeable model and the gate-level implementation, such as the ones shown in Sample 4-2.

Sample 4-1.
RTL coding guidelines to avoid undesirable hardware structures

1. To avoid latches, set all outputs of combinatorial blocks to default values at the beginning of the block.

2. To avoid internal buses, do not assign *reg*s from two separate *always* blocks (Verilog only).

3. To avoid tristate buffers, do not assign the value Z (VHDL) or 1'bz (Verilog).

Sample 4-2.
RTL coding guidelines to maintain simulation behavior

1. All inputs must be listed in the sensitivity list of a combinatorial block.

2. The clock and asynchronous reset must be in the sensitivity list of a sequential block.

3. Use a nonblocking assignment when assigning to a *reg* intended to be inferred as a flip-flop (Verilog only).

The adherence to the synthesizeable subset and proper coding guidelines can be verified easily using a linting tool (more details are in the section titled "Linting Tools" on page 26). After several months of experience, the subset becomes very natural to hardware designers. It matches their mental model of a hardware design: state machines, operators, multiplexers, decoders, latches and clocks etc.

Do not use RTL-like code when writing testbenches.

The synthesizeable subset is adequate for describing the implementation of a particular design. I often claim that VHDL and Verilog are both equally poor at this task. The subset is dictated by the synthesis technology, not by someone with a warped sense of humor playing a practical joke on the entire industry. It is designed to describe hardware structures and logical transformations between registers, matching the capability of logic synthesis technology.

---

1. The IEEE has published standard definitions for RTL coding.

For Verilog, see *"IEEE P1364.1 Standard for Verilog Register Transfer Level Synthesis"* prepared by the Verilog Synthesis Interoperability Working Group of the Design Automation Standards Committee.

For VHDL, see *"IEEE P1076.6 Standard for VHDL Register Transfer Level Synthesis"* prepared by the VHDL Synthesis Interoperability Working Group of the Design Automation Standards Committee.

However, this subset quickly becomes insufficient when writing testbenches that were never intended to be implemented in hardware. HDL and HVL languages have a rich set of constructs and statements. If you have an RTL mindset when writing testbenches and limit yourself to using a coding style designed to describe relatively low-level hardware structures, you will not take full advantage of the language's power. The verification task will be needlessly tedious and complicated.

Contrasting the Approaches

The example below shows a simple handshaking protocol. Your task is to write the VHDL or Verilog code that implements the simple handshaking protocol shown in Figure 4-1. The protocol detects that an acknowledge signal (ACK) is asserted (high) after a requesting signal (REQ) is asserted (high). Once the acknowledge signal is detected, the requesting signal is deasserted, and the protocol then waits for the acknowledge signal to be deasserted.

Figure 4-1.
State diagram
for
handshaking
protocol

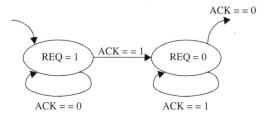

RTL-Thinking Example. A hardware designer, with an RTL mindset, will immediately implement the state machine shown in Figure 4-1. The corresponding VHDL code is shown in Sample 4-3. This relatively simple behavior required 28 lines of code and two processes to describe, and two additional states in a potentially more complex state machine.

Focus on behavior, not implementation.

Behavioral-Thinking Example. A verification engineer, with a behavioral mindset, will instead focus on the *behavior* of the protocol, not its implementation as a state machine. The corresponding code is shown in Sample 4-4. The functionality can be described behaviorally using only four statements.

<table>
<tr><td>

Sample 4-3.
Synthesize-
able VHDL
code for sim-
ple handshak-
ing protocol

</td><td>

```vhdl
type STATE_TYP is (..., MAKE_REQ, RELEASE, ...);
signal STATE, NEXT_STATE: STATE_TYP;
...
COMB: process (STATE, ACK)
begin
   NEXT_STATE <= STATE;
   case STATE is
   ...
   when MAKE_REQ =>
      REQ <= '1';
      if ACK = '1' then
         NEXT_STATE <= RELEASE;
      end if;
   when RELEASE =>
      REQ <= '0';
      if ACK = '0' then
         NEXT_STATE <= ...;
      end if;
   ...
   end case;
end process COMB;

SEQ: process (CLK)
begin
   if CLK'event and CLK = '1' then
      if RESET = '1' then
         STATE <= ...;
      else
         STATE <= NEXT_STATE;
      end if;
   end if;
end process SEQ;
```

</td></tr>
<tr><td>

Sample 4-4.
Behavioral
VHDL code
for simple
handshaking
protocol

</td><td>

```vhdl
process
begin
   ...
   REQ <= '1';
   wait until ACK = '1';
   REQ <= '0';
   wait until ACK = '0';
   ...
end process;
```

</td></tr>
</table>

Behavioral models are faster to write.	Modeling this simple protocol using behavioral constructs should require less than 10% of the time required to model it using synthesizeable constructs. Not only is there less code to write (14%), but also it is simpler, requiring less effort to ensure that it is correct.
Behavioral models simulate faster.	Another benefit of behavioral modeling is the increase in simulation performance. Assuming that there is a long delay between a change in the request and the corresponding acknowledgement, the simulation of the synthesizeable model would still execute the *SEQ* process at every transition of the clock (because that process is sensitive to the clock signal). The process containing the behavioral description would wait for the proper condition of the acknowledge signal, resuming execution only when the protocol is satisfied. If the acknowledge signal replies after a 10 clock-cycle delay, this represents a reduction of process execution from 40 in the synthesizeable version to 2 in the behavioral one, or a 1900% increase in simulation performance.

YOU GOTTA HAVE STYLE!

The synthesizeable subset puts several constraints on the coding style you may use. Even with these restrictions, many less experienced hardware designers manage to write RTL code that is difficult to understand and maintain. There are no such restrictions with behavioral modeling. With this complete and thorough freedom, it is not surprising that even experienced designers produce testbench code that is unmaintainable, fragile and not portable.

A Question of Discipline

| Write maintainable, robust code. | There are no laws against writing bad code. If you do, the consequences do not involve personal fines or prison terms. However, the consequences do involve a real economic cost to your employer. Your code *will* need to be modified: either to fix a functional error, to extend its functionality or to adapt it to a new design. When (not *if*) your code needs to be modified, it will take the person in charge of making that modification more time than would otherwise have been required had the code been written properly the first time. Under extreme conditions, your code may even have to be re-written entirely.[2] |

My first job after graduating from the university was to design and implement a portion of a logic synthesis tool using the C language. In those days, I had been writing code in various languages for over eight years, and I measured my performance as a software engineer by the cleverness of my implementations of algorithms. I felt really proud of myself when I was able to craft a complex computation into a "poetic" one-liner. C is the ultimate software craftsman language!

Invest time now, save support time later.

I soon came to realize the error of my ways. During the eight previous years, I always wrote "disposable" code: The programs were either short-lived (school assignments or personal projects), or they had a narrow audience (utilities for university professors or a learning aid for a particular class). Never had I written a program that would live for several years and be used by dozens of persons, each with their own sophisticated needs and attempting to use my program in ways I had never intended or even thought of. As I found myself having to fix many problems reported by users, I had difficulties understanding my own code written only weeks before. I quickly learned that time invested in writing better code up front would be saved many times over in subsequent support efforts.

Optimize the Right Thing

You should *always* strive for maintainability. Maintainability is important even when writing synthesizeable code. Before optimizing some aspect of your code, make sure it really needs improvement. If your code meets all of its constraints, it does not need to be optimized. Maintainability is the most important aspect of any code you write because understanding and supporting code is the most expensive activity.

Saving lines actually costs money.

There is no economic reason to reduce the number of lines of code. Unless, of course, it also improves the maintainability. Saving one line of code, with an average of 50 characters per line, saves only 50 bytes on the storage medium. With 40GB hard drives costing less than $80 in 2002[3], this represents a savings of one hundred thousandth of one cent ($0.00000001). The time saved in typing,

---

2. Do not think, "*It won't be my problem.*" You may very well be that person and you may not be able to understand your own code weeks later.

3. 93% cheaper in the 3 years since the first edition of this book!

assuming an extremely slow typing speed of one character per second and a loaded labor rate for an engineer at $100,000 a year[4], amounts to $0.69. However, if saving that line reduces the understandability of the code where it will require an additional five minutes to figure out its operation, the additional cost incurred amounts to $4.17. The total *loss* from reducing code by one line equals $3.48. And that is for a single line and a single instance of maintenance.

Optimizing performance costs money.

Similar costs are incurred when optimizing code for performance. These optimizations usually reduce maintainability and must be done only when absolutely required. If the code meets its constraints as is, do not optimize it. That principle applies to synthesizeable code as well. The example in Sample 4-5 is a design example provided in the Vera distribution. It is a synthesizeable description of a 2-bit round-robin arbiter.

RTL code can be too close to schematic capture.

Several aspects of maintainable code were used in Sample 4-5: Identifiers are meaningful and the code is properly indented. However, the continuous assignment statements implementing the combinatorial decoding suggest that the author was thinking in terms of Boolean equations, maybe even working from a schematic design, not in terms of functionality of the design.

This approach simplifies the understanding of the final implementation at the cost of functional understanding. From each concurrent statement, it is easy to figure out the logic gates and flip-flops necessary to implemented it. But try to figure out what happens to the content of the *last_winner* register when there are no requests, or add a third request and grant signal pair. Understanding or modifying the functionality is much more difficult. Other potential problems are the race conditions created by using the blocking assignments in the *always* block (for more details, see "Read/Write Race Conditions" on page 209).

Specify function first, optimize implementation second—and only if needed.

The code shown in Sample 4-6 implements the same function, but it is described with respect to its functionality, not its gate-level implementation. The code sample simplifies the understanding of the function but makes no attempt at describing the final implementation. It is much easier to figure out what happens to the content of

---

4. That, however, is pretty much the same...

Sample 4-5.
Synthesize-
able code for
2-bit round-
robin arbiter

```
/*
################################################
# PROPRIETARY AND CONFIDENIAL               #
# SYSTEMS SCIENCE INC.                       #
# COPYRIGHT (c) 1995 BY SYSTEMS SCIENCE INC. #
################################################
*/
module rrarb(request, grant, reset, clk);
input   [1:0] request;
output  [1:0] grant;
input         reset;
input         clk;
wire          winner;
reg           last_winner
reg     [1:0] grant;
wire    [1:0] next_grant;

assign next_grant[0] =
    ~reset & (request[0] &
             (~request[1] | last_winner));

assign next_grant[1] =
    ~reset & (request[1] &
             (~request[0] | ~last_winner));

assign winner =
    ~reset & ~next_grant[0] &
    (last_winner | next_grant[1]);

always @ (posedge clk)
begin
    last_winner = winner;
    grant = next_grant;
end
endmodule
```

the *last_winner* register when there are no requests or when adding
a new request and grant signal pair. The synthesized results should
be close to that of the previous model. The synthesized results
should not be a concern until it is demonstrated that the results do
not meet area, timing or power constraints. Your primary concern
should be maintainability, unless shown otherwise.

Sample 4-6.
Behavioral
code for 2-bit
round-robin
arbiter

```
module rrarb(request, grant, reset, clk);
input   [1:0] request;
output  [1:0] grant;
input         reset;
input         clk;

reg [1:0] grant;
reg last_winner;
always @ (posedge clk)
begin
    grant <= 2'b00;
    if (reset) last_winner <= 0;
    else begin
        if (request != 2'b00) begin: find_winner
            reg winner;
            case (request)
                2'b01: winner = 0;
                2'b10: winner = 1;
                2'b11: winner = last_winner+1;
            endcase
            grant[winner] <= 1'b1;
            last_winner    <= winner;
        end
    end
end
endmodule
```

Good Comments Improve Maintainability

If reducing the number of lines of code actually increases the overall cost of a design, the same argument applies to comments. One could argue that reducing the number of lines of code can yield a better program, since there are fewer statements to understand and debug. However, the primary purpose of comments is explicitly to improve maintainability of code. No one can argue that reducing their number can lead to better code.

You can write bad comments.

However, just as there is bad code, there are bad comments. Obsolete or outdated comments are worse than no comments at all since they create confusion. Comments that are cryptic or assume some particular knowledge may not be very useful either. One of the most common mistakes in commenting code, illustrated in Sample 4-7, is to describe in written language what the code actually does.

Sample 4-7.
Poor com-
ment in Open-
Vera

```
// Increment addr
addr++;
```

Unless you are trying to learn the language used to implement the model, this comment is self-evident and redundant. It does not add any information. Any reader familiar with the language would have understood the functionality of the statement. Comments should describe the intent and purpose of the code, as illustrated in Sample 4-8. It is information that is not readily available to someone unfamiliar with the design.

Sample 4-8.
Proper com-
ments in
OpenVera

```
// In burst mode, the bytes are written in
// consecutive addresses. Need to access the
// next address to verify that the next byte
// was properly saved.
addr++;
```

Assume an inexperienced audience.

When commenting code, you should assume that your audience is composed of junior engineers who are familiar with the language, but not with the design. Ideally, it should be possible to strip a file of all of its source code and still understand its functionality based on the comments alone.

STRUCTURE OF BEHAVIORAL CODE

This section describes techniques to structure and encapsulate behavioral code for maximum maintainability. Encapsulation can be used to hide implementation details and package reusable code elements.

RTL models require a well-defined structure strategy.

Structuring code is the process of allocating portions of the functionality to different modules or entities. These modules or entities are then connected together to provide the complete functionality of the design. There are many guidelines covering the structure of synthesizeable code. That structure has a direct impact on the ease of meeting timing requirements. The structure of a synthesizeable model is dictated by the limitations of the synthesis tools, often with little regard to the functionality.

Testbenches are structured according to functional needs.

A testbench implemented using behavioral Verilog, VHDL, Open-Vera and *e* code does not face similar tool restrictions. You are free to structure your code any way you like. For maintainability reasons, behavioral code is structured according to functionality or need. If a function is particularly complex, it is easier to break it up into smaller, easier to understand subfunctions. Or, if a function is required more than once, it is easier to code and verify it separately. Then you can use it as many times as necessary with little additional efforts. Table 4-1 shows the equivalent constructs available in each language to help structure code appropriately.

Table 4-1.
Available constructs for structuring code

VHDL	Verilog	OpenVera	*e*
Entity and architecture	Module	Class	Unit, Struct
Function	Function	Function	Method
Procedure	Task	Task	Method, TCM
Package and package body	Module	Class	Unit, Struct
		Inheritance	Inheritance
			Aspect

Encapsulation Hides Implementation Details

Encapsulation is an application of the structuring principle. The idea behind encapsulation is to hide implementation details and decouple the usage of a function from its implementation. That way, the implementation can be modified or optimized without affecting the users, as long as the interface is not modified.

Keep declarations as local as possible.

The simplest encapsulation technique is to keep declarations as local as possible. This technique avoids accidental interactions with another portion of the code where the declaration is also visible. A common problem in Verilog is illustrated in Sample 4-9: Two *always* blocks contain a *for-loop* statement using the register *i* as an iterator. However, the declaration of *i* is global to both blocks. They will interfere with each other's execution and produce unexpected results.

```
integer i;

always
begin
    for (i = 0; i < 32; i = i + 1) begin
        ...
    end
end

always
begin
    for (i = 15; i >= 0; i = i - 1) begin
        ...
    end
end
```

In Verilog, put
local declarations
in named blocks.

In Verilog, you can declare registers local to a *begin/end* block if
the block is named. A proper way of encapsulating the declarations
of the iterators so they do not affect the module-level environment
is to declare them locally in each *always* block, as shown in Sample
4-10. Properly encapsulated, these local variables cannot be accidentally accessible by other *always* or *initial* blocks and create
unexpected behavior.

```
always
begin: block_1
    integer i;
    for (i = 0; i < 32; i = i + 1) begin
        ...
    end
end

always
begin: block_2
    integer i;
    for (i = 15; i >= 0; i = i - 1) begin
        ...
    end
end
```

Verilog *tasks* and
functions can con-
tain local vari-
ables.

Other locations where you can declare local registers in Verilog
include *tasks* and *functions*, after the declaration of their arguments.
An example can be found in Sample 4-11. In VHDL, local variable
declarations can be located before any *begin* keyword.

```
task send;
   input [7:0] data;

   reg           parity;
begin
   ...
end
endtask

function [31:0] average;
   input [31:0] val1;
   input [31:0] val2;

   reg     [32:0] sum;
begin
   sum = val1 + val2;
   average = sum / 2;
end
endfunction
```

Minimize the
scope of local vari-
ables in OpenVera.

In OpenVera, local variable declarations can be located after any open brace. And since curly brackets can be located anywhere in sequential code to create local scope regions, local variables can be created while minimizing their scope and potential undesirable interaction. For example, Sample 4-12 shows how a local iterator variable can be created in the middle of a long sequence of statements by creating a local scope region.

```
function bit mac_frame::compare(mac_frame to)
{
    compare = 0;
    ...
    if (this.data_len !== to.data_len) return;
    {
        integer i;
        for (i = 0; i < this.data_len; i++) {
            if (this.data[i] !== to.data[i])
                return;
        }
    }
    ...
    compare = 1;
}
```

Declarations are
really actions in e.

In e, variable declarations can be located anywhere in sequential code using the *var* action. Since declarations are *actions* (i.e., "sequential statements" in HDL parlance), they can be located anywhere in sequential code sections without requiring a new set of enclosing braces. Declarations will remain visible until the end of the current scope region.

Encapsulating Useful Subprograms

Some functions and procedures are useful across an entire project or between many testbenches. One possibility would be to replicate them wherever they are needed. This obviously increases the required maintenance effort. It also duplicates information that was already captured. VHDL has *packages* to encapsulate any declaration used in more than one entity or architecture. Verilog has no such direct features, but it provides other mechanisms that can serve a similar purpose. OpenVera has *classes* and *e* has *units* and *structs* as encapsulating mechanisms for functions and procedures.

Example: error reporting routines.

One example of procedures that are used by many testbenches are the error reporting routines. To have a consistent error reporting format (which can be parsed easily later to check the result of a regression), a set of standard routines are used to issue messages during simulation. In VHDL, they are implemented as procedures in a package. In Verilog, they are implemented as tasks, with two packaging alternatives. In OpenVera, they are implemented as tasks in a class. In *e*, they are implemented as methods in a predefined *struct*.

Tasks can be packaged in a module and used using a hierarchical name.

Procedures are encapsulated in a container object (module without pins in Verilog, class in OpenVera, struct or unit in *e*) that is then instantiated in constructs that require them. A hierarchical name is used to access the procedure in any encapsulating object instance. The Verilog implementation, shown in Sample 4-13 and used in Sample 4-14, can be compiled on its own since the tasks are contained within a compilation unit. It is also possible to include local variables, such as an error counter. Because all declarations are static in Verilog (unless a task or a function is qualified as *automatic*, see Sample 4-95), and because there is an instance of each variable for each instance of the module, the state of the local variables will be maintained between invocation of the procedures. This is not the case with VHDL where there is a single instance of package variables and subprogram variables, which can have multiple instances. These variables do not survive once the subprogram has completed.

Sample 4-13.
Packaging of
tasks in Ver-
ilog

```
module syslog;

integer warnings;
integer errors;

initial
begin
   warnings = 0;
   errors   = 0;
end

task warning;
   input [80*8:1] msg;
begin
   $write("WARNING at %t: %s", msg);
   warnings = warnings + 1;
end
endtask
   . . .
endmodule
```

Sample 4-14.
Using tasks
packaged
using a mod-
ule in Verilog

```
module testcase;

syslog log();

initial
begin
   . . .
   if (...) log.error("Unexpected response");
   . . .
   log.terminate;
end
endmodule
```

Put global vari-
ables in uninstanti-
ated module.

In the usage example shown in Sample 4-14, there will be an instance of error and warning counters for each instance of the module. In this case, it would be preferable to have a single instance of those counter values. This approach can be accomplished by locating them in an uninstantiated module and referring to them using an absolute path as shown in Sample 4-15. Both modules are located in the same file to hide this implementation detail. In VHDL, *shared variables* in the package can be used with a similar effect. *e* has a similar requirement to Verilog: Global variables must be added to the root *sys* struct.

Sample 4-15.
Global vari-
ables in Ver-
ilog

```
module syslog;

task warning;
    input [80*8:1] msg;
begin
    $write("WARNING at %t: %s", msg);
    syslog_globals.warnings =
        syslog_globals.warnings + 1;
end
endtask

    . . .
endmodule

module syslog_globals;

integer errors;
integer warnings;

initial
begin
    warnings = 0;
    errors   = 0;
end

endmodule
```

This is basic
object-oriented
programming.

Encapsulating procedures and the state variables they operate on in the same object is the primary technique in object-oriented programming. I am *not* claiming that Verilog is object-oriented. The encapsulating module is not an object that can be assigned, copied nor moved around. There is no concept of inheritance and polymorphism. These important object-oriented concepts, present in OpenVera and *e*, will be introduced in "Object-Oriented Programming" on page 166. Sample 4-16 and Sample 4-17 show the same procedures and state variables encapsulated in an object-oriented language. Note how global variables can be created simply by using the *static* attribute in OpenVera. Since this is a true object-oriented programming language, it is necessary to allocate the object instance explicitly using the *new* constructor.

Sample 4-16.
Object-ori-
ented packag-
ing of tasks in
OpenVera

```
class syslog {

    static integer warnings = 0;
    static integer errors   = 0;

    task warning(string msg)
    {
        printf("WARNING at %0d: %s",
                get_time(LO), msg);
        this.warnings++;
    }
    ...
}
```

Sample 4-17.
Using tasks
packaged
using a class
in OpenVera

```
program testcase
{
    syslog log = new;

    ...
    if (...) log.error("Unexpected response");
    ...
    log.terminate;
}
```

Encapsulating Bus-Functional Models

In Chapter 5, I describe how data applied to the design under verification via complex waveforms and protocols can be implemented
using tasks or procedures. These subprograms, called *bus-functional models*, are typically used by many testbenches throughout a
project. If they model a standard interface, such as a PCI bus or a
Utopia interface, they can even be reused between different
projects. Properly packaging these subprograms facilitates their use
and distribution.

Figure 4-2 shows a block diagram of a bus-functional model. On
one side, it drives and samples low-level signals according to a predefined protocol. On the other, subprograms are available to initiate
a transaction with the specified data values. The latter is called a
procedural interface.

In VHDL, use pro-
cedures with *sig-
nal* arguments.

In VHDL, the bus-functional model would be implemented using a
procedure located in a package. For the procedure to be able to
drive the interface signals, they must be passed through the procedure's interface as formals of class *signal*. If the procedure had been

Figure 4-2.
Block diagram
of a bus-
functional
model

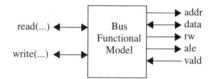

declared in a process, you could drive the signals directly using side effects. It would have been possible in that context since the driver on each signal is clearly identified with the process containing the procedure declaration. Once put in a package, the signals are no longer within the scope of the procedure, nor are the drivers within the procedure attached to any process. Using *signal*-class formals lets a process pass its signal drivers to the procedure for the duration of the transaction. Sample 4-18 shows an example of properly packaged bus-functional models in VHDL.

Sample 4-18.
Encapsulating
bus-functional
models in
VHDL

```
library ieee;
use ieee.numeric_std.all;
package body cpu is

procedure write(variable wadd: in    natural;
                variable wdat: in    natural;
                signal   addr: out   byte;
                signal   data: inout byte;
                signal   rw  : out   std_logic;
                signal   ale : out   std_logic;
                signal   vald: in    std_logic)
is
    constant Tas: time = 10 ns;
begin
    if vald /= '0' then
        wait until vald = '0';
    end if;
    addr <= std_logic_vector(unsigned(wadd, 8));
    data <= std_logic_vector(unsigned(wdat, 8));
    rw   <= '0';
    wait for Tas;
    ale  <= '1';
    wait until vald = '1';
    ale  <= '0';
end write;

end cpu;
```

Task arguments are passed *by value* only.	In Verilog, you might be tempted to implement the bus-functional model using a task where the low-level signals are passed to the tasks, similar to VHDL's procedure. However, Verilog arguments are passed *by value* when the task is called and when it returns. At no other time can a value flow into or out of a task via its interface. For example, the task shown in Sample 4-19 would never work. The assignment to the *bus_rq* variable cannot affect the outside until the task returns. The task cannot return until the *wait* statement sees that the *bus_gt* signal was asserted. But the value of *bus_gt* cannot change from the value it had when the task was called.

Sample 4-19.
Task arguments in Verilog are passed *by value*

```
module arbiter;

task request;
    output bus_rq;
    input  bus_gt;
begin
    // The new value does not "flow" out
    bus_rq <= 1'b1;
    // And changes do not "flow" in
    wait bus_gt == 1'b1;
end
endtask
endmodule
```

Pass signals via module pins in Verilog.	A simple modification to the packaging can work around the problem. Instead of passing the signals through the interface of the task, they are passed through the interface of the module implementing the package, as shown in Sample 4-20. This also simplifies calling the bus-functional model tasks as the (potentially numerous) signals need not be enumerated on the argument list for every call.
Pass signals via a virtual port in OpenVera.	OpenVera has an object designed specifically to represent signal connectivity information: the *virtual port*. An entire signal bundle belonging to a bus-functional model interface can be specified using a single *virtual port*. It can contain signals from different clock domains as well as asynchronous signals. Declaring a *virtual port* creates a new user-defined data type, which can then be used to declare variables and sub-program arguments that will refer to a specific virtual port *binding*. Sample 4-21 shows a class packaging bus-functional model procedure for an ethernet MII interface. Notice how the virtual port is passed as an argument to the constructor and saved in a private data member instead of being passed

Sample 4-20.
Signal inter-
face on Ver-
ilog package

```
module arbiter(bus_rq, bus_gt);
output bus_rq;
input  bus_gt;

task request;
begin
    // The new value "flows" out through the pin
    bus_rq <= 1'b1;
    // And changes "flow" in as well
    wait bus_gt == 1'b1;
end
endtask
endmodule
```

to each procedure. This approach ensures that a single instance of the bus-functional model package operates on a single set of interface signals, maintaining internal state variables belonging to that one interface.

Sample 4-21.
Virtual port
usage in
OpenVera

```
port eth_mii {
    tx_clk, txd, tx_en, rx_clk, rxd, rx_dv,
    col, crs
}

class eth_mii {
    local eth_mii sigs;

    task new(eth_mii sigs) {
        this.sigs = sigs;
    }

    task send(mac_frame frame) {
        ...
    }
    function mac_frame receive() {
        ...
    }
}
```

Use *vrhfix* on gen-
erated header file.

There is one problem with the code in Sample 4-21: When compiled, the generated header file (also known as *vrh* file) will not work. A header file in OpenVera is like a header file in C or a package declaration in VHDL. It is supposed to include all of the external definitions required to use a class. By including a header file, I am supposed to satisfy the compiler and have all of the syntax information to instantiate a class and operate on those instances.

However, the generated header file, generated using either the *-h* or *-H* command-line option, will not include the virtual port declaration. The methodology recommended by Synopsys, as illustrated in Sample 4-22, is to put the virtual port in a separate *interface file* that must be included wherever the virtual port definition is required. I disagree: The virtual port definition is an integral part of a bus-functional model definition and should be located in the same encapsulation file and be part of the same header file, as illustrated in Sample 4-23. The *vrhfix* Perl script[5] will parse the original source file and add any virtual port declaration it finds back into the generated header file.

Sample 4-22.
Encapsulation
of virtual port
declaration
recommended
by Synopsys

In eth_mii.vri:

```
port eth_mii {
    tx_clk, txd, tx_en, rx_clk, rxd, rx_dv,
    col, crs
}
```

In eth_mii.vr:

```
#include "eth_mii.vri"
class eth_mii {
    . . .
}
```

In eth_mii.vrh (generated using "vera -cmp -H"):

```
#include "eth_mii.vri"
extern class eth_mii {
    . . .
}
```

Specify the port
binding when
instantiating the
bus-functional
class.

To create an instance of a bus-functional model, call its constructor. Each instance will be connected to the *port binding* specified in that instance's constructor. Thus, different instances of the same object can be connected to different interfaces. The *binding* may be static (i.e., defined in the interface file) or dynamic (i.e., through the use of the *signal_connect()* task). Sample 4-24 shows an example

5. *vrhfix* fixes other limitations and methodology disagreements that will be discussed throughout the book. Some ideas from the original version were included in the *-H* command-line option, but some problems remain and others were created. All OpenVera examples will assume the use of *vrhfix*. This script is available in the *resources* section of the book's website.

Sample 4-23.
Encapsulation
of virtual port
declaration
using *vrhfix*

In eth_mii.vr:

```
port eth_mii {
    tx_clk, txd, tx_en, rx_clk, rxd, rx_dv,
    col, crs
}

class eth_mii {
    . . .
}
```

In eth_mii.vrh (generated using "vera -cmp -h", then *vrhfix*):

```
port eth_mii {
    tx_clk, txd, tx_en, rx_clk, rxd, rx_dv,
    col, crs
}
extern class eth_mii {
    . . .
}
```

instantiating two instances of the MII bus-functional model, each connected to two different *port bindings*. Sample 4-24 uses static port binding. Static binding should be preferred to dynamic port binding because it allows optimization of the interface between Vera and the HDL simulator that significantly improves performance over dynamic binding. Note that the use of a *single* dynamic binding in an Vera testbench will prevent static binding optimization and overall simulation performance will be reduced.

Sample 4-24.
Instantiating
bus-functional
model class in
OpenVera

```
#include "eth_mii.vrh"
#include "dut.vri"          // Use static bindings

class testbench {
    eth_mii bfm[];

    task new() {
        this.bfm[0] = new(eth_mii_port_0);
        this.bfm[1] = new(eth_mii_port_1);
    }
}
```

Signals are
accessed using full
hierarchical names
in *e*.

e has no concept of ports. Instead, anytime you want to access a signal in the Verilog or VHDL world, you simply refer to it by full hierarchical name (e.g., ~/top/core/mii/txd) anywhere the value of the signal must be sampled or set.

These hardcoded HDL signal names create two problems.

- First, the location of the interface signals in the design hierarchy cannot change. A bus-functional model expecting the interface signals to reside in the instance ~/top/core/mii must be modified to be used in a system-level environment, where the interface signals are now located in the instance ~/top/sys/mii_core/mii.

- Second, the ultimate name of the interface signals cannot be changed. A bus-functional model that refers to signal *txd_0* cannot refer to signal *txd_1* without creating a copy of the source code and modifying the signals references.

Pass signal location by name using *hdl_path()* in *e*.

The first problem is solved by using *units* instead of *structs* to encapsulate bus-functional models. Units are like *structs* except that they are static for the entire duration of the simulation. They are created at initialization time and will be destroyed only at the end of the simulation. Whereas *struct* instances can be dynamically created and destroyed throughout the simulation, *units* cannot. This sounds like an unreasonable limitation until you realize that many things in a design are static as well. For example, the number of pins and interfaces on a design is static. A bus-functional model, connected to these static pins and interfaces should be encapsulated in a *unit* since it too should be static. Each unit can be associated with a particular point in the design hierarchy. The path to that point (for example ~/top/core/mii) is specified by the value of the *hdl_path()* pseudo-method. In a unit, any relative HDL reference (i.e., that does not start with ~/) is assumed to be relative to that unit's *hdl_path()*. The *hdl_path()* is thus similar to a "current working directory" that you "cd" to and from which all relative reference will be made.

Pass signal names by name in *string* variables in *e*.

The second problem can be fixed by using computed signal names. Any portion of a HDL reference specified between parenthesis is evaluated using the *append()* predefined function and inserted in place of the original text, *including* the parenthesis. By using string variables to hold a user-specified signal name, a bus-functional model procedure can refer to the signal by "de-referencing" the string variable, as shown in Sample 4-25. These computed signal names, if ultimately relative, are interpreted relative to the unit's *hdl_path()*. The signal variables remain variables and potentially can be modified at runtime so proper discipline is required. I assume that static unit fields will be introduced in *e* in the not too distant future to prevent this problem and optimize the computation of the HDL reference. Currently, all computed HDL references are re-evaluated each time they are used and their result appended to the *hdl_path()*. If static unit fields were available and used, Specman Elite could pre-compute the HDL references at the beginning of the simulation, similar to what it already does for non-computed signal names, and improve simulation performance.

Sample 4-25.
Packaging
bus-functional
model proce-
dures in *e*

```
unit eth_mii {
    tx_clk: string;
    txd   : string;
    ...
    event tclk is rise('(tx_clk)') @sim;

    sent(mac_frame frame) @tclk is {
        if ('(crs)' != 1'b1) {
            ...
        };
        '(txd)' = ...
    };
};
```

Bind the signal
names using con-
straints.

When instantiating a bus-functional model encapsulated in a *unit*, it will be necessary to specify the location of the signals in the design hierarchy and their names. This location specification is best accomplished using *constraints* on each instance, as shown in Sample 4-26. If the bus-functional model interfaces to signals that tend to have a well-accepted naming convention, it is a good idea to provide (and document!) default signal names in the bus-functional model itself using *soft* constraints as shown in Sample 4-27. That way, users will only need to provide bindings for the names that differ.

Sample 4-26.
Instantiating a
bus-functional
model unit in *e*

```
import eth_mii;

unit testbench {
    bfm[2]: list of eth_mii is instance;
    keep for each in bfm {
        it.hdl_path() == append("~/top.mii_", index);
        it.tx_clk    == "tx_clk";
        it.txd       == "tx_data";
    };
};
```

| **Sample 4-27.** Providing default signal name bindings in *e* | ```
unit eth_mii {
    tx_clk: string;
    txd   : string;
    . . .
    keep soft tx_clk == "tx_clk";
    keep soft txd    == "txd";
    . . .
};
``` |

DATA ABSTRACTION

Synthesizeable models are limited to bits and vectors.

The limitation of logic synthesis technology has forced the synthesizeable subset into dealing only with data formats that are clearly implementable: bits, vectors of bits and integers. Behavioral code has no such restrictions. You are free to use any data representation that fits your need.

Work at the same level as the design under verification.

You must be careful not to let an RTL mindset artificially limit your choice, or to keep you from moving to a higher level of abstraction. You should approach the verification problem at the same level of granularity as the "unit of work" for the design. For an ethernet switch, it is a MAC frame. For an IP cell router, the unit of work is an entire IP packet. For a SONET framer, the unit of work is a SONET frame. For a video compressor, the unit of work is either a video line or an entire frame, depending on the granularity of the compression. The interesting conditions and testcases are much easier to set up at that level than at the low-level bit interface.

Abstracting data in Verilog requires creativity and discipline.

VHDL provides excellent support for abstracting data into high-level representations. Verilog does not have as many features, but with the proper technique and discipline, a lot can be accomplished. The following sections use Verilog to illustrate how various data abstractions can be implemented since it is the more limiting language. Their implementations in VHDL are much easier and you are invited to consult a book on the VHDL language[6] to learn the details. This book also uses OpenVera to show how some of these abstract modeling techniques are used in an object-oriented language.

6. For VHDL, I recommend *VHDL Coding Styles and Methodologies* by Ben Cohen (Kluwer Academic Publisher). Other recommendations for other languages are suggested in the Preface on page xxii.
```

Records

Records are used to represent information composed of various smaller pieces of different types. This section develops a technique for modeling records in Verilog. Records are directly supported by VHDL, OpenVera and *e* (using *record*, *class* and *struct* respectively). Records are ideal for representing packets or frames, where control or signaling information is grouped with user information. The code Sample 4-28 shows the VHDL declaration for a record used to represent an ATM cell. An ATM cell is a fixed-length 53-byte packet with 48 bytes of user data.

Sample 4-28.
VHDL record
for an ATM
cell

```
type atm_payload_typ is array(0 to 47) of
    integer range 0 to 255;

type atm_cell_typ is record
    vpi    : integer range 0 to 4095;
    vci    : integer range 0 to 65535;
    pt     : bit_vector(2 downto 0);
    clp    : bit;
    hec    : bit_vector(7 downto 0);
    payload: atm_payload_typ;
end record;
```

Records can be
faked in Verilog.

Verilog does not support records directly, but they can be faked. Hierarchical names can be used to access any declaration in a module. A module can emulate a record by containing only register declarations. When instantiated, the module instance emulates a record register, with each register in the module becoming a field of the record instance. The record declaration for an ATM cell can be emulated, then used in Verilog, as shown in Sample 4-29. The module containing the declaration for the record can contain instantiation of lower record module, thus creating multi-level record structures.

The faked record
is not a real object.

Although Verilog can fake records, they remain *fakes*. The record is not a single variable such as a register. Therefore, it cannot be assigned as a single unit or aggregate, nor used as a single unit in an expression. For example, Sample 4-30 attempts to compare the content of two cells, assign them and use them as arguments. It would produce a syntax error because the cells are instance names, not variables nor valid expressions.

<table>
<tr><td>

Sample 4-29.
Verilog record
for an ATM
cell

</td><td>

```verilog
module atm_cell_typ;
reg [11:0]  vpi;
reg [15:0]  vci;
reg [ 2:0]  pt;
reg         clp;
reg [ 7:0]  hec;
reg [ 7:0]  payload [0:47];
endmodule

module testcase;

atm_cell_typ a_cell();

initial
begin: test_procedure
   integer i;
   a_cell.vci = 0;
   ...
   for (i = 0; i < 48; i = i + 1) begin
      a_cell.payload[i] = 8'hFF;
   end
end
endmodule
```

</td></tr>
<tr><td>

Sample 4-30.
Verilog
records are not
objects; this is
invalid Verilog

</td><td>

```verilog
module testcase;

atm_cell_typ actual_cell();
atm_cell_typ expect_cell();
atm_cell_typ next_cell();

initial
begin: test_procedure
   ...
   if (actual_cell !== expect_cell) ...
   ...
   expect_cell = next_cell;
   ...
   receive_cell(actual_cell);
   ...
end
endmodule
```

</td></tr>
</table>

Provide conversion function to and from an equivalent vector.

You can work around the limitation of these fake records by using conversion functions between records and equivalent vectors. Following object-oriented practices, these *packing* and *unpacking* functions are located in the record definition module and called using a hierarchical name. The code Sample 4-31 shows how the *tobits* and *frombits* conversion functions can be defined. Sample 4-32 shows how the previously invalid approach of Sample 4-30 becomes usable and by translating ATM cell records to and from bit-vector objects.

Sample 4-31.
Conversion
functions for
Verilog
records for an
ATM cell

```
module atm_cell_typ;
reg [11:0] vpi;
reg [15:0] vci;
...
reg [ 7:0] payload [0:47];

function [53*8:1] tobits;
    input dummy;

    tobits = {vpi, vci, ..., payload[47]};
endfunction

task frombits;
    input [53*8:1] cell;

    {vpi, vci, ..., payload[47]} = cell;
endtask
endmodule
```

Use a symbol to
predefine the size
of the equivalent
record.

One difficulty created by the workaround is knowing, as a user, how big the bit vector representation of the record is. In Sample 4-32, the size of the *bits* temporary reg and the size of the argument of the *receive_cell* task must be known. Should the representation of the record change (e.g., a field is added), all wires and registers declared to carry the equivalent bit vector representation would need to be modified.

Define a symbol
for the vector size
declaration.

The best solution is to define a symbol to hide the size of the corresponding vector from the user as shown in Sample 4-33. This method presents a disadvantage: The symbol must be declared either in a file to be included using the `include directive, or the module defining the record type must be compiled before any module making use of it. The latter also requires that the `resetall directive not be used. Because `define symbols are global to the entire Verilog compilation process, you must make sure that the name will be unique.

Sample 4-32.
Turning Verilog modules into objects

```
module testcase;

atm_cell_typ actual_cell();
atm_cell_typ expect_cell();
atm_cell_typ next_cell();

reg [53*8:1] bits;

initial
begin: test_procedure
    ...
    if (actual_cell.tobits(0) !==
        expect_cell.tobits(0)) ...
    ...
    expect_cell.frombits(next_cell.tobits(0));
    ...
    receive_cell(bits);
    actual_cell.frombits(bits);
end

task receive_cell;
    output [53*8:1] bits;
    ...
endtask
endmodule
```

Records are *classes* in OpenVera and *structs* in *e*.

Records are modeled using *classes* in OpenVera and *structs* in *e*. Because these languages are object-oriented[7], record instances are objects in their own right. As shown in Sample 4-34, they can thus be compared, copied and moved around. But because they are objects, they must be instantiated (or allocated). When declaring an object-type variable, one only declares a *reference* (or a pointer) to an object instance, not to an individual object[8]. Two object-type variables can refer to the same object *instance*. When comparing object references, you are checking whether they refer to the same object, not if the referred objects are identical. A comparison function must be used to compare the content of two object instances.

---

7. Object-oriented (OO) purists could debate to what extent each is truly OO. They are OO enough for this application.

8. For more details on the difference between a reference and an instance, see "Classes" on page 166.

Sample 4-33.
Hiding the
size of the
equivalent
vector

```
module atm_cell_typ;
`define ATM_CELL_TYP [53*8:1]
  ...
endmodule

module testcase;

atm_cell_typ actual_cell();
atm_cell_typ expect_cell();
atm_cell_typ next_cell();

reg `ATM_CELL_TYP bits;

initial
begin: test_procedure
   ...
   if (actual_cell.tobits(0) !==
       expect_cell.tobits(0)) ...
   ...
   expect_cell.frombits(next_cell.tobits(0));
   ...
   receive_cell(bits);
   actual_cell.frombits(bits);
end

task receive_cell;
   output `ATM_CELL_TYP bits;
   ...
endtask
endmodule
```

Conversions func-
tions are automati-
cally provided.

Another advantage of OpenVera and *e* is that, if the fields are
marked as *packed* or *physical* respectively, the conversion functions
to and from arrays or lists of bits are automatically provided and
may not need to be manually implemented. I say "may" because
these implicit conversion functions in real-life *class* and *struct* declarations involving variance, inheritance and extensions usually do
not produce the desired result. However, these implicit conversion
functions remain useful when manually implementing conversion
functions that yield the desired result.

Sample 4-34.
Records in
OpenVera are
objects

```
#include "atm_cell.vrh"

task receive_cell(atm_cell cell)
{
    . . .
}

program testcase
{
    atm_cell actual_cell = new;
    atm_cell expect_cell = new;
    atm_cell next_cell   = new;

    if (!actual_cell.compare(expect_cell)) ...
    . . .
    expect_cell = next_cell;
    . . .
    receive_cell(actual_cell);
}
```

Variant Records

Variant records are
not directly sup-
ported in Verilog,
VHDL or Open-
Vera.

Neither VHDL, Verilog nor OpenVera support variant records. Variant records provide different fields based on the content of another. For example, ethernet frames may or may not contain four additional header bytes carrying virtual LAN information. Variant records automatically handle these two different formats concurrently, converting to and from bits as appropriate.

Variant records
can be defined
manually.

Because variant records are not supported, a record structure that can represent both header formats must contain declarations for the optional fields and control variables indicating which optional fields are active. Any operations on the variant record must check the value of the control field to determine if the optional fields are to be processed. Even in a language where they are directly supported (such as *e*), variant records must be allocated internally to handle the largest possible data structure.

Sample 4-35 shows an OpenVera class modeling a MAC frame with optional virtual LAN tagging. When the *has_vlan* property is non-zero (i.e., true), the *tag*, *priority* and *vlan_id* properties are valid and "exist". When the *has_vlan* property is zero (i.e., false), these subsequent properties are not relevant and do not exist. Tasks and functions in the class, such as *to_bytes* and *from_bytes*, will use the *has_vlan* property to determine whether the other properties exist and process the MAC frame accordingly. The implicit *pack* and *unpack* methods cannot be used because the optional fields would either always be included or excluded, depending on whether they are marked as *packed*. The optional fields remain accessible and the language has no mechanism to prevent their use, so programmer discipline is required. This example also demonstrates another case of record variance: variable-length fields. The *data* property can have a variable number of bytes.

Sample 4-35.
Variant record
in OpenVera

```
class mac_frame {
    rand bit [47:0] da;
    rand bit [47:0] sa;
    rand bit        has_vlan;   // VLAN control
    rand bit [15:0] tag;        // VLAN
    rand bit [ 2:0] priority;   // VLAN
    rand bit        cfi;        // VLAN
    rand bit [11:0] vlan_id;    // VLAN
    rand bit [15:0] len_typ;
    rand integer    len;
    rand bit [ 7:0] data[] assoc_size len;
    rand bit [31:0] fcs;

    function bit [31:0] compute_fcs();

    task to_bytes(bit [7:0] bytes[]) {
        ...
        if (this.has_vlan) {
            ...
        }
        ...
    }
    task from_bytes(bit [7:0] bytes[]);
}
```

e supports variant
records through
when-extensions.

e has a built-in mechanism for supporting variant records. Additional fields (and any other *struct*-level declarations such as methods) can be defined when some other fields have a specific value. They will exist and be accessible only when it will be ascertained that the control field has the required value and the reference to the record instance has been properly typecasted. The implementation of the various methods must manually handle the presence or absence of the optional fields. Testing the value of a control field can be done using the usual == comparison operator then the typecasting can be done using the *as_a( )* operator. However, this is such a frequent operation that *e* provides a shorthand notation with the "*is a*" operator with on-the-fly type casting. See Sample 4-36 for an example.

```
type mac_frame_format: [DIX, IPX, LLC, SNAP];
struct mac_frame {
    format: mac_frame_format;
    da    : uint (bits: 48);
    sa    : uint (bits: 48);
    vlan  : bool;        // VLAN control
    when vlan mac_frame {
        tag     : uint (bits: 16);
        priority: uint (bits:  3);
        cfi     : bit;
        vlan_id : uint (bits: 12);
    };
    len_typ: uint (bits: 16);
    data    : list of byte;
    fcs     : uint;

    to_bytes(): list of bytes {
        ...
        if (me is a vlan mac_frame (vl_me)) {
            result.add(vl_me.tag);
            ...
        };
        ...
    };
    ...
};
```

Optional fields are really there, even if they are not accessible.

The memory for the optional fields is always present in all instances of the variant record, whether they officially exist or not. In fact, it is possible to bypass the type checking mechanism and typecasting requirement by referencing the optional fields using their hidden name. For example, the optional field named *priority* controlled by the value of the *vlan* field is hidden under the name *vlan'property*. Using the hidden name for an optional field is *not* recommended. You may encounter these breaches of elementary programming ethics in code written by engineers who were ignorant of the on-the-fly casting ability of the *is a* operator and were too lazy to perform the explicit type check and casting. You, on the other hand, have no (longer) such excuse.

Could use access variables in VHDL.

A variant record could be emulated in VHDL using an access type for the optional or variable-length fields. However, this would put severe limitations on its usability as signals cannot be of a type containing an access type. Only variables could make use of this variable-length record.

Single inheritance should not be used to model concurrent variances.

You may be tempted to implement variant records using inheritance[9] as shown in Sample 4-37. First, a base class represents the record without any optional field. Then, derived classes (for example, using the *like* inheritance mechanism in *e*) are used to add various optional fields. Although very object-oriented, this approach creates several limitations.

- First, you will not be able to randomly generate a mix of frames, sometimes with the optional fields, sometimes not. Each derivative creates a new data type. Once an instance of a type is created, it cannot be modified. Each time it will be generated randomly, it will produce random values for that type and no other.

- Second, you will not be able to create combinations of optional fields easily. If multiple inheritance is not supported (and OpenVera and *e* do not support it), you cannot recombine multiple derived classes into a single one containing multiple variances. With single inheritance, you would have to create a new type for each combination of optional fields. This quickly grows to a significant number, which grows exponentially when testbench-specific additions must be made to the record type.

Use variance control fields.

If a single class with variance control fields (or *e*'s *when*-inheritance) are used, as in Sample 4-35, a single data type exists. The value of the control fields themselves can be randomized, thus sometimes yielding the presence of the optional field, sometimes yielding their absence. Multiple, orthogonal variances can co-exist in the same class, using different control fields. A new variant only needs to be implemented once, automatically offering all possible combinations with previously implemented orthogonal variances.

---

9. Inheritance will be discussed in more detail in "Inheritance" on page 173.

Sample 4-37.
Variant
records imple-
mented using
single inherit-
ance in Open-
Vera

```
class mac_frame {
    rand bit [47:0] da;
    rand bit [47:0] sa;
    rand bit [15:0] len_typ;
    rand integer    len;
    rand bit [ 7:0] data[] assoc_size len;
    rand bit [31:0] fcs;
    ...
}

class vlan_mac_frame extends mac_frame {
    rand bit [15:0] tag;
    rand bit [ 2:0] priority;
    rand bit [11:0] vlan_id;
    ...
}

class control_mac_frame extends mac_frame {
    rand bit [15:0] opcode;
    ...
}
```

Arrays

Single-dimensional arrays are useful data structures for representing linear information such as fixed-length data sequences, look-up
tables or memories. Two-dimensional arrays are used for planar
data such as graphics or video frames. Three-dimensional arrays are
not frequently used, but they could find an application in representing data for video compression applications such as MPEG. Arrays
with greater numbers of dimensions have rare applications, especially in hardware verification. VHDL and OpenVera support
multi-dimensional arrays for any data type. Verilog-2001 has introduced support for multi-dimensional arrays for *reg* and *wire*. *e* only
supports one-dimensional arrays[10].

e has built-in one-
dimensional
arrays.

As shown in Figure 4-3, array elements are located in consecutive
memory locations. They are accessed by computing their address
using an offset from a base address. Truncation, element replacement and element appending in unused pre-allocated memory at the
end of the array are efficient operations. For example, adding a

---

10. The built-in *list* type in *e* is really an array (see "Lists" on
page 157 for a discussion of the difference).

twelfth element to the array in Figure 4-3 would simply consume one of the unused memory locations. But element insertion, element deleting and array lengthening beyond the amount of pre-allocated memory are expensive operations that require copying of a potentially large number of elements to maintain the integrity of the consecutive memory locations. For example, performing a *pop0()* operation in an *e list*, requires that all other elements in the array be copied or "shifted" down one position.

Figure 4-3.
Array
elements in
memory

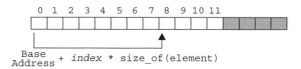

Multi-dimensional arrays must be mapped onto a single dimensional structure.

Even though VHDL, Verilog and OpenVera offer multi-dimensional arrays, they must be mapped to the linear hardware memory of the host computer. That hardware memory is only one-dimensional. Figure 4-4 shows a two-dimensional array mapped into a linear memory. Indexing an element requires knowing the length of each preceding row. To make these (often highly repeated) calculations efficient in large multi-dimensional arrays, it is necessary to have fixed-sized dimensions. Once a VHDL, Verilog or OpenVera array has been declared and allocated, it cannot increase nor decrease its size.

Figure 4-4.
Mapping a
4x4 array to a
linear memory

Use arrays of arrays in *e* and OpenVera.

Multi-dimensional arrays in *e* can be emulated by using an array of *structs* containing an array. Since this multi-dimensional array is composed of independent one-dimensional arrays, the size and number of each dimension can vary. However, looking up one element will require looking up each individual dimension. Sample 4-38 shows an *e* definition, instantiation and reference of a two-dimensional array of RGB values. The same technique can be used in OpenVera, using an associative array of *classes* containing an associative array, if multi-dimensional arrays with variable number of dimensions is required.

Sample 4-38.
Two-dimen-
sional array in
e

```
struct rgb {
    red   : byte;
    green: byte;
    blue  : byte;
};

struct line {
    pixels: list of rgb;
};

struct picture {
    lines: list of line;
};

...
var vga: picture;
gen vga keeping {
    it.lines.size() = 480;
    for each (line) in it.lines {
        line.pixels.size() == 640;
    };
};
var center: rgb = vga.lines[240].pixels[320];
```

Lists

Lists are imple-
mented using
links.

Lists are used to represent ordered linear information and, as such, are very similar to one-dimensional arrays. They fundamentally differ from arrays in their implementation. As shown in Figure 4-5, list elements are located in independent memory locations. The linear and ordered relationship is created by a series of pointers, starting from a *head* pointer, that points to each subsequent element in the list. Lists are frequently doubly-linked in the reverse direction, starting from a *tail* pointer, to facilitate some list operations. *e* lists are implemented using consecutive memory locations, not linked elements. That is why I consider them to be arrays instead of lists.

Figure 4-5.
Doubly-linked
list elements
in memory

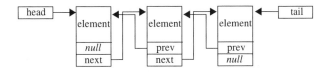

Lists can be more efficient than arrays.

Because of their different implementation, lists are more memory-efficient than arrays if not all locations are used. Arrays must allocate memory for their entire size, whereas the amount of memory used by lists grows and shrinks as the number of elements they contain increases or decreases. Elements can be inserted or deleted anywhere in the list at little cost. All that is required is a re-orientation of the pointer sequence to include or remove the element.

However, while the elements of an array can be accessed randomly, the elements in a list must be accessed sequentially, starting at the head. If the memory usage of your model is of concern, using lists may be the better approach. If access time to various elements is your primary concern, using an array is a more efficient implementation. Sample 4-39 shows an implementation of a doubly-linked list. The Vera *ListMacros.vrh* file provides a predefined doubly-linked list[11] for any user-defined types. The example shown in Sample 4-39 can be turned easily into an *e define as* macro and used to declare lists for any user-defined type[12].

Lists can be used to model large memories.

One of the best applications of a list is to model a large memory. In system-level simulations, you may have to provide a model for a large amount of memory. With the amount of memory available in today's systems, and the overhead associated with modeling them, you may find that you do not have a computer with enough resources to simulate your system-level model efficiently. For example, if you model a memory with 32-bits of addressable bytes using an array of `std_logic_vector` in VHDL, the amount of memory consumed by this array alone exceeds 128Gb (nine logic values requiring 4 bits to represent each `std_logic` bit times 8 bits per byte times 4Gb).

---

11. Double-linked lists create self-referential networks that cause problems in some garbage collection algorithms, such as the one used in Vera. When you are done using a *VeraList* in Vera, be sure to call the *purge()* method to prevent memory leaks.

12. In fact, you can find the double-linked list *e* macro in the *source* section of the book website.

Writing Testbenches: Functional Verification of HDL Models

Sample 4-39.
Doubly-linked
list in *e*

```
struct dl_list {
    head: dl_list_el;
    tail: dl_list_el;

    push(element: my_type) is {
        var link: dl_list_el = new;

        link.element        = element;
        link.previous.next  = link;
        link.next           = null;
        link.previous       = tail;
        tail                = link;
    };
    ...
};

struct dl_list_el {
    element: my_type;
    prev    : dl_list;
    next    : dl_list;
};

...
for (link = a_list.head;
     link != null;
     link = link.next) {
    if (link.element != ...) ...
};
```

Only the sections of the memory currently in use need to be modeled.

In any simulation, it is unlikely that all memory locations are required. Usually, the accesses are limited to a few regions within the memory address space. A list can be used to model a very large memory in a fashion similar to a cache memory. Only regions of the memory that are currently in use are stored in the list. When a particular location is accessed, the list is searched for the region of interest, allocating a new region as necessary. A similar situation occurs when it is necessary to look up packets based on the content of a field, such as the destination address. For example, looking up an ethernet MAC frame based on a 48-bit destination address would require an array with over 281 billion elements. Assuming only minimum-size MAC frames, this array would consume 18 million gigabytes! It is much more efficient to allocate only those locations for which we already have a MAC frame.

This is called a *sparse* memory model.

Figure 4-6 shows a conceptual view of an address space where only the portions that are actively used are physically allocated. This

Figure 4-6.
Sparse
memory
model

4Gb   0

■ = Allocated region

type of partial memory model is called *sparse memory model*. The size of each individual region affects the ultimate performance of the simulation. With a smaller size, the computer's memory is used more efficiently, but more regions are looked up before finding the one of interest. Using larger regions has the opposite effect: More memory usage is traded off for improved look-up efficiency. Sparse memory models are directly supported in VCS by using the *sparse* pragma, in OpenVera with the *associative array* and in *e* with the *keyed list*. All commercial memory models use a sparse memory model. There is a dynamic sparse memory model PLI package provided by Cadence in the NC-Verilog distribution directory. You will find it at:

```
$CDS_HOME/tools/verilog/examples/PLI/damem.
```

A linked list can
be used to model a
sparse memory.

A sparse memory model can be implemented easily using a list of records, where each record represents a region of the memory[13]. The list can grow dynamically by allocating each region on demand, and linking each element in the list to another using access types. The implementation of a sparse memory model using a linked list in VHDL[14] is shown in Sample 4-40. The memory regions are implemented as records: a field for the memory region itself (implemented as an array) and a field for the base address of that region. The record also contains a field to access the next region in the linked list. Because access types and access values are limited to variables, using such an implementation may be impractical if the list needs to be passed through interfaces.

---

13. This statement does not imply that this is how OpenVera's *associative array* or *e*'s *keyed list* are implemented.

14. For a more detailed description and alternative implementation, refer to section 6.1 of "*VHDL Answers to Frequently Asked Questions*", 2nd edition by Ben Cohen (Kluwer Academic Publisher, ISBN 0-7923-8115-7, 1998).

Sample 4-40.
Implementa-
tion of a
sparse mem-
ory using a
linked list in
VHDL

```
process
   subtype byte is std_logic_vector(7 downto 0);
   type region_typ is array(0 to 31) of byte;

   type list_el_typ;
   type list_el_ptr is access list_el_typ;
   type list_el_typ is record
      base_addr  : natural;
      region     : region_typ;
      next_region: list_el_ptr;
   end record;

   variable head: list_el_ptr;

   -- See Sample 4-41 for continuation
begin
   ...
end process;
```

In Sample 4-41, a procedure is implemented to locate and return the section of the memory containing any address of interest. It starts at the head of the list, looking at every element of the list. If the end of the list is reached without finding the required region, a new region is allocated and prepended to the head of the list.

In Sample 4-42, a procedure is provided to read a single memory location. After locating the proper section of memory, it simply returns the content of the appropriate location in the section. There is also a procedure used to assign to a memory location. It works like the procedure to read a location, except that a new value is assigned.

Provide operators for the data struc-ture.

Lists are most useful when they come with a rich set of operators, such as appending or prepending to a list, removing the element at the head or tail of the list, finding out its length or iterating over all of its elements. These operators should be provided in the same package as the data structure. In an object-oriented language, these are *methods* of the list object. The list macros supplied with Open-Vera contain definitions for many list and iterator operators. Refer to Appendix B of the Vera User's Manual for more details. The built-in *list* type in *e*, although not technically a list, comes with a rich set of list operations and iteration. The double-linked list exam-ples shown in Sample 4-39 shows a single *push()* operator to append a new element at the end of the list. A complete list imple-

Sample 4-41.
Looking up a
sparse mem-
ory model in
VHDL

```
process
   -- See Sample 4-40 for declarations

   procedure get_region(addr: in  natural;
                         here: out list_el_ptr)
   is
      variable element: list_el_ptr;
   begin
      element := head;
      -- Have we reached the end of the list?
      while (element /= null) loop
         -- Is the address of interest in this
         -- list element?
         if (element.base_addr <= addr and
             addr < element.base_addr +
                  element.region'length) then
            here := element;
            return;
         end if;
         element := element.next_region;
      end loop;
      element := new list_el_typ;
      element.base_addr :=
         addr / element.region'length;
      element.next_region := head;
      head := element;
      here := element;
   end get_region;

   -- See Sample 4-42 for continuation
begin
   ...
end process;
```

mentation would come with many more useful operators. OpenVera
also provides a *mailbox*, which is a simple FIFO. Although it has
none of the usual list operators, it is suitable for simple queue applications where data is removed from the front of the list and is added
to the end of the list.

<table>
<tr>
<td>

Sample 4-42.
Reading and writing a location in a sparse memory model

</td>
<td>

```
process
   -- See Sample 4-41 for declarations

   procedure lookup(addr :  in  integer;
                          value: out byte) is
      variable element: list_el_ptr;
   begin
      get_region(addr, element);
      value := element.region(addr -
                              element.base_addr);
   end lookup;

   procedure set(addr:  in integer;
                     value: in byte) is
      variable element: list_el_ptr;
   begin
      get_region(addr, element);
      element.region(addr - element.base_addr) :=
         value;
   end set;

   variable val: byte;
begin
   set(10000, "01011100");
   lookup(10000, val);
   assert val = "01011100";
end process;
```

</td>
</tr>
</table>

Files

External input files complicate configuration management.

Personally, I prefer to avoid using external input files for testbenches. Configuration management of the testbench and the design under verification is complex enough. Without good practices, it is very difficult to make sure that you are simulating the right version of the correct model together with the proper implementation of the right testbench. If you must add to the mix making sure you have the right version of input files, often generated by scripts from some other format of some other files, configuration management grows exponentially in complexity. For example, many use files to initialize Verilog memories, as shown in Sample 4-43.

Understanding the implementation of the testcase now requires looking at two files and understanding their interaction. If the file always contains the same data for the same testcase, it can be replaced with an explicit initialization of the memory in the Verilog

Sample 4-43.
Initializing a
Verilog mem-
ory using an
external file

```
module testcase;

reg [7:0] pattern [0:55];

initial $readmemh(pattern, "pattern.memh");

endmodule
```

code, as shown in Sample 4-44. Now, only a single file needs to be managed and understood. In some cases, using external files is unavoidable, such as when using input data that was produced by an external program or recorded from actual data streams.

Sample 4-44.
Explicitly ini-
tializing a Ver-
ilog memory

```
module testcase;

reg [7:0] pattern [0:55];

initial
begin
    pattern[0]  = 8'h00;
    pattern[1]  = 8'hFF;
    . . .
    pattern[55] = 8'hC0;
end

endmodule
```

Files can program
bus-functional
models.

Programmable testbenches are architected around programmable bus-functional models and checkers, and can be programmed via an external input file. The "program" can be as simple as a sequence of data patterns or as complex as a pseudo assembly language with opcodes and operands interpreted by an engine implemented in Verilog or VHDL. However, this approach to programmable bus-functional models makes it extremely difficult to coordinate or synchronize a particular operation of the bus-functional model with some external events or other bus-functional model. It requires the introduction of synthetic synchronization instructions.

If a new synchronization mechanism is required, a new instruction must be added. It is easier to "program" a bus-functional model using the testbench language (which is a rich high-level language) with calls to the procedural interface of the bus-functional model. The program, being part of the testbench code, has visibility over the necessary states of the design or other bus-functional model to coordinate and synchronize them effectively. A program can also be a set of limit values for constraints in a randomly-driven bus-functional model. But this limits programs to modifying constraint boundary conditions. Programs cannot add entirely new constraints or relax existing ones.

External files can eliminate recompilation. Using external input files can save a lot of compilation time if you use a compiled simulator such as NC-Verilog, VCS or any VHDL simulator. If you can modify your testcase by modifying external input files, it is not necessary to recompile the model of the design under verification nor the testbench. For large designs, this compilation time can be significant, especially for a gate-level design with SDF back-annotation. However, with HVLs such as OpenVera and *e,* which co-simulate with the HDL model, simulating a different testbench only requires loading a different set of HVL objects or source files. If the design and the interface of the HVL to the design does not change (i.e., the *shell* or *stub* files), it is not necessary to recompile the Verilog or VHDL code either. With Specman Elite, it is even possible to load a new testbench without even quitting the simulation. Using external files to control a testbench is no longer necessary to avoid re-compilation if HVLs are used.

Mapping High-Level Data Types to Physical Interfaces

It is very unlikely that high-level data types are directly usable by any device that must be verified. Any complex data structure has to be mapped to bits, bytes, addresses and registers. They are sent to or received from the design using a physical-level interface using a more basic data representation, such as a bit, byte or word, usually including synchronization, framing or handshaking signals. In Chapter 5, I show techniques using bus-functional models for applying high-level data to a design via a low-level physical interface (and vice-versa on the output side).

OBJECT-ORIENTED PROGRAMMING

OpenVera and *e* are object-oriented.

Object-oriented programming is a methodology that has been used with great success in software engineering for many years. After structured programming (i.e., the removal of the "go to" statement and introduction of subprograms and control structures), it is the next evolutionary step in language design. Object-oriented used to be one of those buzzwords used to describe almost everything. In this book, object-oriented is used to identify a methodology that makes use of (and a language that supports) classes, inheritance and polymorphism. Both OpenVera and *e* meet this definition of object-oriented.

Classes

Objects are data and procedures together.

Classes are a collection of variables and subprograms that create object types. In OpenVera, a class is defined using the *class* statement. In *e*, a class is defined using the *struct* or *unit* statement. A class defines an object's state as a collection of data members. A class also defines all possible operations on the object using methods. In OpenVera, data members are called *properties* and methods are *functions* and *tasks* declared within the class. Sample 4-35 shows an example of a *class* declaration in OpenVera. In *e*, data members are called *fields* and methods are called *methods* and *time-consuming methods (TCMs)*. Sample 4-36 shows an example of a *struct* declaration in *e*.

Everything is modeled as an object.

Packets, frames and cells are modeled as objects. Their various fields are data members. Methods exist to calculate and check the value of any CRC or error protection field, translate the field values to and from a sequence of physically transmitted bytes and segment a large object into a list of smaller ones. Processor instructions are modeled as objects. Their opcode and various operands are data members. Methods exist to produce the object code value, relocate branch destination values and display the current value as an assembly code statement. Floating point values—which are not directly supported by either language—are modeled as objects. Data members represent their integer and fractional parts. Methods perform arithmetic and logical operations, translate to and from a fixed-point value or display the floating-point number using a user-specified format.

Bus-functional models are objects. Their interface signals are data members. Methods implement the procedural interface. A design configuration is modeled as an object. The routing table, coefficient array or interface configuration parameters are modeled using data members. Methods are used to download the configuration specified in the data members in the design, or generate a random design configuration. A scoreboard (see "Scoreboarding" on page 348 for more details on scoreboards) is modeled as an object. The lists of expected output data sequences are data members. Methods exist to transform a new input data value according to the current design configuration, compare an output data value against expected ones and check the scoreboard for any losses at the end of the simulation. An entire testbench is an object in which the bus-functional models and scoreboards sub-objects are data members. Methods implement the reset procedure, main testcase sequence and termination procedures.

Data members consume memory. Methods do not.

By default, each data member is local to each object instance. Methods, however, are global to the class. For example, if there are 1,000 instances of the *mac_frame* class (see Sample 4-35), there are 1,000 instances of the *da* property but only a single instance of the code implementing the *to_bytes* function. The memory required to implement the data members will be replicated for each object instance. Methods will not consume more memory, whether an object is instantiated only once or 1,000 times. If memory consumption starts to be a problem, focus on the data members of the objects with the most instances.

Data members can be global.

It is often necessary to have information that will be shared among all instances of an object. For example, if each object has its own instance of an error counter, it would be difficult to determine the number of error messages that were produced during a simulation. If that error counter is global, that task becomes much easier. Another example is automatically assigning unique ID numbers to each object instance. A global ID number source is the only way to ensure uniqueness. In OpenVera, a data member global to the class is declared using the *static* attribute, as shown in Sample 4-45. *e* has no concept of data members global to a *struct*, but they can be emulated by putting them in a scope common to all instances, such as the global *sys* struct, as shown in Sample 4-46. Unlike static data members that remain in their individual class name space, there is only one *sys* name space global to the entire simulation. It is necessary to use some naming convention or encapsulation in another object to prevent name collision with other global variables added to *sys* by other testbench components.

Sample 4-45.
Global data members in OpenVera

```
class sim_status {
    static integer n_errors = 0;
}

class mac_frame {
    sim_status status;
    ...
    if (this.compute_fcs() !== this.fcs) {
        printf("Bad FCS");
        this.status.n_errors++;
    }
}

class mii {
    sim_status status;
    ...
    if (col === 1'b1 && crs !== 1'b1) {
        printf("Collision without carrier\n");
        this.status.n_errors++;
    }
}
```

Sample 4-46.
Global data members in *e*

```
extend sys {
    mac_frame_next_id: uint;
};

struct mac_frame {
    id: uint;
    pre_generate() is also {
        me.id = sys.mac_frame_next_id;
        sys.mac_frame_next_id += 1;
    }
};
```

Verilog's *module* and VHDL's *package* are not objects.

You may be tempted to conclude that Verilog is object-oriented since modules, used to emulate records as shown in "Records" on page 146, can contain both variables (data members) and functions and tasks (methods). Verilog is not object oriented because modules are not true objects. Modules are hierarchical constructs, not data structures. They cannot be assigned. nor passed to tasks or functions as arguments. They cannot be compared or used in expressions. VHDL packages may also look like objects: They can contain data (shared variables) and procedures. In addition to presenting all of the same non-object behaviors that Verilog modules do, VHDL packages cannot be instantiated more than once. Even though a package is used in many library units, a single instance exists for the entire elaborated simulation model. And if that weren't enough, neither language can emulate inheritance nor polymorphism, which are other key aspects of object-orientedness.

Objects have public and private declarations.

The concept behind objects is to encapsulate data and its transformation operations to present to the user a coherent and stable interface. As the object evolves or is modified, the interface visible to the user should remain constant. To help enforce this, objects usually have two separate declaration spaces: *public* and *private* declarations, similar to VHDL's separate package and package body concept. Public declarations are accessible from the outside of the object (i.e., by the users), whereas private declarations are only accessible from within the object (i.e., by the implementer).

If the public interface is never modified (or modified in such a way as not to impact the user), the entire private implementation can be modified or re-written without affecting the users. By default, declarations are public. In OpenVera, to make a declaration private, use the *local* attribute, as shown in Sample 4-47. Any *local* data members or methods will not be found in the class declaration in the generated header file, since it is not publicly accessible. *e* has a different data protection granularity. Any declaration marked *package*, as shown in Sample 4-48, is only accessible from other package members. Since any source file can declare itself as a member of the package, data protection is controlled by the user, not the author. *Package* members are visible to all classes in the package and is closer, as a concept, to C++'s *friend* classes.

Keep non-randomized data members private.

Once a data member has been made public, you can count on other objects to make direct use of it. It is not possible to control its access, nor ensure that its value remains consistent with other data members. For example, the *length* property in Sample 4-49 can be modified independently of the *data* property. It is very easy for a user to corrupt the internal state of the object by directly operating on the data members. Traditional object-oriented practice commands that all accesses to an object be done through methods. These methods can ensure that the value of the properties are coherent at all times, as shown in Sample 4-50. However, if you end up

Sample 4-47.
Local class
declarations in
OpenVera

```
class sim_status {
   static integer n_errors = 0;
}

class mac_frame {
   local sim_status status;
   ...
   if (this.compute_fcs() !== this.fcs) {
      printf("Bad FCS");
      this.status.n_errors++;
   }
}

class mii {
   local sim_status status;
   ...
   if (col === 1'b1 && crs !== 1'b1) {
      printf("Collision without carrier\n");
      this.status.n_errors++;
   }
}
```

Sample 4-48.
Package pro-
tection in *e*.

```
package mac_frame;

extend sys {
   package mac_frame_next_id: uint;
   keep max_frame_next_id == 0;
};

struct mac_frame {
   !id: uint;
   pre_generate() is also {
      me.id = sys.mac_frame_next_id;
      sys.mac_frame_next_id += 1;
   };
   compute_fcs(): uint (bits: 32) is {
      ...
   };
};
```

with a pair of *set_data()* and *get_data()* methods for each data
member, then internal coherency is probably not required, or you
are providing methods with too low a level of abstraction. You
might as well make the data members public. OpenVera and *e* place
an additional requirement for making a data member public. It is
not possible to externally constrain a private data member since it is
not accessible. All randomized data members must be public to

allow them to be constrained. More on constraints and randomization will be discussed in "Random Stimulus" on page 354.

Sample 4-49.
Unsafe object
state

```
class byte_list {
    integer length = 0;
    bit [7:0] data[];
}

program ignoramus_user {
    byte_list my_list = new;

    my_list.length = 100;
    my_list.data[0] = 1;
    // List is corrupted: array has 1 element,
    // NOT 100.
}
```

Sample 4-50.
Safe object
state

```
class byte_list {
    local integer length = 0;
    local bit [7:0] data[];
    task resize(integer length);
}

program ignoramus_user {
    byte_list my_list = new;

    my_list.length = 100;   // Syntax error!
    my_list.resize(100);
    // List now has 100 element,
}
```

Reference and instance are different things.

When declaring an object variable, all you are declaring is a *reference* (or a pointer) to an object of a specific class. By default, object variables do not refer to any object (Figure 4-7(a)). They must first be initialized by allocating an *instance* of an object created using *new* or *gen*[15]. When assigning an object variable to another, all you are doing is making a copy of the reference, not a copy of the object instance (Figure 4-7(b)). If one of these two references modifies the object, it is modified for both variables since they both refer to the same object. A common mistake is to put references to objects in a

---

15. The *gen* action is unique to *e*. Furthermore, *e* will recursively allocate all object-type data members in a newly allocated instance. In Open-Vera, this must be done explicitly in the constructors.

list, but keep using the same reference to generate new values, as shown in Sample 4-51. Since there is a single call to *new*, a single instance of the ATM cell object exists. The list ends up containing several references to the last value of the cell, instead of to 10 random cells as expected. Similarly, when comparing two object variables, you are comparing their reference, not their content. If both variables refer to the same object instance, the comparison will be true (Figure 4-7(b)). If both objects refer to two different objects (even if they have identical content), the comparison will be false (Figure 4-7(c)).

Figure 4-7.
Object
reference vs.
object instance

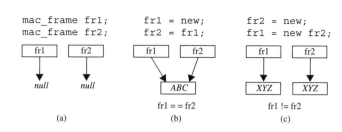

Sample 4-51.
Common mistake with object reference

```
var cell: atm_cell = new;
var cells: list of atm_cell;

while (cells.size() < 10) {
    cells.push(cell);
}
```

Comparison and copying can be shallow or deep.

Because copying or comparing object variables only deals with the references, methods must be used to perform comparison or copy functions. If the object instances being compared also contain references to other objects, how are these handled? If only the references are used, then the operation is said to be *shallow* (Figure 4-8(a)). If the operation is applied recursively down the hierarchy of objects, the operation is said to be *deep* (Figure 4-8(b)). In *e*, every *struct* includes the implicit *copy* method to perform a shallow copy operation and the *deep_copy()* and *deep_compare()* predefined routines can perform a deep copy or compare operation on any instance of a *struct* or list. In OpenVera, the *new* constructor can be used to perform a shallow copy operation, but comparison and deep copy methods must be written manually for each *class*.

Standard methods must be provided.

When defining a new object type, you should always provide the methods usually required for using an object. You should try to pro-

Figure 4-8.
Shallow and
deep
operations

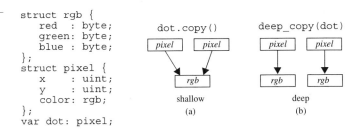

```
struct rgb {
    red   : byte;
    green : byte;
    blue  : byte;
};
struct pixel {
    x     : uint;
    y     : uint;
    color : rgb;
};
var dot: pixel;
```

vide these methods using a consistent name and argument to avoid dealing with meaningless differences between various classes. For example, methods to display the object in a meaningful human-readable format always prove indispensable. Both *e* and OpenVera come with predefined object print methods or statements, but they usually produce the information in a format that is too detailed, verbose or not intuitive. Other methods that should be included are deep-copy, deep-compare (OpenVera only), packing to and from a list of bytes, words or quads, calculating and checking error-protection fields, checking the internal consistency of data members and converting to and from other object types.

Inheritance

Objects can build upon other objects.

What if there is a class that does almost everything you need, but it is missing only that one little feature, what do you do? The traditional approach would dictate that you make a copy of the useful code, call it something else, and make the necessary modifications. But this has just created additional code that must be maintained and understood. With inheritance, your needs can be built upon existing objects—even those you do not have source code for—by only specifying the desired difference in behavior. Any unchanged behavior is automatically *inherited* from the original object. Any changes made to the original class are also automatically inherited by the new class, reducing maintenance efforts. The original class is called the *parent* or *base* class. The new class inheriting from the base class is called a *derived* class. OpenVera's inheritance and *e*'s *like* inheritance are traditional object-oriented inheritance mechanisms. *e* possesses an additional inheritance mechanism: *when* inheritance, which was used to model variant records (see Sample 4-36). Refer to section 4.10 of the *e Language Reference Manual* for a detailed discussion of the differences between *when* and *like* inheritance.

Derived objects can overload parent members and methods.

The difference in functionality between a derived class and a parent class can be expressed by adding new data members and methods, adding to the parent's methods or replacing data members and methods with new ones. For example, the verification of your design requires that you inject MAC frames with corrupted FCS values. But the MAC frame class shown in Sample 4-35 always has a good FCS value. You can create a new MAC frame object that can have a bad FCS value, based on the value of a control property,[16] using inheritance, as shown in Sample 4-52. Notice how the new version of the *compute_fcs()* method adds functionality to the parent methods. In OpenVera, you can refer to overloaded data members and methods in the parent class using the *super* prefix. In *e*, the same would be accomplished by extending the *compute_fcs()* method using *is also*, where the call to the original method is implicit.

Sample 4-52. Adding functionality through inheritance

```
class mac_frame_may_be_bad extends mac_frame {
    rand bit is_bad;

    function bit [31:0] compute_fcs() {
        compute_fcs = super.compute_fcs();
        if (this.is_bad) {
            bit [4:0] i = random;
            compute_fcs[i] ^= 1;
        }
    }
}
```

Children take after their parent.

Because derived classes are extensions of their base class, they remain valid instances of their base class. As shown in Sample 4-53, they can be assigned to base class variables without any type conversion (i.e., automatic downcasting). From that point on, they will be viewed as if they were an instance of the base class. An

16. Why not make this derived class always a bad frame? Because generating a stream containing a mix of good and bad frames would require instantiating a mix of different classes. This way, only one class needs to be instantiated. The class will decide on its own whether the frame is good. And this approach is easier to constrain.

instance of a derived class, referred to as a base class, can be assigned back to a derived class variable with explicit upcasting.

This assignment makes all code and models that already operate on the base class available to operate on the derived class—without their knowledge. For example, as shown in Sample 4-54, the derived MAC frame class with potential bad FCS values can be downcasted to the base class and sent through the existing MII bus-functional model from Sample 4-21. Upcasting an instance of the base class or of a derived class on a different inheritance branch (or *lineage*) causes an error. To prevent runtime errors when you cannot rely on implicit knowledge an object's lineage, it is always possible to test the lineage of an object. In OpenVera, use the *cast_assign()* task with a *CHECK* argument. In *e*, use the *is a* conditional operator.

Sample 4-53.
Downcasting and upcasting an inheritance tree in *e*

```
struct instruction ...;
extend ARITHMETIC instruction ...;
extend BRANCH instruction ...;
extend CONDITIONAL BRANCH instruction ...;
...
var instr:                       instruction;
var arith: ARITHMETIC            instruction;
var brch : BRANCH                instruction;
var br_if: CONDITIONAL BRANCH instruction;
...
instr = arith;     // OK: downcasting
arith = br;        // ERROR: different lineage
instr = br_if;     // OK: downcasting

br = instr.as_a(BRANCH instruction); // upcasting

arith = instr.as_a(ARITHMETIC instruction);
         // ERROR: instance on wrong lineage

if instr is a ARITHMETIC instruction (ai) {
   arith = ai; // OK: Exec'd if correct lineage
};
```

Sample 4-54.
Using code written for the base class with a derived class

```
eth_mii               bfm   = new(...);
mac_frame_may_be_bad tx_fr = new;

bfm.send(frame);
```

Declarations can
be semi-private.

Derived classes cannot access any private declaration in the base class. It is often necessary to let a derived class have more intimate knowledge of the internal state and implementation of a base class than what is visible through the public interface. Declarations can have the *protected* attribute. This makes them visible *only* to the base class and any derived classes. Protected data members and methods are "semi-private" declarations shared only between a base class and its derived classes. When implementing a derived class, it is assumed that you have a more intimate knowledge of a base class (how it works, what it depends on) than a casual user. That is why it is possible to gain greater visibility into a base class (if allowed by the base class author). As a user of a class, do not casually create derived classes simply to get at the protected members and methods. Usually, these are not safeguarded as well as the public interface and are more subject to being modified. *e* provides a further restriction: members marked as *private* are only visible to the base class and derived classes in the same package.

Inheritance and
instance are differ-
ent things.

When building a new class upon an existing class, should you inherit from it or simply instantiate it as a data member in the new class? That depends on the relationship between the two classes. If the new class is a different way of looking at the older class, but fundamentally represents the same object, then inheritance should be used. The corruptible MAC frame in Sample 4-52 is the perfect example. The objective of this new class is not to create a completely separate and new object. It is to add a new capability to the existing one. Whether good or bad, this new MAC frame version can still be viewed and treated like the old one—albeit with a small, hidden difference.

The other example is the pixel and RGB objects shown in Figure 4-8. If you already have a *color* object and want to build a *pixel* object that has a color specification, you should instantiate the color object as a data member of the pixel object, not derive from it. Why? Because a pixel is not a color. A color is only an attribute of a pixel, just like its position on the screen.

Polymorphism

Polymorphism means multiple forms.

The term "polymorphism" means to have many forms. The concept of polymorphism was hinted at when I talked about the automatic downcasting of a derived class when using existing methods that deal with the base class, as shown in Sample 4-54. Any object has the ability to take the form of an instance of its base class. You can create an entire genealogy of objects. Objects on different branches can be treated as if they were the same object, when viewed as a common base class. If all objects are derived from a single root object, they can all be viewed as instances of that root base class. For example, all *struct* definitions in *e* are automatically derived from the *any_struct* object class. Polymorphism lets you write generic methods that can deal with any objects.

A class can be designed to be used only as a base.

Polymorphism does not happen by accident. You have to plan your class genealogy to isolate common and useful functions in base classes. Sometimes, the common information exists, but does not make sense as a complete object. It would be a mistake to create an instance of such a base class because the base class is not designed to represent an object on its own. Rather, the base class is designed to take advantage of polymorphism and create a single set of generic operations.

For example, ATM cells come in two flavors: UNI and NNI. Both have the same size, both have a large number of common fields. They only differ by their interpretation of the first 12 bits, as shown in Figure 4-9. To take advantage of the polymorphism between the two ATM cell flavors, you can create an *any_atm_cell* class that contains all the common fields. Derived classes will be used to add the fields unique to each flavor. But the class *any_atm_cell* should not be allowed to be instantiated on its own: it is *not* a valid ATM cell! A user must always use one of the derived classes. It is possible, in OpenVera, to have the compiler enforce this usage by declaring the base class as *virtual*. As shown in Sample 4-55, attempting to "new" an instance of a virtual class is an error.

Which method is called?

Dealing with the object reference and type casting is relatively simple. The value of the reference stays the same while the casting operation lets the compiler provide runtime type checking. The bigger question is, when a method is overloaded or extended in a derived class, like the *compute_fcs()* method in Sample 4-52, which

Figure 4-9.
Differences in
UNI and NNI
ATM cells

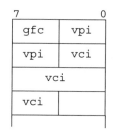

UNI ATM Cell

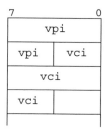

NNI ATM Cell

Sample 4-55.
Using *virtual*
classes

```
virtual any_atm_cell {
    bit [11:0] vpi;
    ...
}
class uni_atm_cell extends any_atm_cell {
    bit [3:0] gfc;
    bit [7:0] vpi;
}
class nni_atm_cell extends any_atm_cell {
    bit [11:0] vpi;
}
...
any_atm_cell a_cell;
uni_atm_cell uni_cell = new;
nni_atm_cell nni_cell = new;

a_cell = new;          // ERROR
a_cell = uni_cell;
a_cell = nni_cell;
cast_assign(nni_cell, a_cell);
cast_assign(uni_cell, a_cell);   // Runtime error
```

version is called when an instance of a derived class is referred to as an instance of its base class?

A parent can act like a child.

In OpenVera, the original method in the base class is called by default. If the methods are declared as *virtual* in OpenVera, the *overloaded method* is called. In *e*, methods are always virtual. Sample 4-56 shows how a method in the base class (which is never overloaded) makes use of a virtual method. The header error check (HEC) computation requires operating on all header bytes—which vary depending on the cell's flavor. When the *compute_hec()* method invokes the (always virtual in *e*) *pack_header()* method, what is called is the extended method found in the particular extension corresponding to the current value of the *flavor* field. The list of bits returned by *pack_header()* will always contain all of the header bits, regardless of the actual flavor of the ATM cell instance. In OpenVera, most methods should be declared virtual to give the possibility of being adapted to the extensions of each derived class and maintain the behavior expected by existing code that uses the original base class.

Sample 4-56.
Using *virtual*
methods

```
struct atm_cell {
    flavor: [UNI, NNI];
    ...
    compute_hec(): uint (bits: 8) is {
        var header: list of bit;
        header = pack_header();
        ...
    };
    pack_header(): list of bit is {
        result.add(%{vpi, clp, pt});
    };
};
extend UNI atm_cell {
    ...
    pack_header(): list of bit is first {
        result = %{gfc, vci};
    };
};
extend NNI atm_cell {
    ...
    pack_header(): list of bit is first {
        result = vci;
    };
};
```

Randomization is
a virtual process.

Although they are not explicitly documented as such, the predefined method *randomize()* and constraint blocks are virtual in OpenVera. This means that, when randomizing a base class variable that refers to a derived class instance, the derived class is being randomized, subject to the constraints and overloaded *pre_randomize()* and *post_randomize()* methods in the derived class. In *e*, since all extensions are virtual, generating an instance of a base class will always generate all of its *like* extensions and relevant *when* extensions, based on the (possibly randomly-generated) control field values for the variant parts.

Limitations of OpenVera and *e*'s OOP Implementation

No multiple inheritance.

Classes in OpenVera can be derived from a single base class only. Similarly in *e*, structs can be *like* extended from a single base struct only. It is not possible to have a class be derived from more than one base class, nor is it possible to recombine different derivatives of a common base class into a single class that encompasses both extensions. *e's when* extension can be used to emulate multiple inheritance by using different control fields. But all *when* extensions must originate from a single class. It is not possible to combine two separate base classes into a single class.

No static data members in *e*.

e does not have the concept of static data members. A static field would have a single instance and be shared by all instances of the struct that contains it. Instead, a global data space such a *sys* must be used. See Sample 4-46 for an example.

No class-level enumerals in OpenVera.

Although OpenVera has an enumerated type, the enumerals in each type share a common name space. You cannot use the same enumeral in different enumerated types, like you can in VHDL. C++'s solution is to offer class-level enumerated types. But those do not exist in OpenVera. A solution is to use static properties to define symbolic values, as shown in Sample 4-57. But type checking is lost since all symbolic values are now integers.

Sample 4-57. Emulating class-level enumerals in OpenVera

```
class atm_cell {
    bit flavor;
    static bit UNI = 0;
    static bit NNI = 1;
    ...
}
...
atm_cell cell = new;
cell.flavor = cell.UNI;
```

No truly private data protection in *e*.

There is no concept of the traditional private interface to a class in *e*. Access to class members can be restricted to other objects declared in the same package or to class extensions. But since any source file can declare itself as a member of a package, and anyone can declare a class extension, anyone can gain access to the restricted declarations.

ASPECT-ORIENTED PROGRAMMING

AOP is the next step after object-oriented programming.

Aspect-Oriented Programming (AOP), also called Subject-Oriented Programming, has been the subject of research in the software engineering community for quite some time. Several aspect- and subject-oriented extensions of popular programming languages exist. *Aspectj*[17] (an aspect-oriented extension to Java) and *AspectC++*[18] (you guessed it: an aspect-oriented extension to C++), as well as new languages designed around AOP (such as *e*) are examples. More details on the aspect-oriented programming movement can be found at:

```
http://aosd.net
```

AOP fits the verification requirements.

AOP is a powerful mechanism when applied to verification. The task of verification is to use things in ways that were not expected. It is easy to model something that does the right thing—and that's what you want to use most of the time. But verification requires breaking things to verify how a design will react to the bad data. Verification is also concerned about different things at different times. One testcase verifies that the parity error on a packet is properly detected and dealt with. Another testcase deals with violating the timing on the processor interface and verifying that the value of internal registers are not adversely affected. Both are different concerns on the same environment. One should not be forced to execute with the other and run the risk of interference. AOP is a mechanism that can separate these concerns into separate aspects of the verification environment.

The Problem with Object-Oriented Programming

OOP was thought to solve software engineering problems.

When Object-Oriented Programming (OOP) came along, it was perceived as the solution to the maintenance and evolution problem of large, complex software systems. It was thought that the strong encapsulation of the data into objects and the tight control over its interface through methods would allow the creation of objects that could be evolved and maintained without affecting other objects. Reality proves to be a little different.

---

17. See `http://www.aspectj.org` for more details.

18. See `http://www.aspectc.org` for more details.

Functional aspects involve many objects.

The inheritance and polymorphism mechanisms in object-oriented systems force all of the object classes into a strict hierarchy. Classes are clearly parents, children or siblings of each other. The strong encapsulation forces each class to be treated separately. The problem is that a new feature that must be added to the software system never fits into this neat hierarchical structure. This new feature (or "aspect") involves modifying several classes, which are likely located in separate files. The new aspect is thus diluted over several programming structures and many different files. As more aspects are added, they complicate the maintenance problem and the addition of future aspects. Furthermore, it is not possible to make aspects optional. Once an aspect is implemented, it cannot be removed without some premeditated runtime option to disable it. Disabling an aspect can only *logically* remove it; the aspect remains there *physically*.

Derived objects create new data types.

I can already hear some of you asking: Why not implement an aspect by deriving from the necessary classes and locate all of those derivatives into a single file? Wouldn't that centralize the implementation of an aspect and make it easy to remove it? Unfortunately, no. The problem with creating a derived class is that it creates a new type, separate from (but related to) the original class. All of the existing code in the system still refers, uses and instantiates the original class. You would therefore have to modify all of those references to use the new derived classes to include the additional aspect.

For example, assume you have a class that models a FIFO (let's call it "fifo"). You add a "flag" aspect to the FIFO: flags that will indicate when the FIFO is almost empty or almost full. You create a derived class called "fifo_with_flags". Although you have not modified the original FIFO model, you now have to modify every bit of code that used the FIFO model to use your additional aspect (i.e., replace every instance of "fifo" with "fifo_with_flags"). This entire process will have to be repeated when you add the "programmable-almost-empty-and-almost-full-levels" aspect on top of the "flags" aspect.

Exponential growth of derived class combinations.

There is another problem created with implementing aspects by deriving classes. Most languages (and specifically *e* and OpenVera) do not support multiple inheritance. Using the FIFO model example again, assume you have a class that modifies the original FIFO with synchronous read and write clocks to add an "asynchronous" aspect. How do you deal with the two orthogonal flag-related aspects? You can implement each aspect in a linear fashion (i.e., flags, then programmable levels, then asynchronous clocks or asynchronous clocks, then flags, then programmable levels). But this approach would force every FIFO used in the system to include those aspects it does not need. Or you can create a derived class for every possible combination of aspects (i.e., flags, flags + programmable levels, asynchronous clocks, asynchronous clocks + flags, asynchronous clocks + flags + programmable levels). But this is going to create a large number of classes that grows exponentially with the number of orthogonal aspects.

AOP is about "patching" code.

The solution to the problem is to be able to modify the original object classes without modifying the original source code (which is "golden"). Aspect-oriented languages offer a well-defined mechanism for adding declarations and inserting or replacing code from the outside of a class, without actually editing the original class implementation. I can see the hair raising on the neck of many readers. I thought the same thing when I first saw this mechanism. "*Evil!*" I said. But once you understand the concept and the methodology that goes with it, I am sure you will change your mind, just as I did.

Variant Data with Variant Code

AOP is creating variant records, after the fact.

In "Variant Records" on page 151, I showed how variant records should be implemented in OpenVera and *e*. In that particular instance, as the author of the original *mac_frame* class, I had prior knowledge that the virtual LAN tagging variance was required and modeled the object accordingly. I can be even more forward-looking and add variant fields for 802.2 LLC and SNAP information. Verification exists because it is impossible for anyone to think of everything beforehand. What if I suddenly need to support control frames in my verification environment? With the aspect-oriented features of *e*, this can be done after the fact without modifying any existing code, simply by loading the additional code shown in Sample 4-58. Note how enumerated types can be extended with new enumerals as needed.

If data members can vary, so should the code.

Aspects involve more than adding data members or constraints. Aspects usually require modification to methods so they can

Sample 4-58.
Adding a control frame aspect to Sample 4-36

```
extend mac_frame_format: [CONTROL];

type opcodes: [PAUSE = 1] (bits: 16);
extend CONTROL mac_frame {
    opdcode: opcodes;
    keep len_typ  == 0x8808;
    keep data_len <= 44;
};
extend PAUSE CONTROL mac_frame {
    keep da == 0x0180C2000001;
    quantas: uint (bits: 16);
    keep data_len == 42;
    keep for each in data {
        it == 0x00;
    };
};
```

include the new data members in their operation. For example, adding the "bad HEC" aspect to the ATM cell object, requires adding a data member to control whether to corrupt the cell (having a cell that is always corrupted would not be very useful). It also requires modifying the HEC computation method to return an invalid value. Sample 4-59 shows the bad HEC aspect that can be added to the ATM cell object. Notice how code is inserted at the end of the existing *compute_hec()* method to corrupt the (correct) calculated HEC value by replacing it with a random—but different—value. A *soft* constraint is used on the variance control field *bad_hec* to maintain the default behavior of the original *atm_cell* by default. When adding a new aspect, the usage of the aspect should be made optional by encapsulating all of the extensions into a variances of the original object.

Sample 4-59.
Adding a HEC corruption aspect to an ATM cell object

```
extend atm_cell {
    bad_hec: bool;
    keep soft not bad_hec;
};
extend bad_hec atm_cell {
    compute_hec(): byte is also {
        gen result keeping {
            it != result;
        };
    };
};
```

What is an aspect? The aspects used in examples so far were pretty simple. They involved extending a single class—which could have been relatively simple in a strict object-oriented hierarchy. But aspects usually have wider implications, even the simple example aspects used so far. For example, the *control frame* aspect added to the ethernet MAC frame also has implications on the MAC-layer bus-functional model. This bus model must be able to generate control frames. It must also be able to respond to PAUSE control frames and stop the flow of frames for the specified number of quantas. This aspect also affects any ethernet interface checkers. As the checker sees PAUSE control frames flow to a MAC-layer interface, it must verify that the interface pauses for the specified time period. An aspect is a complete implementation of a specific function in a system.

Aspects can be simple functions, entire testcases or composed of sub-aspects. An aspect could be the handling of corrupted data objects (which affects the object itself, the object generators, the bus-functional models and the self-checking structure). An aspect could be the introduction of a new category of instruction opcodes in a processor (affecting the opcode object, the code generator and the trace predictor). An aspect can be a constrained testcase that adds constraints on several generators and adds some synchronization mechanism to create a particularly interesting corner case. An aspect could be a directed debug testcase, limiting the generators to generating a single object and the design configuration to a straightforward one. An aspect could be the addition of functional coverage measurement to an existing verification environment, involving additions to the generators, self-checking structures, bus-functional models or white-box monitors into the design. An aspect could be a set of aspects, designed to re-run a troublesome testcase, involving setting the seed to a specific value and loading all other aspects that were used in the original simulation plus additional debugging aspects that provide useful trace information.

Aspects must be managed. However complex or simple an aspect is, regardless of how many objects are affected by it, they are implemented in separate files that can be added to an environment as needed. You only need to include in an environment those aspects that are needed. Managing aspects becomes an important part of simulation management. Which aspects are orthogonal? Which ones are optional? In what order should they be loaded? Which aspects depend on what other aspects?

Limitations of *e*'s AOP Implementation

More discipline is required.

Everything can be extended: enumerated types, object classes, methods, temporal definitions of events and functional coverage groups. And when I say everything, I mean everything: every type, every class and every method of every class. There is no mechanism to control the extension points, what can be extended and what shouldn't. On the other hand, the existence of such a mechanism would be of questionable value: Isn't AOP about being able to modify an existing implementation to add stuff the original author didn't think about? If we are limited to what the author has decided to let us extend, aren't we back to square one? *e* allows complete freedom.

But freedom requires self-discipline. In a large system, it is often difficult to know all of the impacts of including many different aspects, created by different persons. There has to be a discipline to the various extensions. A few experienced individuals should be responsible for directing how and where various aspects should be inserted. The system must be designed to allow for insertions. Users should use aspects-within-aspects to limit the scope of their extensions. The topic of aspect-oriented coding methodology goes far beyond this simple overview section. Some coding guidelines are available in Appendix A. But just as many books were written on structured programming, then later on object-oriented programming, many books are (and will be) written on aspect-oriented programming.

Individual extensions for individual values.

When extending an object definition using a variance control field, the extensions can be specified for only a single value of the control field. For example, an "instruction" object is being extended into an "add" and a "sub" object. If there is any additional function that is common to both extensions, it must be replicated for each instruction—once for the control field being equal to "add", another time for the control field being equal to "sub". It is not possible to extend an object based on a control field being equal to this or that value. This limitation is usually handled by declaring an additional virtual field to "merge" the separate values into one, as shown in Sample 4-60. But care must be exercised to maintain the coherency between the new virtual field and the original variance control field.

Fields can only be appended.

When extending an object class definition with additional fields, the fields are physically appended to the existing fields, as if they

Sample 4-60.
Adding virtual field to control common extensions

```
extend instruction {
   aos: bool;
   keep aos == opcode in [ADD, SUB];
   when aos instruction {
      // Common extensions to ADD and SUB
      ...
   };
};
```

had been declared after them. Most of the time, this is not a problem. It only becomes a problem when using the built-in *pack* and *unpack* methods. If the additional fields should be physically packed in the middle of the existing fields (such as the virtual LAN fields in Sample 4-36 or the opcode and operand fields in Sample 4-58), the resulting default packing and unpacking processes will yield an incorrect result. Some potential data extension points are known a priori, such as before and after the user payload in a packet. In that case, you can introduce a placeholder physical field at the appropriate location of the *struct* to be packed. Initially empty, it can then be extended to contain additional fields, as illustrated in Sample 4-61 and Sample 4-62. However, the new data fields are located in a sub-object and require an additional component in their name when referenced.

Sample 4-61.
Placeholders for data extensions

```
type mac_frame_type: [RAW];

struct mac_frame_pre_data {
   kind: mac_frame_type;
};

struct mac_frame {
   kind: mac_frame_type;
   ...
   len_typ: uint (bits: 16);
   %pre_data: mac_frame_pre_data;
   %data: list of byte;
   %fcs: uint (bits: 32);
   keep pre_data.kind == me.kind;
};
```

Methods can only be appended, prepended or replaced.

When extending an object class implementation with additional sequential code in existing methods, the additional code can be appended or prepended only to the existing code, or the additional code can completely replace the original code. It is not possible to add code at arbitrary points in the original code sequence (i.e., there

Sample 4-62.
Extending a
struct physi-
cal field in the
"middle"

```
extend mac_frame_type: [VLAN];

extend VLAN mac_frame_pre_data {
    %priority: uint (bits:  3);
    %cfi       : bit;
    %vlan_id : uint (bits: 12);
    %len_typ : uint (bits: 16);
};

extend VLAN mac_frame {
    keep len_typ == 0x8100;
};
```

is no *is middle* extension mechanism). If the aspect requires additional functionality in the body of a *for-loop*, you are out of luck.
When writing code designed to be extended, it is a good idea to add
calls to empty methods (called *hook* methods) at judicious locations
in the code sequence to let an aspect add functionality wherever
appropriate. For example, Sample 4-63 shows how hook methods
open up the possibility of adding code to the body of nested *for-
loops*.

Sample 4-63.
Hook meth-
ods let you
add aspects
where you
need them

```
pre_packet_parity(pkt: ip_pkt) is empty;
pre_data_parity(data: *byte) is empty;
post_packet_parity(parity: *byte) is empty;
...
for each (pkt) in packets {
    var parity: byte = 0x00;
    me.pre_packet_parity(pkt);
    for each (db) in pkt.data {
        me.pre_data_parity(db);
        parity ^= db;
    };
    me.post_packet_parity(pkt, parity);
};
```

Aspects are order-
dependent.

When extending an object class implementation with additional
sequential code in existing methods, the new code is inserted in or
replaces the existing code, including any previously loaded aspects.
Therefore, the order of execution of aspect-specific code will
depend on the order in which the individual aspects were loaded.
For example, if two aspects extend a method using *is first*, the code
related to the aspect loaded *last* will be executed first. If a third
aspect extends the same method using *is only*, the code from the
previous two aspects, as well as the original method body, will be

effectively wiped out and will never be executed. Loaded in a different order, (e.g., #3, then #2 then #1), the two *is first* extensions could execute in reverse order, before expected code that no longer exists. Similarly, the *return* action will abort the execution of a method and all of the previously loaded extensions. But the *return* action will continue with the execution of any *is also* extensions loaded after the extension (or original implementation) containing the *return* action.

THE PARALLEL SIMULATION ENGINE

C and C++ lack essential concepts for hardware modeling.

Why hasn't C been used as a hardware description language instead of creating Verilog, VHDL, SystemVerilog, Jeda, OpenVera, *e* and many others? Because the basic C language lacks three fundamental concepts necessary to model hardware designs: connectivity, time and concurrency. Basic C++ also lacks the necessary features to support the HVL productivity cycle: randomization, constrainability and functional coverage measurement.

Connectivity, Time and Concurrency

Connectivity is the ability to describe a design using simpler blocks then connecting them together. Schematic capture tools are perfect examples of connectivity support.

Time is the ability to represent how the internal state of a design evolves over time and to control its progression and rate. This concept is different from *execution time* which is a simple measure of how long a program runs.

Concurrency is the ability to describe actions that occur at the same time, independently of each other.

C and C++ can be extended.

Many extensions and coding styles for C or C++ exist that include some or all of these concepts. SystemC is a set of C++ classes to introduce the concept of connectivity, time and concurrency. The SystemC Verification Library is a set of C++ classes that provides support for randomization, constraints and temporal expressions. OpenVera feels like a hybrid between Verilog and C++.

Connectivity, Time and Concurrency in HDLs and HVLs

Each language implements these concepts in different ways.

The connectivity, time and concurrency concepts are very important to understand when learning to model using a modeling language. Each language implements them in a different fashion, some easier to understand than the others.

For example, connectivity in Verilog is implemented by directly instantiating modules within modules, and connecting the pins of the modules to wires or registers. Understanding why registers cannot be used in some circumstances requires understanding the concept of concurrency. Concurrency is described in detail in the following sections.

In VHDL, connectivity is implemented with entities, architectures, components and configurations. The mechanics of connectivity in VHDL require a lot of statements and apparent duplication of information and is often one of the most frustrating concepts to learn in VHDL.

In OpenVera and *e* the only connectivity mechanism provided is to connect the testbench to the design under verification. They both interface to the HDL worlds and operate through an automatically generated *shell* file. The content of the *shell* file reflects the properties of the interface signals and must be re-generated every time you modify the specification of the interface between OpenVera or *e* and Verilog or VHDL.

Time can be implemented as absolute or unit-less relative values.

The concept of time is also implemented differently. Verilog uses a unit-less time value. The time values from multiple modules are correlated using a scale factor specified using the `timescale compiler directive. In VHDL, all time values are absolute, with their units clearly stated. OpenVera and *e*, time is specified as a unit-less value. The actual unit of time is based on the timescale or resolution of the Verilog or VHDL *shell* or *stub* file.

Simulators differ most in their implementation of concurrency.

The implementation of concurrency is where languages and simulators differ the most. Although they all use an *event-driven* simulation process, they differ in the granularity of their concurrency, and in the timing and focus of assignments between concurrent constructs. To write VHDL, it is necessary to understand the implementation of this concept because of the restrictions concurrency imposes on the use of the language. Other languages put very few restrictions on their use. Verilog, *e* and OpenVera rely on the author to use concurrency appropriately. If you are limited to a certain coding style, such as the synthesizeable subset, you can write functional Verilog code without having to understand its implementation of the concept of concurrency.

You write better
testbenches when
you understand
concurrency.

When writing testbenches, you are not confined to such coding styles. It becomes necessary to understand how concurrency is implemented and how concurrency affects the execution of the various components of the testbench.

Many testbenches are written with a severe lack of understanding of this important concept. In the best case, the execution and overall control structure of the testbench code is difficult to follow and maintain. In the worst case, the testbench fails to execute properly on a different simulator, on different versions of the same simulator or when using different command-line options. The understanding of concurrency is often what separates the experienced designer from the newcomers.

The Problems with Concurrency

There are two problems with concurrency. The first one is in describing concurrent systems. The second is executing them.

Concurrent systems are difficult to describe.

Since computers were created, computer scientists have tried to figure out a way to take advantage of the increased performance offered by multi-processor machines. They are relatively easy to build and many parallel architectures have been designed. However, they proved much more difficult to program. I do not know if that difficulty originated with the mindset imposed by the early Von Neumann architecture still used in today's processors, or by an innate limitation of our intellect.

Concurrent systems are described using a hybrid approach.

Human beings are adept at performing relatively complex tasks in parallel. For example, you can drive in heavy traffic while carrying a conversation with a passenger. But it seems that we are better at describing a process or following instructions in a sequential manner. For example, a recipe is always described using a sequence of steps. The description of concurrent systems has evolved into a hybrid approach. Individual processes running in parallel with each other are themselves described using sequential instructions. For example, a dessert recipe includes instructions for the cake and the icing as separate instructions that can be performed in parallel, but the instructions themselves follow a sequential order.

VHDL and Verilog models are concurrent processes described sequentially.

A similar principle is used in both VHDL and Verilog. In VHDL, the concurrent processes are the *process* statements (all concurrent statements are simple shorthand forms for processes). In Verilog, the concurrent processes are the *always* and *initial* blocks and the *continuous signal assignment* statements. In *e*, concurrent processes are *time-consuming method threads* started using the *start*, *all of* and *first of* actions, temporal expressions and *on* blocks. In OpenVera, concurrent processes are the *program thread* and *execution threads containing timing control statements* started using the *fork/join* statement The exact behavior of each concurrent construct, in all languages, is described individually using sequential statements.

Every *process* in a VHDL model, every *always* and *initial* block in a Verilog model and every *thread* in *e* and OpenVera execute in parallel with each other, but internally each executes sequentially. It is a common misconception that Verilog's *initial* blocks mean "initialize". Unlike VHDL, there is *no* initialization phase in Verilog. Everything is implicitly initialized to x and *initial* blocks are identical to *always* blocks except that they execute only once. They are removed from the simulation once the last statement in the *initial* block executes.

Emulating Parallelism on a Sequential Processor

Concurrent systems must be executed on single processor machines.

If you look inside the workstation that you use to simulate your model, you will see that there is a single processor. Even if you have a multi-processor machine, you can always write a model with one more parallel construct than you have processors available. How do you execute a parallel description on a single processor, which is itself a sequential machine?

Multi-tasking operating systems are like simulators.

If you use a modern computer, you probably have a windowing graphical interface. During normal day-to-day use, you are very likely to have several windows open at once, each of them running a different application. On multi-user machines, there may be several others running a similar environment on the same computer. The applications running in all of these windows appear to work all in parallel even though there is a single sequential processor to execute them. How is that possible? You probably answered *time-sharing*. With time-sharing, each application uses the entire processor for small portions of time. Each application has its turn according to priority and activity. If the performance of the processor and operating system is high enough, the interruptions in the execution of a program are below our threshold of detection: It appears as if each program runs smoothly 100% of the time, in parallel with all the others.

Simulators are time-sharing engines.

A simulator works using the same principle. Each *process*, or *always* and *initial* block or *thread* has the simulation engine for some portion of time. Each *appears* to be executing in parallel with the others when, in fact, they are each executed sequentially, one after another. There is one important difference in the time-sharing process of a simulator. Unlike a multi-tasking operating system, the simulator assumes that the various parallel constructs cooperate to obtain fair access to the simulation resources.

Simulators do not have time slice limits.

In an operating system, every process has a limit on the amount of processor time it can have during each execution slice. Once that limit is exhausted, the process is kicked out of the processor to be replaced by another. There is no such limit in a simulator. Any process keeps executing until it explicitly requests to be kicked out. Thus, it is possible in a simulation to have a process grab the simulation engine and never let it go. Ensuring that the parallel constructs properly cooperate in a simulation is a large part of understanding how concurrency is implemented.

Processes simulate until they execute a timing statement.

In VHDL, a process simulates, and keeps simulating, until a *wait* statement is executed. When the *wait* statement is executed, the process is kicked out of the simulation engine and replaced by another one. This process remains "out of circulation" until the condition it is waiting for is realized. Verilog has a similar model: *always* and *initial* blocks simulate and keep simulating until a @, # or a blocking assignment is executed. It also stops executing if a *wait* statement whose condition is currently *false* is executed. In OpenVera, threads may be interrupted only when a signal drive, @, *delay, wait_var, sync* or *suspend_thread* statement is executed. In *e*, threads may be interrupted only when a *wait*, *sync* or TCM call action is executed. If a concurrent process does not execute some form of an active timing statement[19], it remains in the simulation engine, locking all other processes out.

The Simulation Cycle

HDL simulators execute processes at the current time, then assign zero-delay future values.

Figure 4-10 shows the VHDL and Verilog simulation cycle. For a given timestep, the simulation engine executes each of the parallel processes that must be executed. While executing, these processes may perform assignments of future values using signal assignments in VHDL or nonblocking assignments in Verilog. Once all processes are executed (i.e., they are all waiting for something), the simulator assigns any future values scheduled for the current timestep (i.e. zero-delay assignments). Processes sensitive to the new values are then executed at the delta-cycle. This cycle continues until there are no more processes that must be executed at the current timestep and there are no more zero-delay future values.

Figure 4-10. VHDL and Verilog simulation cycle

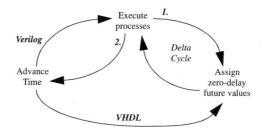

Simulators then advance time or starve.

If there is nothing left to be done at the current time, there *must* be either:

1. A process waiting for a specific amount of time

2. A future value to be assigned after a non-zero delay

If one of the conditions is present, the simulator advances time to the next time period where there is useful work to be done. The simulator then assigns a future value, which causes processes sensitive to the signals assigned these values to be executed, or execute processes that were waiting. If neither of the conditions are true, then the simulation stops on its own, having reached a quiescent state and suffering from event starvation.

---

19. Some timing control statements can be inactive if the condition they are supposed to wait for is already true. They include (but are not limited to) Verilog's *wait* statement, OpenVera's *sync* statement and *e*'s *sync* action.

Simulators do not increment time step by step.

The simulator does *not* increment time by a basic time unit, timestep or time increment. Regardless of the simulation resolution, the simulation advances time as far as necessary, in a single step, to the next point in time where there is useful work to do. Usually, that point in time is the delay in the clock generator. Increasing the simulation time resolution should not significantly decrease the simulation performance of a behavioral or RTL model.[20]

Zero-delay cycles are called delta cycles.

The state of the simulation progresses along two axes: zero-time and simulation time. As processes are simulated and new values are assigned after zero delays, the state of the simulation evolves and progresses, but time does not advance. Since time does not advance, but the state of the simulation evolves, these zero-delay cycles where processes are evaluated and zero-delay future values are assigned are called *delta-cycles*. The simulation progresses first along the delta axis then along the real-time axis, as shown in Figure 4-11. It is possible to write models that simulate entirely along the delta-time axis. It is also possible to write models that are unintentionally stuck in delta cycles, preventing time from advancing.

Figure 4-11.
Time progression along two axis

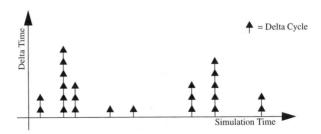

VHDL and Verilog behave differently after advancing time.

In Figure 4-10, you will notice that the VHDL and Verilog simulation cycles differ after time advances. In VHDL, future values are assigned before the execution of processes. In Verilog processes are executed first. Given the choice between executing a process or assigning a new value at the exact same point in time in the future, VHDL assigns the new value, while Verilog executes the process first. This may produce different simulation results between apparently identical VHDL and Verilog models, such as those shown in Sample 4-64 and Sample 4-65. A message is displayed in the Ver-

---

20. This is the case for some Verilog simulators.

ilog version, but not in the VHDL one. It may also affect the behavior in a co-simulation environment when a new value crosses the VHDL/Verilog boundary.

Sample 4-64.
Verilog model apparently identical to a VHDL model

```
module testcase;

reg R;

initial
begin
    R = 1'b0;
    R <= #10 1'b1;
    #10;
    if (R !== 1'b1) $write("R is not 1\n");
end

endmodule
```

Sample 4-65.
VHDL apparently identical to a Verilog model

```
entity case is
end case;

architecture test of case is
    signal R: bit := '0';
begin
    process
    begin
        R <= '1' after 10 ns;
        wait for 10 ns;
        assert R = '1'
            report "R is not 1"
            severity NOTE;
        wait;
    end process;
end test;
```

The Co-Simulation Cycle

HVLs co-simulate with HDLs.

Figure 4-12 shows the co-simulation cycle between OpenVera or *e* and VHDL or Verilog. After completing all of the delta cycles at the current simulation time, instead of advancing time, the *e* or Open-Vera simulator may be invoked. It performs as many delta cycles (in Specman Elite they are called *simulation ticks*) as it can. Then control is returned to the HDL simulator. If assignments were made from OpenVera or *e* to the HDL world at the current time, they are performed, which may cause additional HDL and HVL delta cycles. If there is nothing to be done at the current simulation time, the HDL simulator advances time.

Figure 4-12.
HDL and
HVL co-
simulation
cycle

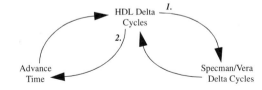

The HDL simula-
tor controls time.

Notice how time is never advanced within OpenVera or *e*. They always execute in zero-time. Time is controlled by the associated HDL simulator. That is the reason why it is more efficient to generate clocks in the HDL side than in the HVL side. Vera and Specman Elite come with a stand-alone simulator that can simulate your code without a Verilog or VHDL simulator. These stand-alone simulators have no notion of real time and will simply increment the simulation time by one after every set of delta cycles.

Parallel vs. Sequential

Use sequential
descriptions as
much as possible.

As explained earlier, humans can understand sequential descriptions much easier than concurrent descriptions. Anything that is described using a single sequence of statements is easier to understand and maintain than the equivalent behavior described using parallel constructs. The independence of their location and ordering in the source file adds to the complexity of concurrent descriptions. A concurrent description that would be relatively easy to understand can be obfuscated by simply separating the pertinent concurrent statements with a few other unrelated concurrent constructs. Therefore, functionality should be described using sequential constructs as much as possible.

A frequent misuse of sequential constructs in Verilog involves the initialization of registers. For example, Sample 4-66 shows a clock generator implemented using two concurrent constructs: an *initial* and an *always* block.

However, generating a clock is an inherently sequential process: It starts at one value then toggles between one and zero at a constant rate. A better description, using a single concurrent construct, is shown in Sample 4-67.

Sample 4-66.
Misuse of concurrency in Verilog

```
reg clk;
initial clk = 1'b0;
always #50 clk = ~clk;
```

Sample 4-67.
Proper use of concurrency in Verilog

```
reg clk;
initial
begin
   clk = 1'b0;
   forever #50 clk = ~clk;
end
```

Deterministic sequential behavior does not need concurrency.

Another less obvious case of misused concurrency happens when the behavior of the various processes is deterministically sequential because of the data flow. For example, Sample 4-68 shows a VHDL process labeled *P2* that can execute only once the process labelled *P1* triggers the signal *do*. The *P1* process then waits for the completion of process *P2* before resuming its execution. The sequence of execution cannot be other than the first half of *P1*, *P2*, then the second half of *P1*.

Sample 4-68.
Deterministic sequential execution in VHDL

```
architecture test of bench is
   signal do, done: boolean;
begin
   P1: process
   begin
      -- First half of P1
      ...
      do <= not do;
      wait on done;
      -- Second half of P1
      ...
   end process P1;

   P2: process
   begin
      wait on do;
      -- All of P2
      ...
      done <= not done;
   end process P2;
end test;
```

The implementation in Sample 4-69 shows the equivalent functionality, implemented using a single process. Not only is the execution

Sample 4-69.
Simplified
sequential exe-
cution in
VHDL

```
architecture test of bench is
begin
    P1_2: process
    begin
        -- First half of P1
        ...
        -- All of P2
        ...
        -- Second half of P1
        ...
    end process P1_2;
end test;
```

flow easier to follow, but also it does not require the control signals *do* and *done*.

Fork/Join Statement

Control flow may alternate between sequential and concurrent regions.

The overall control flow for a testcase often involves a sequence of sequential steps followed by concurrent ones. For example, verifying a configuration of a design may require configuring the device through several consecutive reads and writes via the CPU interface, then concurrently sending and receiving data. This process is then repeated for another configuration. Figure 4-13 shows a control flow diagram of such a control structure.

Figure 4-13.
Series of
sequential and
concurrent
control flows

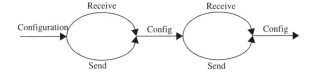

Implement using a *fork/join* statement in Verilog.

The easiest way to implement this type of control flow structure is to use a *fork/join* statement in Verilog and OpenVera and an *all of* action in *e*. This statement dynamically creates concurrent processes within a region of sequential code. The sequential execution resumes after the fork/*join* statement, once all the concurrent regions are complete as illustrated in Figure 4-14(a). For example, the code in Sample 4-70 waits for the maximum of Ta, Tb and Tc.

The *join* condition can have many flavors.

In *e* and OpenVera, the join condition may have different flavors. By default, the code after a *fork/join* statement resumes only once all of the branches have completed their respective execution.

Figure 4-14.
Execution
threads in
fork/join
statements

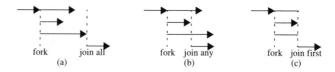

fork   join all   fork   join any   fork   join first
(a)   (b)   (c)

Sample 4-70.
Example of
using the *fork/*
join statement
in Verilog

```
initial
begin
    ...
    fork
        #(Ta);
        #(Tb);
        #(Tc);
    join
    ...
end
endmodule
```

Sometimes it may be useful to continue execution as soon as one of
the branches completes its execution, as illustrated in Figure 4-
14(c). Sample 4-71 shows how the *first of* action in *e* is used to generate an event at regular intervals until it is eventually acknowl-
edged.

Sample 4-71.
Variant of the
fork/join
action in *e*

```
...
first of {
    {
        wait cycle @ack;
    };
    {
        while TRUE {
            emit req;
            wait delay(10);
        };
    };
};
...
```

OpenVera's *join*
any does not ter-
minate other
branches.

The equivalent OpenVera code is shown in Sample 4-72. Notice the
presence of the *terminate* statement after the *join any*. In OpenVera,
unlike *e*, the other branches of the *fork/join any* statement keep executing after execution resumes after the *join*, as illustrated in
Figure 4-14(b). Your functionality may require that they be allowed
to complete in parallel. But if they must be aborted, as is the case in

Sample 4-72, they must be aborted explicitly using the *terminate* statement.

```
. . .
fork
    sync(ALL, ack);
    while (1) {
        trigger(req);
        delay(10);
    }
join any
terminate
. . .
```

The *fork/join* statement can be disabled in Verilog.

e's *first of* action and OpenVera's terminated *join any/terminate* statement can be emulated in Verilog by disabling the named *fork/join* statement from within the execution branches. For example, the code in Sample 4-73 detects and reports a time-out if the *posedge* on *gt* is not received within *Tmax* time units.

Sample 4-73.
Example of
disabling the
fork/join state-
ment in Ver-
ilog

```
initial
begin
    . . .
    fork: wait_for_gt
        @ (posedge gt) disable wait_for_gt;
        #(Tmax) begin
            $write("Time-out on gt\n");
            disable wait_for_gt;
        end
    join
    . . .
end
endmodule
```

VHDL has no *fork/join* construct.

Unfortunately, VHDL does not have a *fork/join* statement or its equivalent. It is necessary to emulate this behavior using separate processes and controlling their execution via another process. Emulating the functionality of the *fork* is simple: An event on a single signal can be used to trigger the execution of the concurrent regions. Emulating the functionality of the *join* is more complicated. You could use an event on a signal for each branch of the join, but this would require a signal for every branch. Adding a new branch would require adding a new signal and modifying the *wait* statement implementing the *join*.

```
package fork_join is

type join_ctl_typ is (join, fork, run);
type branches_typ is
    array(integer range <>) of join_ctl_typ;

function join_all(branches: branches_typ)
    return join_ctl_typ;
function join_one(branches: branches_typ)
    return join_ctl_typ;

subtype fork_join_all is join_all join_ctl_typ;
subtype fork_join_one is join_one join_ctl_typ;

end fork_join;
```

Emulate the *join*
statement using a
resolution func-
tion.

Resolution functions provide a simpler mechanism for handling an arbitrary number of branches. A different resolution function can be used to implement a *join-all* (Figure 4-14(a)) or a *join-any* (Figure 4-14(b)) functionality. Sample 4-74 and Sample 4-75 show the implementation of the *join* resolutions functions, while the code in Sample 4-76 shows how to use it. My prayers to the VHDL gods for a *fork/join* statement remain, to this day, unanswered.

Sample 4-75.
Implementation of the *fork/join* emulation in VHDL

```
package body fork_join is

function join_all(branches: branches_typ)
    return join_ctl_typ is
begin
    for i in branches'range loop
        if branches(i) = fork then
            return fork;
        end if;
        if branches(i) = run then
            return run;
        end if;
    end loop;
    return join;
end join_all;

function join_one(branches: branches_typ)
    return join_ctl_typ is
begin
    for i in branches'range loop
        if branches(i) = fork then
            return fork;
        end if;
        if branches(i) = join then
            return join;
        end if;
    end loop;
    return run;
end join_one;

end fork_join;
```

Sample 4-76.
Using the
emulation of
the *fork/join*
statement in
VHDL

```
use work.fork_join.all;
architecture test of bench is
    signal fk_jn1: fork_join_all;
begin
    process
    begin
        -- Fork
        fk_jn1 <= fork;
        wait until fk_jn1 = fork;
        fk_jn1 <= run;
        -- Branch #0
        ...
        fk_jn1 <= join;
        -- Join
        wait until fk_jn1 = join;
    end process;

    branch1: process
    begin
        fk_jn1 <= fork;
        wait until fk_jn1 = fork;
        fk_jn1 <= run;
        ...
        fk_jn1 <= join;
        wait;
    end process branch1;

    branch2: process
    begin
        fk_jn1 <= fork;
        wait until fk_jn1 = fork;
        fk_jn1 <= run;
        ...
        fk_jn1 <= join;
        wait;
    end process branch2;
end test;
```

The Difference Between Driving and Assigning

Assignments write a value to a memory location.

Regular programming languages provide variables that can contain arbitrary values of the appropriate type. They are implemented as simple memory locations. Assigning to these variables is the simple process of storing a value into that memory location. VHDL variables and Verilog *reg* operate in the same way. When an assignment is completed, whether blocking or nonblocking, the newly assigned value overwrites any previous value in the memory location. Previous assignments have no effects on the final result. Regular assignments behave like a multiplexer. A single value from all of the potential contributors is somehow selected.

The last assignment determines the value.

For example, in Sample 4-77, the value of the register R goes from x to 5 to 4 to 3 to 2 to 1, then finally to 0. Since R is a variable shared by all three concurrent blocks, a single memory location exists. Whatever value was assigned last by a concurrent block, is the value stored in the variable. This is where the keyword *reg* comes from for Verilog variables. Registers—or flip-flops—retain whatever value was last loaded into them, without regard to the previous values or other concurrent sources.

Sample 4-77.
Assignments to a shared variable in Verilog

```
module assignments;
integer R;

initial R <= #20 3;

initial
begin
   R = 5;
   R = #35 2;
end

initial
begin
   R <= #100 1
   #15 R = 4;
   #220;
   R = 0;
end

endmodule
```

Hardware description languages need the concept of a wire.

The variable is sufficient for ordinary sequential programming languages. When describing hardware, a construct that can describe the behavior of a wire used to connect multiple devices together must be provided. Figure 4-15 shows a wire, presumably part of a data bus, connected to several devices. Each device, using a tristate driver, can drive a value onto the wire. The final logic value on the wire depends on *all* the individual values being driven, not just the last one, like a variable.

Figure 4-15.
Multiple drivers on a wire

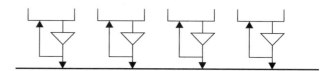

Individual values from connected devices must be driven continuously onto the wire.

To model connectivity via a wire properly, any value driven by a device must be driven continuously onto that wire, in parallel with the other driving values. The final value on that wire depends on all of the continuously driven individual values.

For example, on a tristate wire, the individual driven values of z, 1, weak-0 and z would produce a final result of 1. Figure 4-16 shows the implementation of the wire driver in Verilog and VHDL.

In Verilog, this continuous drive is implemented using a continuous assignment while the final value is determined by the type of wire being used (*wire*, *wor*, *wand* or *trireg*) and the strength of the individual driven values.

In VHDL, the continuous drive is implemented in each process that assigns a signal while the final value is determined by the user-defined resolution function.

Each concurrent construct has its own, single driver.

Parallel drivers on a wire require concurrent constructs to describe them. Many inexperienced engineers, when learning to code for synthesis try to implement the design shown in Figure 4-17 using the code shown in Sample 4-78. Unfortunately, since a single register is used with variable assignments in sequential code, a *multiplexer* is synthesized instead of the expected parallel drivers. The proper solution requires three concurrent constructs, one for each driver, and is shown in Sample 4-79.

Figure 4-16.
Implementa-
tion of
continuous
drive in
Verilog and
VHDL

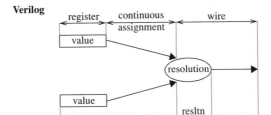

Figure 4-17.
Simple design
with three
tristate drivers

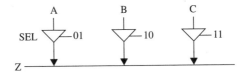

Sample 4-78.
Implementa-
tion using a
multiplexer

```
module simple(A, B, C, SEL, Z);
input       A, B, C;
input   [1:0] SEL;
output      Z;

reg Z;
always @ (A or B or C or SEL)
begin
    case (SEL)
    2'b00: Z = 1'bz;
    2'b01: Z = A;
    2'b10: Z = B;
    2'b11: Z = C;
    endcase
end
endmodule
```

Sample 4-79.
Implementa-
tion using
three tristate
drivers

```
module simple(A, B, C, SEL, Z);
input       A, B, C;
input   [1:0] SEL;
output      Z;

assign Z = (SEL == 2'b01) ? A : 1'bz;
assign Z = (SEL == 2'b10) ? B : 1'bz;
assign Z = (SEL == 2'b11) ? C : 1'bz;

endmodule
```

HVLs have no concept of continuous drivers.

e and OpenVera only deal with variables. They have no concept equivalent to Verilog's *wire* or VHDL's *signal*. In *e*, assignments to HDL references are, by default, assumed to be to HDL variables. If separate drivers are required, the HDL signal must be qualified using a *verilog* or *vhdl* statement. In OpenVera, each output interface signal is a single instance of a variable, even if it is bound to different virtual port instances or used in separate threads. That is why Vera will detect and flag conflicting assignments to the same HDL signal, made at the same simulation time, even if they are made through different virtual port bindings or execution threads. If separate drivers are required, another interface signal, bound to the same HDL node, must be used.

RACE CONDITIONS

The simulation cycle creates race conditions.

If you refer to Figure 4-10 and the section titled "Emulating Parallelism on a Sequential Processor" on page 192, you will see that parallel threads are executed one after another, during the same timestep. The order in which the threads are executed is *not deterministic*. Race conditions exist when multiple concurrent threads compete for the same shared resource over the same time period.

RTL coding guidelines hide race conditions

Race conditions are conveniently eliminated when limiting yourself to writing synthesizeable code. But once you start using all the features of the language, you may find yourself with code that is not portable across different simulators, different versions of the same simulator or by using different command-line arguments. Any change in the simulation algorithm that causes concurrent threads to be executed in a different order will yield different simulation results.

Shared variables can create race conditions.

All variables in Verilog, *e* and OpenVera are shared among concurrent threads within their scope (except for *shadow* variables in OpenVera). VHDL was initially designed to make race conditions impossible to implement. However, with the introduction of shared variables in the 1993 version of the standard, race conditions are now just as easily introduced in VHDL as in Verilog. Although all of the examples are shown in Verilog, if you share variables between concurrent execution threads in VHDL, *e* or OpenVera, pay close attention to the race conditions described below.

Read/Write Race Conditions

A *read/write* race condition happens when two concurrent threads attempt to read and write the same shared variable in the same timestep. If you look at the code in Sample 4-80, you will notice that the first *always* block assigns the variable *count* while the second one displays it. But *both threads execute at the rising edge of the clock.*

Sample 4-80.
Example of a read/write race condition

```
module rw_race(clk);
input clk;

integer count;

always @ (posedge clk)
begin
    count = count + 1;
end

always @ (posedge clk)
begin
    $write("Count is equal to %0d\n", count);
end

endmodule
```

The execution order determines the final result.

Let's assume that the current value of *count* is 10. If the first block is executed first, the value of *count* is updated to 11. When the second block is executed, the value 11 is displayed. However, if the second block executes first, the value of 10 is displayed, the value of *count* being incremented only when the first block executes later.

Some read/write race conditions can be solved by using nonblocking assignments.

This type of race condition can be solved easily by using a nonblocking assignment in Verilog or a signal in VHDL, such as shown in Sample 4-81. Referring again to Figure 4-10: When the first block executes, the nonblocking assignment *schedules* the new value of 11, with a delay of zero, to the next timestep. When the second block executes, the value of *count* is *still* 10. The new value is assigned to *count* only when all blocks executing at this timestep are executed, creating a delta cycle.

Prefer sequential over parallel code.

Using a nonblocking or signal assignment resolves the race condition by introducing an infinitesimal delay between the *write* and the *read* operation. The unspoken assumption is that a nonblocking or signal assignment is available. In OpenVera, nonblocking assign-

Sample 4-81.
Avoiding a
read/write race
condition
using a non-
blocking
assignment

```verilog
module rw_race(clk);
input clk;

integer count;

always @ (posedge clk)
begin
    count <= count + 1;
end

always @ (posedge clk)
begin
    $write("Count is equal to %0d\n", count);
end

endmodule
```

ments can be made only to interface signals. They are not available when assigning to a variable. In *e*, nonblocking assignments do not exist at all. You should avoid creating parallel threads when a single sequential thread would do the same job, as shown in Sample 4-82.

Sample 4-82.
Avoiding a
read/write race
condition
using sequen-
tial code

```verilog
module rw_race(clk);
input clk;

integer count;

always @ (posedge clk)
begin
    $write("Count is equal to %0d\n", count);
    count <= count + 1;
end

endmodule
```

Continuous
assignments cre-
ate races.

A more insidious read/write race condition can occur in Verilog between *always* or *initial* blocks and continuous assignments. Examine the code in Sample 4-83 closely. What value of *out* will be displayed? The answer depends on the simulator and the command line you are using. Without using any command-line options, Verilog-XL says that *out* is "xxxxxxxx". VCS says that *out* is "00000001". Why the difference of opinion?

Sample 4-83.
A Verilog riddle

```
module rw_race;

wire [7:0] out;
assign out = count + 1;

integer count;
initial
begin
    count = 0;
    $write("Out = %b\n", out);
end

endmodule
```

Verilog-XL does not interrupt blocks to execute continuous assignments.

The difference comes from their interpretation of the simulation cycle. When the *initial* block assigns a new value to *count*, Verilog-XL schedules the execution of the continuous assignment for the next timestep, since it is sensitive to *count*. The execution of the *initial* block is not interrupted and the value of *out* displayed is the one it had after initialization, since the continuous assignment has not yet been executed.

VCS does.

VCS, on the other hand, executes the continuous assignment *as soon as count is assigned* in the *initial* block. The execution of the *initial* block is interrupted after the assignment to *count* while the continuous assignment is executed. The execution of the *initial* block resumes immediately afterward. The immediate propagation of events through continuous assignments is one of the techniques VCS implementers have used to speed-up simulation, unfortunately at the price of incompatibility with Verilog-XL.

This type of race condition cannot be avoided easily.

Unfortunately, this type of error condition is not as easy to avoid or eliminate as the one between two blocks. When writing behavioral code, you must be careful about the timing between assignments to registers in the right-hand side of a continuous assignment and reading the wire driven by it. To make matters worse, the race condition may involve non-zero delays as well as multiple continuous assignment statements, such as in Sample 4-84. A *read/write* race condition occurs if the delay between the time the right-hand side of a continuous assignment is updated, and the time any wire on the left-hand side is read, is equal to the propagation delay of all intervening continuous assignments. Figure 4-18 illustrates the timing of these race conditions. The only way to avoid such race condi-

---

tions is to avoid using continuous assignments for internal decoding logic.

Sample 4-84.
Another read/
write race con-
dition

```
module rw_race;

wire [7:0] out, tmp;
integer count;
assign #1 out = tmp - 1;
assign #3 tmp = count + 1;

initial
begin
   count = 0;
   #4;
   // "out" will be 0 or x's.
   $write("Out = %b\n", out);
end

endmodule
```

Figure 4-18.
Timing of a
read/write
race condition

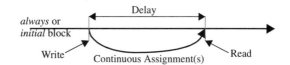

Write/Write Race Conditions

A *write/write* race condition occurs when two concurrent threads write to the same register at the same timestep. If you look at the code in Sample 4-85, you will notice that both *processes* assign the variable *flag* under different conditions and both blocks execute at any change of the clock. This setup creates a *write/write* race condition if both conditions are true.

The execution
order determines
the final result.

If you refer one more time to Figure 4-10 and the section titled "Emulating Parallelism on a Sequential Processor" on page 192, you will see that both processes are executed one after another, during the same timestep. Again, the order in which the processes execute is *not deterministic*. Let's assume that both conditions are true. If the first process is executed first, the value of *flag* is updated to FALSE. When the second process is executed, the value of *flag* is updated to TRUE. However, if the second process executes first,

Sample 4-85.
Example of a
write/write
race condition

```
architecture test of bench is
    shared variable flag: boolean;
begin
    process (clk)
    begin
        if (<cond1>) then
            flag := FALSE;
        end if;
    end process;

    process (clk)
    begin
        if (<cond2>) then
            flag = TRUE;
        end if;
    end process;
end test;
```

the value of *flag* is updated to TRUE, then it is updated to FALSE
when the first process executes later.

Sample 4-86.
Another exam-
ple of a write/
write race con-
dition

```
module ww_race(clk);
input clk;

reg flag;

always @ (posedge clk)
begin
    if (<cond1>) flag <= 0;
end

always @ (posedge clk)
begin
    if (<cond2>) flag <= 1;
end

endmodule
```

Nonblocking
assignments do not
solve the problem.

You might be tempted to use the same solution to eliminate the race
condition as was used to eliminate the *read/write* race condition, as
shown in Sample 4-86. Using a signal in VHDL creates multiple
drivers, one per thread, which eliminates the race condition. Using
nonblocking assignments in Verilog simply moves the *write/write*
race condition from the register assignment to the scheduling of the
future value. If the first block executes first, the future value 0 is
scheduled for the next timestep. When the second block executes,

the future value 1 is also scheduled for the next timestep, overwriting the previously scheduled value of 0. If the blocks execute in the opposite sequence, the scheduled value of 0 overwrites the previously scheduled value of 1.

There is no way out of this one.

There is no mechanism to prevent this type of race condition. The logic of your model must make sure that both conditions are never true at the same time. It would be a good practice to put an assertion in your model to verify that it is indeed always the case.

Pop quiz!

Why can't you have a *write/write* race condition on a Verilog wire?[21]

Initialization Races

There is no initialization phase in Verilog.

The most frequent race conditions can be found at the beginning of the simulation, when all blocks are executed for the first time. Unlike VHDL, Verilog has *no* initialization phase. Everything is initialized to x, then the simulation starts normally. Even the in-line reg initialization introduced in Verilog-2001 is internally translated into an *initial* block. It is a common misconception that *initial* blocks are used to initialize variables. *Initial* blocks are *identical* to *always* blocks, except that they execute only once, whereas *always* blocks execute forever, as if they were stuck in an infinite loop.

Initial blocks are not executed first.

When the simulation is started, the *initial* and *always* blocks are executed one after another, in *any order*. The *initial* blocks are *not* executed first—although doing so would not be illegal and some simulators, such as Silos III, do just that. Most simulators, for no other reason than to be compatible with Verilog-XL and legacy code containing race conditions, first execute blocks in the same order as they are specified in the file[22]. But subsequent execution order is not so deterministic.

When simulating the code in Sample 4-87 using an XL-compliant simulator, the first *always* block would be executed and suspended immediately, waiting for the rising edge of the clock. The *initial*

---

21. Because wires are driven, not assigned. The value from each parallel construct would contribute to the final logic value on the wire, without overwriting the other.

22. You should avoid depending on this behavior.

Writing Testbenches: Functional Verification of HDL Models

block is executed next, assigning the new value of 1 to the register named *clk*, which was previously initialized to x. A transition from x to 1 being considered a rising edge, the first *always* block sees the event and is scheduled to be executed again at the next timestep. However, since the last *always* block was *not* yet executed, and thus is not waiting for the rising edge of the clock, it does not see this edge. When the last block is finally executed, it is also immediately suspended, waiting for the *next* rising edge on *clk*. An XL-compliant simulator would therefore execute the body of the first *always* block, but not the second. However, that is not a requirement. If a simulator chooses to execute the *initial* block first, the body of neither block would execute at time 0.

Sample 4-87.
Race condition at simulation startup

```
module init_race;
reg clk;

always @ (posedge clk)
begin
    $write("Block #1 at %t\n", $time);
end

initial clk = 1'b1;

always @ (posedge clk)
begin
    $write("Block #3 at %t\n", $time);
end

endmodule
```

There is a co-simulation initialization race too.

The Vera and Specman Elite simulators are slaves to the Verilog simulator. They are started when the appropriate call is made when executing an *initial* block in the *shell* file. There is no way to predict nor enforce the execution order of the HVL initialization statement. This inability creates a race condition between the initialization of Verilog variables and the HVL code. Therefore, you cannot rely on Verilog variables being appropriately initialized when the *e* or OpenVera code first executes. See "Random Generation of Reference Signal Parameters" on page 239 for a more detailed discussion and example.

Guidelines for Avoiding Race Conditions

Race conditions can be avoided if you follow strict coding guidelines, which effectively restricts the usage of Verilog, OpenVera and *e* to what is automatically enforced by VHDL without the use of shared variables. These guidelines differ from typical RTL coding guidelines because of the stricter rules on using blocking vs. nonblocking assignment or the use of continuous assignment statements. RTL coding guidelines are designed to fit the model to the inferred hardware structure. Testbenches use the full language, and as such require guidelines designed to fit the model to the underlying simulation engine.

1. If a variable is declared outside of a concurrent thread structure, assign to it using a nonblocking assignment (when available). Reserve the blocking assignment for variables local to the thread.

2. Assign to a variable from a single concurrent thread.

3. In Verilog, use continuous assignments to drive inout pins only. Avoid using them to model internal combinatorial functions. Prefer sequential code in a large *always* block to several continuous assignments.

4. Do not assign any value at time 0.

Semaphores

Use semaphores.

The problem with guidelines is that there is no way to ensure that everyone follows them. Competing access to shared resources by concurrent threads is an old problem with an equally old solution: the semaphore. When traffic coming from multiple directions (the concurrent thread) need to cross an intersection (the shared resource), a traffic light (the semaphore) is used to make sure that only one direction of traffic gets to cross the intersection at the same time. A semaphore can be used to ensure that only one thread executes the portion of their code that can potentially create a race condition.

A semaphore is a write/write race condition put to good use.

A semaphore is a shared variable that is set by a single execution thread only if the semaphore is currently cleared. That thread is then responsible for clearing the semaphore after completing its access to the shared resource. Sample 4-88 shows an implementation of a semaphore[23] in Verilog. The *in_use* variable indicates whether the semaphore is set. If the *lock* task is invoked while the *in_use* variable is set to 1, the thread waits until the *lock* task is eventually cleared. The *unlock* task clears the semaphore. A similar implementation can be done in VHDL using a shared variable.

Sample 4-88.
A Verilog
semaphore

```
module semaphore;

reg in_use;

task lock;
begin
   while (in_use === 1'b1) wait (in_use !== 1'b1);
   in_use = 1'b1;
end
endtask

task unlock;
   in_use = 1'b0;
endtask
endmodule
```

This would not work on a true parallel system.

The key to the proper operation of the semaphore implementation shown in Sample 4-88 is the *while* loop in the *lock* task. Let's assume that three concurrent threads are vying for the semaphore by calling the *lock* task at the exact same simulation cycle. Because concurrent threads are really executed sequentially, one at the time, one of these threads (lets call it #1) will execute first. The *in_use* register being equal to 1'bx, the condition of the *while* loop will be false and it will set the semaphore and return from the *lock* task. Threads #2 and #3 run, one after another, and get to the *while* loop. Because *in_use* is now set to 1'b1, they enter the *while* loop and wait for *in_use* to be not identical to 1'b1. Eventually thread #1 will release the semaphore by calling the *unlock* task. This will wake up threads #2 and #3. One of them (let's pick #2) will run first, find the condition of the *while* loop false, set the semaphore then return from the *lock* task. When thread #3 runs, it finds the condition of the *while* loop still true (because of thread #2) and waits again.

This semaphore may not be fair.

This simple semaphore implementation relies on the ordering of thread execution to ensure that access is fair. If the Verilog simula-

---

23. Computer scientists have a very narrow definition of a semaphore that is probably not met by this implementation. However, it is good enough for Verilog.

tor implements a first-in-first-out execution order on the *wait* statement (i.e. the thread that has been waiting the longest is run first), then the semaphore will be fair. If it uses a last-in-first-out execution order (i.e. the thread with the most recent invocation of the *wait* statement executes first), then this semaphore will be completely unfair. This simple implementation would also not work if a thread does not suspend its execution between the *unlock* and *lock* task calls: Since *in_use* had just been cleared by the *unlock* task, the *while* loop would not be entered and the thread would acquire the semaphore again.

Semaphores are built-in in Open-Vera and *e*.

e has a predefined semaphore object called *locker*. OpenVera comes with predefined semaphore tasks *semaphore_get()* and *semaphore_put()*. OpenVera also comes with a special kind of semaphore it calls *regions*. Unlike regular semaphores, which protect the execution of critical code sections, regions protect access to critical data sections. With regions, concurrent threads can reserve the exclusive use of particular data values (e.g., addresses, identifiers, opcodes). It is like applying a semaphore on a set of data values instead of a section of code. Unless a thread releases a value in a region, other threads cannot use it.

VERILOG PORTABILITY ISSUES

Two compliant simulators can produce different results.

In my many years of consulting in design verification, I have yet to see a *single* testbench that simulates with identical results on Verilog-XL and VCS. Half the time, these same testbenches can produce different results by using different command-line options or use a different version of the same simulator! Yet, both Verilog simulators are fully compliant with the IEEE standard. Most of the time, the differences are due to race conditions (see "Race Conditions" on page 208). Sometimes, the differences are due to different interpretations of the Verilog standard: Many implementation details were left unspecified or existing discrepancies between simulators were also declared "unspecified". Simulator vendors are thus free to implement these unspecified portions of the standard any way they want, yielding different simulation results.

The primary cause is the author's lack of experience.

The primary cause of the simulation differences are the authors. Verilog *appears* easy to learn because it produces the expected response rather quickly. Making sure that the results are reproducible under different conditions is another matter. Learning the idiosyncrasies of the language are what takes time and differentiates an experienced modeler from a new one. It is possible to write testbenches that will simulate with identical results on all simulators and with all command-line options.

Events from Overwritten Scheduled Values

If a scheduled value is overwritten by another scheduled value, can the original value cause an event? The answer to that question is left undefined by the Verilog standard. If you look at the code in Sample 4-89, will anything be displayed at time 10?

Sample 4-89.
Overwriting
scheduled values in Verilog

```
module events;

reg stobe;

always @ (strobe)
begin
    $write("Stobe is %b\n", strobe);
end

initial
begin
    strobe = 1'b0;
    strobe <= #10 1'b1;
    strobe <= #10 1'b0;
end

endmodule
```

Overwriting a scheduled value may generate an event.

Figure 4-19 shows the queue of scheduled future values for register *strobe* just before the last statement of the *initial* block is about to execute. After executing that last statement, and scheduling the new value of 0 after 10 time units in the future, what happens to the previously scheduled value of 1? Is it removed? Is it left there? If so, which value will be assigned to *strobe* 10 time units from now? Only 0 (and thus not generating an event on *strobe*) or both in zero-time (and generating an event)? The answer to this question is simulator dependent. In asynchronous descriptions, avoid overwriting previously scheduled values using nonblocking assignments.

Figure 4-19.
Event queue
on *strobe*

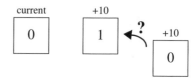

Disabled Scheduled Values

disable statements
can be used to con-
trol loops.

The *disable* statement is great for modeling reset conditions (see
"Modeling Reset" on page 383 for more details) or loop control to
emulate the behavior of VHDL's *next* and *exit* statements. The code
in Sample 4-90 shows how a loop can be controlled using the *disable* statement.

Sample 4-90.
Loop control
using the *dis-
able* state-
ment in Ver-
ilog

```
module loop_control;

initial
begin
    ...
    begin: exit_label
        while (...) begin: next_label
            ...
            // Force a new iteration
            if (...) disable next_label;
            ...
            // Break out of the loop
            if (...) disable exit_label;
            ...
        end // next_label
    end // exit_label
    ...
end

endmodule
```

Nonblocking
assignment values
may be affected by
the *disable* state-
ment.

The Verilog standard does not specify what happens to still-pending
values that were scheduled using a nonblocking assignment within
a block that is disabled. Consider the code in Sample 4-91. When a
reset condition is detected, the *always* block modeling the CPU
interface is disabled to restart it from the beginning. What should
happen to the various values assigned to the CPU interface signals
data and *dtack* using nonblocking assignments, but that may not
have been assigned to the registers yet? Depending on the simulator
you are using, these values may be removed from the scheduled
value queue and never make it to the intended registers, or they may

<table>
<tr>
<td>

Sample 4-91.
Nonblocking
assignments
potentially
affected by a
disable state-
ment

</td>
<td>

```verilog
module cpuif(...);

always
begin: if_logic
    ...
    data  <= #(Ta) read_val;
    dtack <= #(Tack) 1'b1;
    @ (negedge ale);
    data  <= #(Thold) 32'bz;
    dtack <= #(Thold) 1'b0;
    ...
end

always wait (reset == 1'b1)
begin
    disable if_logic;
    wait (reset != 1'b1);
end
endmodule
```

</td>
</tr>
</table>

remain unaffected by the *disable* statement. Avoid disabling a block
where nonblocking assignments are performed.

Output Arguments on Disabled Tasks

Output values may
not make it out of
disabled tasks.

Another area where the behavior of Verilog is left unspecified is the
value of output arguments in disabled tasks. Look at the code in
Sample 4-92. The *read* task has an output argument returning the
value that was read. Within the task, a *disable* statement is used to
abort its execution at the end of the read cycle. Because the entire
task was disabled, whether the value of *rdat* is copied out into the
register *actual* used to invoke the task is not specified in the Verilog
standard.

Disable the inner
block instead of
the task.

In some simulators, the value of *actual* is updated with the value of
rdat, effectively completing the read cycle. In some others, the
value of *actual* remains unchanged, leaving the read cycle incomplete. This unspecified behavior can be avoided easily by disabling
the internal *begin/end* block inside the task instead of the task itself,
as shown in Sample 4-93.

Sample 4-92.
Unspecified
behavior of
disabled tasks

```
task read;
    input   [7:0] radd;
    output [7:0] rdat;
begin
    . . .
    if (valid) begin
        rdat = data;
        disable read;
    end
    . . .
end
endtask

initial
begin: test_procedure
    reg [7:0] actual;

    read(8'hF0, actual);
    . . .
end
```

Sample 4-93.
Avoiding
unspecified
behavior of
disabled tasks

```
task read;
    input   [7:0] radd;
    output [7:0] rdat;
begin: read_cycle
    . . .
    if (valid) begin
        rdat = data;
        disable read_cycle;
    end
    . . .
end
endtask
```

Non-Re-Entrant Tasks

This is not an
unspecified behav-
ior.

Unless a task is declared as *automatic*, they are not re-entrant. Non-re-entrant tasks are not really an unspecified behavior in Verilog. All simulators have non-re-entrant tasks because every declaration in a Verilog model is static. By default, no declaration is dynamically allocated upon invocation of a subprogram or entry into a block of code.

The same memory space is used for all invocations of a task.

When you declare a task or a function, the memory space for its arguments and all other locally declared variables is allocated at compile time. There is a single location for the subprogram and all of its local variables. The memory is not allocated at runtime each time the task or function is invoked. Every time a subprogram is invoked, the *same* memory space is used. This reuse of memory space does not cause problems in functions or in tasks that do not include @, # or *wait* statements because the local data space is used in a single invocation. The memory space is no longer in use by the time a second invocation is made. However, if a task contains delay control statements, it may still be active when a second invocation is made.

A second invocation clobbers the data space of an active prior invocation.

Examine the code in Sample 4-94. The task named *write* contains delay control statements and is invoked from two different *initial* blocks. In Figure 4-20(a), the content of the arguments, local to the task, is shown after the invocation from the first *initial* block. While this first invocation is waiting, the second *initial* block is executed[24] and invokes the *write* task again, setting its local arguments to the values shown in Figure 4-20(b). When the first invocation resumes, it continues its execution, using the arguments provided by the second invocation: Its data space was overwritten. The first invocation goes on to write the value 8'h34 at address 8'h5A.

Sample 4-94.
Non-re-entrant task

```
task write;
    input [7:0] wadd;
    input [7:0] wdat;
begin
    ad_dt <= wadd;
    ale   <= 1'b1;
    rw    <= 1'b1;
    @ (posedge rdy);
    ad_dt <= wdat;
    ale   <= 1'b0;
    @ (negedge rdy);
end
endtask

initial write(8'h5A, 8'h00);
initial write(8'hAD, 8'h34);
```

24. This specific execution order is only an example. The *initial* blocks could execute in reverse order with equally catastrophic results.

Figure 4-20.
Task data
space

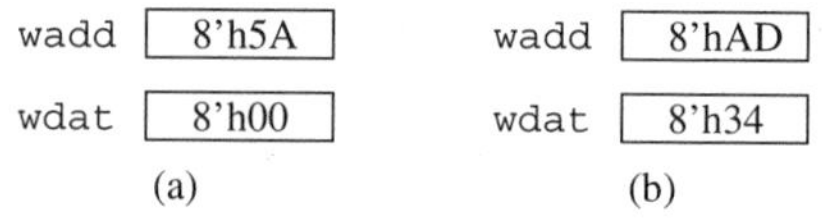

| wadd | 8'h5A | | wadd | 8'hAD |
| wdat | 8'h00 | | wdat | 8'h34 |

          (a)   (b)

Concurrent task
activations may
not be so obvious.

The concurrent invocation of the same task in Sample 4-94 is pretty obvious. But most of the time, the conditions where a task is activated more than once are much more obscure. In a large verification environment, with numerous tasks invoked under a complex control structure, it is very easy to concurrently activate a task and corrupt an entire testcase without you, or Verilog, being aware of it.

automatic tasks
are re-entrant.

In Verilog-2001, tasks can be made re-entrant by declaring them *automatic*. This causes the arguments and local variables to be dynamically created upon invocation of the task. Because these variables are no longer static, they cannot be referred to externally using a hierarchical name, nor can they be displayed on a waveform viewer. Furthermore, making a task re-entrant only solves the local data part of the problem. Separate threads still exist with the potential for race conditions to shared variables. For example, if the task in Sample 4-94 were made re-entrant by adding the keyword *automatic*, would the problem be solved? No. Even though each thread would have their correct respective values of the address and data to write, both will assign to the *same shared variables ad_dt, ale* and *rw*, creating write/write race conditions (see "Write/Write Race Conditions" on page 212).

Use a semaphore
to detect concur-
rent task activa-
tion.

The best approach to avoid this fatal condition is to use a semaphore to detect concurrent activation or protect against the write/write race condition. When using *automatic* tasks, the semaphore shown in Sample 4-88 will ensure proper operation of the task, as shown in Sample 4-95.

Sample 4-95.
Using a sema-
phore in a re-
entrant task

```
semaphore sem();

task automatic write;
    input [7:0] wadd;
    input [7:0] wdat;
begin
    sem.lock();
    ...
    sem.unlock();
end
endtask
```

A semaphore does not help non-re-entrant tasks.

Semaphores can only help protect shared resources if they are used *before* the shared resource is accessed. The solution shown in Sample 4-95 would not work for a non-re-entrant task because the data space of the task was already corrupted. It is too late. One solution would be to use the semaphore before the non-re-entrant task is invoked, as shown in Sample 4-96. What if someone forgets to use the semaphore before calling the task?

Sample 4-96.
Using a semaphore with a non-re-entrant task

```
semaphore sem();

task write;
...
endtask

initial
begin
    sem.lock();
    write(8'h5A, 8'h00);
    sem.unlock();
end

initial
begin
    sem.lock();
    write(8'hAD, 8'h34);
    sem.unlock();
end
```

You can detect concurrent task activation.

A modified version of the semaphore can be used to detect concurrent activation of a non-re-entrant task. As shown in Sample 4-97, the *in_use* register indicates whether the task is currently activated. If the task is invoked while the *in_use* register is set to 1, the simulation is terminated. Because the data space of the task has already been clobbered, it is not possible to recover from the error. Terminating the simulation is the only option. The problem must be fixed by retiming the access to the task (usually through a semaphore) to ensure that no concurrent invocation takes place.

Back when I was still using Verilog, I put a guard around any non-re-entrant task. This let my model tell me immediately if I misused it. I could immediately fix the problem, without having to diagnose a testbench failure back to a concurrent task activation. The time invested in adding the simple guard was well worth it. If the task I wrote was to be used by others, the message produced by the concurrent activation detection specifically stated that the error was not

```
task write;
   input [7:0] wadd;
   input [7:0] wdat;

 · reg in_use;
begin
   if (in_use === 1'b1) $stop;
   in_use = 1'b1;

   ad_dt <= wadd;
   ale   <= 1'b1;
   rw    <= 1'b1;
   @ (posedge rdy);
   ad_dt <= wdat;
   ale   <= 1'b0;
   @ (negedge rdy);

   in_use = 1'b0;
end
endtask
```

in my task code, but in *their* use of it and to go look for a concurrent activation. This has saved me many technical support calls.

SUMMARY

When writing testbenches, think function, not implementation. Abandon the RTL coding mindset. Do not think in terms of logic, registers and state machines. Think in terms of data transformation, program state and execution flow.

Your first objective is to write maintainable code. Write relevant comments that describe your intent, not the code. Optimize for performance only when necessary.

Minimize the scope of your variables as much as possible. Declare local variables in the scope where they are needed.

Package reusable subprograms and bus-functional models in suitable constructs to facilitate their reuse. Maintain separate name spaces as much as possible. Make sure that it is possible to have multiple instances of a bus-functional model connected to different interface signals without interference or collisions.

Use data abstraction. Collect related data into records, arrays and lists.

Separate public interfaces from private implementation. Plan your class inheritance and take advantage of polymorphism to create generic bus-functional models and utility subprograms.

Understand the concurrency model used in simulating HDLs and HVLs. It will help write more efficient models and avoid race conditions. Use semaphores to protect shared resources.

Understand the unspecified portion of the Verilog standard. This portion is a source of non-portability between different Verilog simulators, versions and command-line options.

CHAPTER 5   STIMULUS AND RESPONSE

The purpose of writing testbenches is to apply stimulus to a design and observe the response. That response must then be compared against the expected behavior.

Generating stimulus is the process of providing input signals to the design under verification as shown in Figure 5-1. From the perspective of the stimulus generator, every input of the design is an output of the generator.

Figure 5-1.
Stimulus
generation

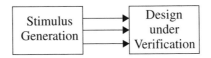

Monitoring is the process of observing output signals from the design under verification as shown in Figure 5-2. From the perspective of the response monitor, every output of the design is an input of the monitor.

Figure 5-2.
Response
monitoring

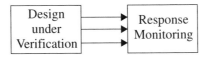

Stimulus and Response

This chapter shows how to apply stimulus and observe response.	In this chapter, I show how to generate stimulus and observe responses. I also show how to abstract data flowing to and from the design from a physical level composed of 1's, 0's and elapsed time to a transaction level composed of data objects and procedures. The greatest challenge with stimulus is making sure it is an accurate representation of the environment, not just a simple case. When monitoring responses, one has to be careful not to miss any data and detect as many errors as possible.
The next chapter shows how to structure a test-bench.	In the next chapter, I show how to best structure the stimulus generators and response monitors to create a transaction-layer self-checking environment. Constrainable random generation is then added on top of the stimulus generators and response monitors. If you prefer a top-down perspective, I recommend you start with the next chapter then come back to this one.

REFERENCE SIGNALS

Clock signals must be generated with care.	Because a clock signal has a very simple repetitive pattern, it is one of the first and most fundamental signals to generate. It is also the most critical signal to generate accurately. Many other signals use clock signals to synchronize themselves. When using OpenVera or *e*, I recommend generating all clock signals from VHDL or Verilog.
Explicitly assign 1 and 0.	The behavioral code to generate a 50% duty-cycle 100MHz clock signal is shown in Sample 5-1. To produce a more robust clock generator, use explicit assignments of values 0 and 1. Using a statement like `clk = ~clk` would depend on the proper initialization of the clock signal to a value different than the default values of 1'bx or U. Assigning explicit values also provides better control over the initial phase of the clock; you control whether the clock is starting high or low. The same style also can be used to model clocks with different duty cycles (i.e., different duration of the high and low phases).

Sample 5-1.
Generating a
50% duty-
cycle clock

```
reg clk;
always
begin
    #5;
    clk = 1'b0;
    #5;
    clk = 1'b1;
end
```

Any repetitive waveform is easy to generate.	Waveforms with deterministic edge-to-edge relationships with an easily identifiable period also are easy to generate. It is a simple process of generating each edge in sequence, at the appropriate time. For example, Figure 5-3 outlines an apparently complex waveform. However, Sample 5-2 shows that it is simple to generate.

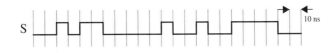

Figure 5-3.
Apparently
complex
waveform

Sample 5-2.
Generating a
deterministic
waveform

```
process
begin
    S <= '0'; wait for 20 ns;
    S <= '1'; wait for 10 ns;
    S <= '0'; wait for 10 ns;
    S <= '1'; wait for 20 ns;
    S <= '0'; wait for 50 ns;
    S <= '1'; wait for 10 ns;
    S <= '0'; wait for 20 ns;
    S <= '1'; wait for 10 ns;
    S <= '0'; wait for 20 ns;
    S <= '1'; wait for 40 ns;
    S <= '0'; wait for 20 ns;
    ...
end process;
```

Time Resolution Issues

Integer division may speed-up the clock.	When generating waveforms in Verilog, you must select the appropriate timescale and precision to properly place the edges at the correct offset in time. When using an expression, such as `cycle/2`, to compute delays, you must make sure that integer operations do not truncate a fractional part. For example, the clock generated in Sample 5-3 produces a period of 14 ns because of truncation caused by the integer division.
The Verilog timescale may affect the timing of edges.	If the precision of the currently active timescale is not sufficiently high, delay values are rounded up or down. When this happens to the delay values of clock signals, it shifts the relative position of the clock edges. For example, the clock generated in Sample 5-4 produces a period of 16 ns because of rounding the result of the real division to an integer value.

Sample 5-3.
Truncation
errors in stim-
ulus genera-
tion

```
'timescale 1ns/1ns
module testbench;
...
reg clk;
parameter cycle = 15;
always
begin
    #(cycle/2);   // Integer division
    clk = 1'b0;
    #(cycle/2);   // Integer division
    clk = 1'b1;
end
endmodule
```

Sample 5-4.
Rounding
errors in stim-
ulus genera-
tion

```
'timescale 1ns/1ns
module testbench;
...
reg clk;
parameter cycle = 15;
always
begin
    #(cycle/2.0);   // Real division
    clk = 1'b0;
    #(cycle/2.0);   // Real division
    clk = 1'b1;
end
endmodule
```

Because the timescale in Sample 5-5 offers the necessary precision
for a 7.5 ns half-period, only this signal generates a 50% duty-cycle
clock signal with a precise 15 ns period.

Sample 5-5.
Proper preci-
sion in stimu-
lus generation

```
'timescale 1ns/100ps
module testbench;
...
reg clk;
parameter cycle = 15;
always
begin
    #(cycle/2.0);
    clk = 1'b0;
    #(cycle/2.0);
    clk = 1'b1;
end
endmodule
```

Aligning Signals in Delta-Time

Delta delays are functionally equivalent to real delays.

In the specification shown in Figure 5-4, the transition of *clk2* is aligned with a transition on *clk*. There are many ways of generating these two signals. Depending on the approach used, these aligned transitions may occur in the same delta cycle, or in different delta cycles. Although delta-cycle delays are considered zero-delays by the simulator, functionally they have the same effect as real delays.

A derived waveform, such as the one shown in Figure 5-4, apparently is easy to generate. A simple process, sensitive to the proper edge of the original signal as shown in Sample 5-6, and voila! Even the waveform viewer shows that it is right!

Figure 5-4.
Derived
waveform
specification

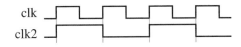

Sample 5-6.
Improperly
generating a
derived wave-
form

```
clk2_gen: process(clk)
begin
    if clk = '1' then
        clk2 <= not clk2;
    end if;
end process clk2_gen;
```

Watch for delta delays in derived waveforms.

The problem is not apparent visually. Because of the simulation cycle (See "The Simulation Cycle" on page 194), there is a delta cycle between the rising edge of the base clock signal and the transition on the derived clock signal, as shown in Figure 5-5. Any data transferred from the base clock domain to the derived clock domain goes through this additional delta cycle delay. In a zero-delay simulation, such as a behavioral or RTL model, this additional delta-cycle delay can have the same effect as an entire clock cycle delay.

Figure 5-5.
Delta delay in
derived
waveform

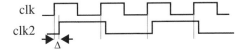

Propagation delays make it work in the real world.

Why is it that generating divided clocks in simulation the same way it is done in the real world does not work? Because, in synchronous designs, there is always a race condition between the *clk*-to-D-input path and the *clk*-to-*clk2* path of adjacent flip-flops. This constant race condition is solved by making sure that the delay through the *clk*-to-D path is always longer than the delay through the *clk*-to-*clk2* path. In the real world, these signal propagation delays will never be zero. Device physics and clock skew management provides a simple solution. In zero-time behavioral or RTL models, propagation delays are composed of delta cycles. If the number of delta cycles in the *clk*-to-D path is smaller than the number of delta cycles in the *clk*-to-*clk2* path, an entire clock cycle delay will be lost.

Align derived signals in delta-time.

The solution is surprisingly similar to that used in the real world. It is necessary to minimize the delta-cycle skew between the base and derived signals. This skew can be completely eliminated by aligning their respective edges in delta time. The only way to perform this task is to re-derive the base signal through a divide-by-1 operation, as shown in Sample 5-7 and illustrated in Figure 5-6. The base signal is never used by other processes. Instead, they must use the divide-by-1 signal.

Figure 5-6.
Generation of aligned derived signals

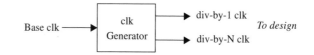

Sample 5-7.
Properly generating a derived waveform

```
derived_gen: process(clk)
begin
    clk1 <= clk;
    if clk = '1' then
        clk2 <= not clk2;
    end if;
end process derived_gen;
```

Differential data signals need not be aligned.

When generating a differential data signal pair, it is not necessary to align both polarities in the same delta cycle. Adding an inversion delay in one of the phase signals only adds to the clock-to-D-input path delay. This technique goes in the right direction to solve the race condition. As shown in Sample 5-8, the inversion of the *d* signal in the connection to the *dn* pin may introduce an additional delta cycle in the *d*-to-*dn* path compared to the *d*-to-*dp* path.

Sample 5-8.
Generating
differential
data signals

```
wire [15:0] d;

bfm cpu(..., .d(d), ...);
design dut (..., .dp(d), .dn(~d), ...);
```

Clock Multipliers

Implemented
using PLLs.

Many designs have a very high-speed front-end interface that is driven using a multiple of a recovered clock or the lower-frequency system clock. This clock multiplication is performed using an internal or external PLL (phase locked loop). PLLs are inherently analog circuits. They are very costly to simulate in a digital simulator. When an internal PLL is used, the analog component that implements the PLL is often modeled as an empty module or entity. It is up to the testbench to create an appropriate clock multiple signal in a behavioral fashion.

The reference
clock could
become the
derived clock.

A simple strategy is to reverse the role of the reference and derived clock. Since clock dividers are so easy to model, you could generate the high-frequency clock then use it to derive the lower-frequency system-clock. Sample 5-9 shows the Verilog model for a multiply-by-4 clock generator using the divide-by-4 strategy. But this only works under two conditions: The reference clock is also an input to the design, and the frequency of the reference clock is known and fixed.

Sample 5-9.
Generating
clock multiples by division
sion

```
initial
begin
    clk4 <= 1'b0;
    clk1 <= 1'b0;
    forever begin
        repeat (4) begin
            #10;
            clk4 = ~clk4;
        end
        clk1  = ~clk1;
    end
end
```

Synchronize the
multipled clock to
the reference
clock.

The first condition can be eliminated by synchronizing the multi-plied clock signal with the reference clock. It will be possible to generate the multiplied clock signal even if the reference clock is supplied by the design. Sample 5-10 shows the VHDL model of a multiply-by-4 clock generator, synchronized with an input refer-ence clock. But the problem of the hard-coded multiplied clock period remains. This model assumes a reference clock with an 80 ns period. What if the reference clock has a different frequency in a different simulation run? How can we generalize this model into a generic clock-multiplier PLL model?

Sample 5-10.
Synchroniz-ing multiplied
clock to input
reference
clock

```
process (clk)
begin
   for I from 1 to 4 loop
      clk4 <= not clk4;
      wait for 10 ns;
   end loop;
end
```

Measure the
period of the refer-ence clock.

Why not let the model learn the period of the reference signal? You can measure the time difference between two consecutive edges, divide this value by 4, and voila! A generic PLL model. Sample 5-11 shows a PLL model with a continuous measure of the reference signal. As the frequency of the reference clock signal changes, the frequency of the multiplied clock will adapt.

Watch that time-scale!

When using Verilog, the usual words of caution (see "Time Resolu-tion Issues" on page 231) apply regarding the precision of the timescale. The computed period of the multiplied signal is a real value that will likely have a fractional part. The actual delay value between two consecutive edges of the multiplied clock will be the computed value rounded to the current timescale precision. If the size of this error is small enough compared to the period of the ref-erence signal, this should not cause a problem.

Asynchronous Reference Signals

Figure 5-7 shows a specification for two unrelated clock signals. They are used by two separate clock domains in the design under verification. *clk100* is a 100 MHz signal while *clk33* is a 33 MHz signal. You could be tempted to model these two clock signals as shown in Sample 5-12, using the higher frequency signal as a refer-ence to generate the lower-frequency one with a divide-by-3 strat-

Sample 5-11.
Generic clock
multiplier
model

```
module pll(input  wire ref_clk,
                  output reg  out_clk);
parameter FACTOR = 4;

initial
begin: adaptive
    real stamp, period;

    out_clk <= 1'b0;
    @(ref_clk);
    stamp = $realtime;

    forever begin
        @(ref_clk);
        period = ($realtime - stamp)/FACTOR;
        stamp = $realtime;

        repeat (FACTOR) begin
            out_clk = ~out_clk;
            #(period);
        end
    end
end
endmodule
```

egy. This approach will indeed generate the waveforms shown in
Figure 5-7. But that is only one of the possible solutions, and one
that may not highlight some classes of problems.

Figure 5-7.
Unrelated
waveform
specification

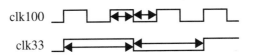

clk100

clk33

Sample 5-12.
Improperly
generating
unrelated
waveform

```
clk33_gen: process(clk100)
    variable count: integer := 0;
begin
    count := count + 1;
    if count = 3 then
        clk33 <= not clk33;
        count := 0;
    end if;
end process clk33_gen;
```

Alignment on
paper is not a spec-
ification.

The problem comes from the *inference* that the waveforms are
aligned simply because they are aligned in the figure. There are no
explicit or implicit timing relationships between the two signals as

there is no timing arrow going from an edge in one waveform to an edge in the other waveform. Drawing tools have a grid system that facilitates drawing straight lines. But they also have the side effect of aligning objects. When writing a specification, you must be careful that these implicit alignments do not create the illusion of a relationship. When reading a specification, do not assume a relationship unless it is explicitly stated.

Generate unrelated signals in separate processes.

Sample 5-13 shows a better way to generate these unrelated clock signals. Since they are not synchronized in any way, they are generated using separate concurrent processes. This separation will make it easy to modify the frequency of one signal without affecting the frequency of the other. Also, notice how each signal is explicitly skewed at the begining of the simulation to avoid having the edges aligned at the same simulation time. This approach is a good practice to highlight potential problems in the clock-domain crossing portion of the design. By varying these initial signal skew values, it will be possible to verify the correct functionality of the design across different asynchronous clock relationships.

Sample 5-13.
Generating
unrelated
waveforms

```
initial
begin
    clk100 <= 0;
    #2;
    forever begin
        #5;
        clk100 = ~clk100;
    end
end

initial
begin
    clk33 <= 0;
    #5;
    forever begin
        #15;
        clk33 = ~clk33;
    end
end
```

Writing Testbenches: Functional Verification of HDL Models

Random Generation of Reference Signal Parameters

All signals are related in simulation.

In the previous section, I explained why unrelated signals should be modeled as separate processes and skewed with respect to each other to avoid creating an implicit relationship that does not exist between them. The truth is: There is no way to accurately model unrelated signals. Each waveform is described with respect to the current simulation time. Because all waveforms are described using the same built-in reference, they are all implicitly related. Even though I made my best effort to avoid modeling any relationship between the two clock signals generated in Sample 5-13, they *are* related because of the deterministic nature of the simulator. Unless I manually modify one of the timing parameters, they will maintain the same relationship in all simulations.

Asynchronous means random.

When we say that two signals are asynchronous to each other, we are saying that they have a random phase relationship. That phase relationship will be different every time and cannot be predicted. When I specified explicit skew delay values in Sample 5-13, I introduced certainty where there wasn't any. These delay values should be generated randomly to increase the chances that, over the thousands of simulation runs the design will be subjected to, any problem related to clock skews will be highlighted.

Randomize in OpenVera or e.

One solution would be to call $random or using a random number generation package in VHDL[1] to generate the delay values. If you are using an HDL, that is indeed your only solution. Even though signals are modeled in HDL, their parameters may be generated using an HVL. You can generate signal parameters in the HVL side[2] at the begining of the simulation, transfer those parameters over to the HDL side, then use the randomly generated values in your HDL model. This procedure requires that there be a place for OpenVera or *e* to deposit those randomly-generated signal parameters. Sample 5-14 uses variables as a placeholder for the timing parameters. It also provides the variables with default values just in

---

1. References to random number generation and linear-feedback shift register packages in VHDL can be found in the *resources* section of:

   http://janick.bergeron.com/wtb

2. See "Random Stimulus" on page 354 for a discussion of why all random generation should be performed in the HVL side.

Writing Testbenches: Functional Verification of HDL Models

case the random parameter generator is not included in the HVL side—which it has no control over.

Sample 5-15 shows the corresponding signal parameter generator on the HVL side. Because neither *e* nor OpenVera support real numbers, the timing parameters must be supplied as integer values. To enable the randomly-generated values to cover the entire precision of the clock generator, these integer values must be interpreted as in number of *precision* units and manually scaled to the *scale* unit.

Sample 5-14.
Allowing for randomly generated parameters

```
`timescale 1ns/100ps
...
integer skew   = 0;
integer period = 100;
initial
begin
   clk <= 0;
   #(skew/10.0);
   forever begin
      #(period/20.0);
      clk = ~clk;
   end
end
```

Sample 5-15.
Generating waveform parameters

```
unit clk_parameters {
   skew  : uint;
   period: uint;

   keep skew < 2 * period;
   keep period in [18..22];

   run() is also {
      'skew'   = me.skew;
      'period' = me.period;
   };
};
```

Wait before using values from the HVL side.

There is a potential problem with the code in Sample 5-14. OpenVera and *e* are slave processes to the simulator. They are started by invoking a PLI task or foreign procedure. It is necessary for the Verilog or VHDL model to let the HVL simulator get started, run, generate random values then assign them back to the HDL side before they can be used.

There is a race condition between the process that starts the HVL simulator and *initial* blocks in the clock generator. Remember that explicit variable initialization in Verilog-2001 creates implicit *initial* blocks. Therefore, there are three races: the assignment of *skew* with the default value, getting to the "#(skew)" statement and the assignment of *skew* with a random value by the HVL. The race between the HDL simulator and the #(skew) statement is resolved easily by adding an additional real delay to ensure that the HVL simulator has completed its assignment of the timing parameters. The race between the signal initialization and the HVL assignment of random value (which would not exist in VHDL) is solved by initializing the value of the parameter variables procedurally if they contain any unknowns.

Sample 5-16.
Generating waveforms with randomly generated parameters

```
integer skew;
integer period;
initial
begin
    #2;
    clk = 0;
    if (^skew === 1'bx)    skew = 0;
    if (^period === 1'bx) period = 100;
    #(skew/10.0);
    forever begin
        #(period/20.0);
        clk = ~clk;
    end
end
```

Applying Reset

Synchronized signals must be properly modeled.

The first signal to be generated after the clock signals is the hardware reset signal. The reset signal must be shaped properly to reset the design correctly. The generation of a synchronous reset signal should also reflect its synchronization with any clock signal. For example, consider the specification for a reset signal shown in Figure 5-8. The code in Sample 5-17 shows how such a waveform is generated frequently.

Figure 5-8.
Reset waveform specification

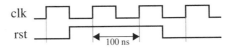

Sample 5-17.
Improperly
generating a
synchronous
reset

```
always
begin
   #50 clk = 1'b0;
   #50 clk = 1'b1;
end

initial
begin
   rst = 1'b0;
   #150 rst = 1'b1;
   #200 rst = 1'b0;
end
```

Race conditions
can be created eas-
ily between syn-
chronized signals.

There are two problems with the way these two waveforms are generated in Sample 5-17. The first problem is functional: There is a
race condition between the *clk* and *rst* signals. At simulation time
150, and again later at simulation time 350, both registers are
assigned at the same timestep. Because the *blocking* assignment is
used for both assignments, one of them is assigned first. A block
sensitive to the falling edge of *clk* may execute before or after *rst* is
assigned. From the perspective of that block, the specification
shown in Figure 5-8 could appear to be violated. The race condition
can be eliminated by using *nonblocking* assignments, as shown in
Sample 5-18. Both *clk* and *rst* signals are assigned between
timesteps when no blocks are executing. If the design under verification uses the falling edge of *clk* as the active edge, *rst* is
already—and reliably—assigned.

Sample 5-18.
Race-free gen-
eration of a
synchronous
reset

```
always
begin
   #50 clk <= 1'b0;
   #50 clk <= 1'b1;
end

initial
begin
   rst = 1'b0;
   #150 rst <= 1'b1;
   #200 rst <= 1'b0;
end
```

Lack of maintain-
ability can intro-
duce functional
errors.

The second problem, which is just as serious as the first one, is
maintainability of the description. You could argue that the first
problem is more serious, since it is functional. The entire simulation can produce the wrong result under certain conditions. Main-

tainability has no such functional impact. Or has it? What if you made a change as simple as changing the phase or frequency of the clock. How would you know to change the generation of the reset signal to match the new clock waveform?

Conditions in real life are different than those in this book.

In the context of Sample 5-18, with Figure 5-8 nearby, you would probably adjust the generation of the *rst* signal. But outside of this book, in the real world, these two blocks could be separated by hundreds of lines, or even be in different files. The specification is usually a document one inch thick, printed on both sides. The timing diagram shown in Figure 5-8 could be buried in an anonymous appendix, while the pressing requirements of changing the clock frequency or phase was stated urgently in an email message. And you were busy debugging this other testbench when you received that pesky email message! Would you know to change the generation of the reset signal as well? I know I would not.

Model the synchronization within the generation.

Waiting for an apparently arbitrary delay cannot guarantee synchronization with respect to the delay of the clock generation. A much better way of modeling synchronized waveforms is to include the synchronization in the generation of the dependent signals, as shown in Sample 5-19. The proper way to synchronize the *rst* signal with the *clk* signal is for the generator to wait for the significant synchronizing event, whenever it may occur. The timing or phase of the clock generator can be modified now, without affecting the proper generation of the *rst* waveform. From the perspective of a design sensitive to the falling edge of *clk*, *rst* is reliably assigned one delta-cycle after the clock edge.

Reset may need to be applied repeatedly during a simulation.

There is a problem with the way the *rst* waveform is generated in Sample 5-19. The *initial* block runs only once and is eliminated from the simulation once completed. There is no way to have it execute again during a simulation. What if it were necessary to reset the device under verification multiple times during the same simulation? An example is the "hardware reset" testcase that verifies proper reset operation: After setting some internal register, the hardware reset must be applied to verify that these registers return to their reset value. Having control when reset is applied is also very useful. This control lets testbenches perform preparatory operations before resetting the design and starting the actual stimulus. Furthermore, when using Specman Elite, it is possible to wipe out the current testbench and load a new one without exiting the simulation. This unloading and loading requires that the model of the

Sample 5-19.
Proper genera-
tion of a syn-
chronous reset

```
always
begin
    #50 clk = 1'b0;
    #50 clk = 1'b1;
end

initial
begin
    rst = 1'b0;
    wait (clk !== 1'bx);
    @ (negedge clk);
    rst <= 1'b1;
    @ (negedge clk);
    @ (negedge clk);
    rst <= 1'b0;
end
```

design be reset in what Specman Elite perceives to be the middle of a very long simulation to have the new testbench start with the design in a known state.

Generate reset
from within a sub-
program.

The proper mechanism to encapsulate statements that you may need to repeat during a simulation is to use a *task* or a *procedure* as shown in Sample 5-20. To repeat the reset signaling, simply call the subprogram. To maintain the behavior of using an *initial* block to reset the device under verification automatically at the beginning of the simulation (which may or may not be desirable), simply call the task in an *initial* block.

Sample 5-20.
Encapsulating
the generation
of a synchro-
nous reset

```
always
begin
    #50 clk <= 1'b0;
    #50 clk <= 1'b1;
end

task hw_reset
begin
    rst = 1'b0;
    wait (clk !== 1'bx);
    @ (negedge clk);
    rst <= 1'b1;
    @ (negedge clk);
    @ (negedge clk);
    rst <= 1'b0;
end
endtask
initial hw_reset;
```

Tasks and procedures can be called from OpenVera or *e*.

If you are using OpenVera or *e*, it is possible to invoke the reset task or procedure in the HDL world from the HVL world. The testbench is thus able to reset the design any time the reset is required. An alternative would be to put the reset subprogram in an HVL method, as shown in Sample 5-21. Notice how the extension to the *run()* method is used to initialize the output reset signal automatically at the beginning of the simulation. An automatic resetting of the design at the beginning of the simulation can be implemented by starting an invocation of the *hw_reset()* time-consuming method in that same method. In OpenVera, this would be done in the class constructor (i.e., the *new()* task).

Sample 5-21.
Synchronous reset in *e*.

```
unit hw_reset {
    clk: string;
    rst: string;

    event active_edge is fall('(clk)') @sim;

    hw_reset() @active_edge is {
        '(rst)' = 1'b1;
        wait [2] * cycle;
        '(rst)' = 1'b0;
    };
    run() is also {
        '(rst)' = 1'b0;
    };
};
```

Reset signal can be generated from HVLs.

Reset generation and control can be performed from the HVL testbench. This feature gives the testbench better control over the reset parameters and its coordination with other device stimuli, not just the clock.

Are you paying attention?

Pop quiz: What is missing from the *hw_reset* task in Sample 5-20 and Sample 5-21? The answer can be found in this footnote.[3]

3. The task *hw_reset* contains delay control statements. These statements should contain a semaphore to detect concurrent activation. You can read more about this issue in "Non-Re-Entrant Tasks" on page 222.

SIMPLE STIMULUS

In this section, I explain how to generate deterministic waveforms. Various techniques are developed to generate stimulus signals in the best way. I also demonstrate how to encapsulate and package signal generation operations using simple bus-functional models.

Applying Synchronous Data Values

There is a race condition between the clock and data signal.

Sample 5-22 shows how you could generate a zero-delay synchronous data waveform. This approach is identical to the way flip-flops are inferred in an RTL model. As illustrated in Figure 5-9, there is a delay between the edge on the clock and the transition on *data,* but the delay is a single delta cycle. In terms of simulation time, there is no delay. For RTL models, this infinitesimal clock-to-Q delay is sufficient to model the behavior of synchronous circuits properly. However, this delay assumes that all clock edges are aligned in delta time (see "Aligning Signals in Delta-Time" on page 233). If you are generating both clock and data signals from the outside of the model of the design under verification, you have no way of ensuring that the total number of delta-cycle delays between the clock and the data is maintained and that the data signal will arrive before the clock.

Sample 5-22.
Zero-delay
generation of
synchronous
data

```
sync_data_gen: process
begin
    wait until clk = '0';
    data <= ...;
    wait until clk = '0';
    data <= ...;
    ...
    wait;
end process sync_data_gen;
```

Figure 5-9.
Synchronous
data
waveforms

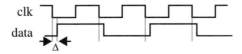

The clock may be delayed more than the data.

For many possible reasons, the clock signal may be delayed by more delta cycles than its corresponding data signal. These delays could be introduced by using different I/O pad models for the clock and data pins. They also could be introduced by the clock distribution network, which does not exist on the data signal. If the clock signal is delayed more than the data signal, even in zero-time as shown in Figure 5-10, the effect is the same as removing an entire clock cycle from the data path.

Delay the data from the active clock edge.

Interface specifications never specify zero-delay values. A physical interface always has a real delay between the active edge of a clock signal and its synchronous data. When generating synchronous data, always provide a real delay between the active edge and the transition on the data signal, as shown in Sample 5-23, or synchronize the data signal with the inactive edge of the clock.

Figure 5-10.
Delta delays in
clock path

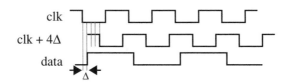

Sample 5-23.
Non-zero-
delay genera-
tion of syn-
chronous data

```
sync_data_gen:
begin
    wait until clk = '0';
    data <= ... after 1 ns;
    wait until clk = '0';
    data <= ... after 1 ns;
    ...
    wait;
end process sync_data_gen;
```

Encapsulating Waveform Generation

A subprogram can be used to apply data properly.

Just as we encapsulated the application of the hardware reset signal (see Sample 5-20), we can encapsulate the application of synchronous input data. As illustrated in Figure 5-11, data signals must be applied with a proper setup and hold time—but no more—to meet the input timing constraints. Instead of repeating the synchronization for each input value, a subprogram can be used for synchronization with the input clock. It would also apply the data received as input argument. The code in Sample 5-24 shows the implementation and use of a *task* applying a user-specified value to input signals according to the specification in Figure 5-11. Notice how the input is set to unknowns after the specified hold time to stress the timing of the interface. This technique is also known as "surround by X". Leaving the input to a constant value would not detect cases where the device under verification does not meet the maximum hold requirement. Note that functional verification is not the ideal vehicle for verifying setup and hold times and is only applicable to gate-level simulations. Verification of setup and hold times is better accomplished using a static timing analyzer.

Figure 5-11.
Input data
waveform
specification

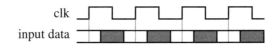

Sample 5-24.
Encapsulating
the applica-
tion of input
data values

```
task apply_values;
    input [...] values;
begin
    inputs <= values;
    @(posedge clk);
    #(Thold);
    inputs <= 'bx;
    #(cycle - Thold - Tsetup);
end
endtask

initial
begin
    hw_reset;
    apply_values(...);
    apply_values(...);
    ...
end
```

Abstracting Waveform Generation

Input vectors are
difficult to write
and maintain.

Using synchronous test values, also known as test vectors, to verify a design is rather cumbersome. They are hard to interpret and difficult to specify correctly. For example, using cycle-by-cycle values to verify a synchronously resettable D flip-flop with a 2-to-1 multiplexer on the input, as shown in Figure 5-12, could be stimulated using the vectors shown in Sample 5-25.

Use subprograms
to encapsulate
operations.

It would be easier if the operation accomplished by the test vectors were abstracted. The device under verification can perform only two things:

Figure 5-12.
2-to-1 input
sync reset D
flip-flop

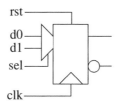

rst
d0
d1
sel

clk

Sample 5-25.
Test vectors
for 2-to-1
input sync
reset D flip-
flop

```
initial
begin
    // Values: rst, d0, d1, sel
    apply_vector(4'b1110);
    apply_vector(4'b0100);
    apply_vector(4'b1111);
    apply_vector(4'b0011);
    apply_vector(4'b0010);
    apply_vector(4'b0011);
    apply_vector(4'b1111);
    . . .
end
```

- A synchronous reset

- Load from input *d0* or *d1*

Instead of providing test vectors to perform these operations repeatedly, why not provide subprograms that perform these operations? All that will be left is to call the subprograms in the appropriate order with the appropriate data.

Try to apply the worst possible combination of inputs.

The subprogram to perform the synchronous reset is very simple. It needs to assert the *rst* input, then wait for the active edge of the clock. But what about the other inputs? You could decide to leave them unchanged, but is that the worst possible case? What if the reset was not functional and the device loaded one of the inputs and that input was set to 0? It would be impossible to differentiate the wrong behavior from the correct one. To create the worst possible condition, both *d0* and *d1* inputs must be set to 1. The *sel* input can be set randomly, since either input selection should be functionally identical. An implementation of the *reset* procedure is shown in Sample 5-26.

Pass input values as arguments to the subprogram.

The second operation this design can perform is to load input *d0* or *d1*. The *task* to perform this operation is shown in Sample 5-27. Unlike resetting the design, loading data can have different input

Sample 5-26.
Abstracting
the synchro-
nous reset
operation

```
procedure reset(signal rst: out std_logic;
                signal d0 : out std_logic;
                signal d1 : out std_logic;
                signal sel: out std_logic;
                signal clk: in  std_logic);
begin
   rst <= 1'b1;
   d0  <= 1'b1;
   d1  <= 1'b1;
   sel <= random();
   wait until clk = '1';
   wait for Thold;
   rst <= 'X';
   d0  <= 'X';
   d1  <= 'X';
   sel <= 'X';
   wait for Tcycle - Thold - Tsetup;
end procedure reset;
```

values: It can load either a 1 or a 0. The value of the input to load is passed as an argument to the task, as well as a specification of which input to use, *d0* or *d1*. The worst condition is created when the other input is set to the complement of the input value. If the device is not functioning properly and is loading from the wrong input, then the result will be clearly wrong.

Stimulus gener-
ated with
abstracted opera-
tions is easier to
write and main-
tain.

Once operation abstractions are available, providing the proper stimulus to the design under verification is easy to write and understand. Notice how a 16-bit variable is used to pass a two-character string identifying the name of the input pin to use. Using a name-based specification makes it easier to understand which pin is being loaded. Compare the code in Sample 5-28 with the code of Sample 5-25. If the polarity of the *rst* input were changed, which verification approach would be easiest to understand and modify?

Sample 5-27.
Abstracting
the load opera-
tion

```
task load;
   input data;
   input [2*8:1] which;
begin
   rst <= 1'b0;
   if (which == "d0") begin
      sel <= 1'b0;
      d0 <=  data;
      d1 <= ~data;
   end
   else if (which == "d1") begin
      sel <= 1'b1;
      d1 <=  data;
      d0 <= ~data;
   end
   else begin
      $write("Invalid input \"%s\".\n", which);
      $finish;
   end
   @ (posedge clk);
   #(Thold);
   {rst, d0, d1, sel} <= 4'bxxxx;
   #(cycle - Thold - Tsetup);
end
endtask
```

Sample 5-28.
Verifying the
design using
operation
abstractions

```
initial
begin
   sync_reset;
   load(1'b1, "d0");
   sync_reset;
   load(1'b1, "d1");
   load(1'b0, "d0");
   load(1'b1, "d1");
   sync_reset;
   ...
end
```

SIMPLE OUTPUT

Generating stimulus is only half of the job. Actually, it is more like 25% of the job. The other parts, verifying that the output is as expected and collecting functional coverage measurements, is much more time consuming and error prone. There are various ways the output can be checked against expectations. The outputs have varying degrees of applicability and repeatability. In this section, I will review techniques, some good, some not so good, for verifying simple responses.

Visual Inspection of Response

Results can be printed.

The most obvious method for verifying the output of a simulation is to inspect the results visually. The visual display can be an ASCII printout of the input and output values at specific points in time, as shown in Sample 5-29.

Sample 5-29.
ASCII view of simulation results

```
              r  s
              sddeqq
     Time  t011 b
     ----------
     0100  1110xx
     0105  111001
     0200  010001
     0205  010010
     0300  111110
     0305  111101
     0400  001101
     0405  001110
     0500  001010
     0505  001010
     0600  001110
     0605  001110
     0700  111110
     0705  111101
     . . .
```

Producing Simulation Results

To print simulation results, you must model the signal sampling.

The specific points in time that are significant for a particular design or testbench are always different. Which signals are significant is also different and may change as the simulation progresses. If you know which time points and signals are significant for determining the correctness of the simulation results, you have to be able to model that knowledge. Producing the proper simulation results involves modeling the behavior of the signal sampling.

Many sampling techniques can be used.
There are many sampling techniques, each as valid as the other. The correct sampling technique depends on your needs and on what makes the simulation results significant. Just as you have to decide which input sequence is relevant for the functionality you are trying to verify, you must also decide on the output sampling that is relevant for determining the success or failure of the function under verification.

You can sample at regular intervals.
The simplest sampling technique is to sample the relevant signals at a regular interval. The interval can be an absolute delay value, as illustrated in Sample 5-30, or the interval can be a reference signal such as the clock, as illustrated in Sample 5-31. Note how the *$strobe* statement is used in Verilog instead of *$write* and the use of a *postponed* process in VHDL. Then note how the recorded values are the final, stable values for the current simulation cycle are ensured.

Sample 5-30.
Sampling at a delay interval

```
parameter INTERVAL = 10;
always
begin
    #(INTERVAL);
    $strobe(...);
end
```

Sample 5-31.
Sampling based on a reference signal

```
postponed process (clk)
    variable L: line;
begin
    if clk'event and clk = '0' then
        write(L, ...);
        writeline(output, L);
    end if;
end process;
```

You can sample based on a signal changing value.
Another popular sampling technique is to sample a set of signals whenever one of them changes. This technique is used to reduce the amount of data produced during a simulation when signals do not change at a constant interval.

To sample a set of signals, simply make a *process* or *always* block sensitive to the signals whose changes are significant, as shown in Sample 5-32. The set of signals displayed and monitored can be different. Verilog has a built-in task, called *$monitor*, to perform this sampling when the set of displayed and monitored signals are identical.

An example of using the *$monitor* task is shown in Sample 5-33. Its behavior is different from the VHDL sampling process shown in Sample 5-32: Changes in values of signals *rst*, *d0*, *d1*, *sel*, *q*, and *qb* cause the display of simulation results, whereas only changes in *q* and *qb* trigger the sampling in the VHDL example. Note that Verilog simulations are limited to a single active *$monitor* task. Any subsequent call to *$monitor* <u>replaces</u> the previous monitor.

Sample 5-32.
Sampling based on signal changes

```
process (q, qb)
    variable L: line;
begin
    write(L, rst & d0 & d1 & sel & q & qb);
    writeline(output, L);
end process;
```

Sample 5-33.
Sampling using the *$monitor* task

```
initial
begin
    $monitor("...", rst, d0, d1, sel, q, qb);
end
```

Minimizing Sampling

To improve simulation performance, minimize sampling.

The use of an output device on a computer slows down the execution of any program. Therefore, recording simulation output reduces the performance of the simulation. To maximize the speed of a simulation, minimize the amount of simulation output produced during its execution.

In Verilog, an active *$monitor* task can be turned on and off by using the *$monitoron* and *$monitoroff* tasks, respectively. If you are using an explicit sampling *always* block or are using VHDL, you should include sampling minimization techniques in your model, as illustrated in Sample 5-34. A very efficient way of minimizing sampling is to have the stimulus turn on the sampling when an interesting section of the testcase is entered, as shown in Sample 5-35.

Sample 5-34.
Minimizing
sampling
	``
	process
	begin
	    wait until <interesting_condition>;
	    while (<interesting_condition>)loop
	        wait on q, qb;
	        write(l, rst & d0 & d1 & sel & q & qb);
	        writeline(output, l);
	    end loop;
	end process;
	``

Sample 5-35.
Controlling
the sampling
from the stim-
ulus
	``
	initial
	begin
	    $monitor("...", rst, d0, d1, sel, q, qb);
	    $monitoroff;
	    sync_reset;
	    load(1'b1, "d0");
	    sync_reset;
	    $monitoron;
	    load(1'b1, "d1");
	    load(1'b0, "d0");
	    load(1'b1, "d1");
	    sync_reset;
	    $monitoroff;
	    ...
	end
	``

Visual Inspection of Waveforms

Results are better
viewed when plot-
ted over time.

Waveform displays usually provide a more intuitive visual representation of simulation results. Figure 5-13 shows the same information as Sample 5-29, but using a waveform view. The waveform
view has the advantage of providing a continuous display of many
values over the entire simulation time, not just at specific time
points as in a text view. Therefore, you need not specify or model a
particular sampling technique. The signals are continuously sampled, usually into an efficient database format. Sampling for waveforms must be turned on explicitly. It is a tool-dependent process
that is different for each language[4] and each tool.

---

4. Verilog has a standard waveform database called the VCD file.
   Although all waveform viewers can display simulation results from a
   VCD file, all of the more advanced viewers use their own proprietary
   database to store additional signal information.

Figure 5-13.
Waveform
view of
simulation
results

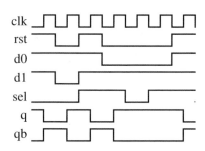

Minimize the number and duration of sampled signals.

The default behavior is to sample all signals during the entire simulation. The waveform sampling process consumes a significant portion of the simulation resources. Reducing the number of signals sampled, or reducing the duration of the sampling, increases the simulation performance. However, it is a trade-off with running a simulation multiple times to obtain traces of signals that were found to be necessary for diagnosing the cause of a functional error in a previous iteration. During a typical verification process, all signals should be sampled at the beginning, when the number of bugs is significant and their location is unknown. As the code stabilizes and simulations move to greater levels of integration, less and less signals are sampled. During regression runs, no signals are sampled. The rule of thumb is: If you expect the simulation to fail, sample a lot of signals; if you expect it to pass, don't sample any.

Self-Checking Testbenches

Visual inspection is not acceptable.

The model of the D flip-flop with a 2-to-1 input mux being verified has a functional error. Can you identify it using either views of the simulation results in Sample 5-29 or Figure 5-13? How long did it take to diagnose the problem?[5]

Code the response with the stimulus.

This example was for a very simple design, over a very short period of time, and for a very small number of signals (and you knew there was a bug). Imagine visually inspecting simulation results spanning hundreds of thousands of clock cycles, and involving hundreds of input and output signals. Then imagine repeating this visual inspection for every testbench and for every simulation of every testbench. The probability that you will miss identifying an error is

---

5. The logic value on input *d0* is ignored and a 1 is always loaded.

Writing Testbenches: Functional Verification of HDL Models

equal to one. You must automate the process of comparing the simulation results against the expected outputs.

Input and Output Vectors

Specify the expected output values for each clock cycle.

The first step in automating output verification is to include the expected output with the input stimulus for every clock cycle. The vector application task in Sample 5-24 can be easily modified to include the comparison of the output signals with the specified output vector, as shown in Sample 5-36. The testcase becomes a series of input/output test vectors, as shown in Sample 5-37.

Sample 5-36.
Application of input and verification of output data vectors

```
task apply_vector;
    input [...] in_data;
    input [...] out_data;
begin
    inputs <= in_data;
    @(posedge clk);
    fork
        begin
            #(Thold);
            inputs <= ...'bx;
        end
        begin
            #(Td);
            if (outputs !== out_data) ...;
        end
        #(cycle - Thold - Tsetup);
    join
end
endtask
```

Sample 5-37.
Input/output test vectors for 2-to-1 input sync reset D flip-flop

```
initial
begin
    // In: rst, d0, d1, sel
    // Out: q, qb
    apply_vector(4'b1110, 2'b00);
    apply_vector(4'b0100, 2'b10);
    apply_vector(4'b1111, 2'b00);
    apply_vector(4'b0011, 2'b10);
    apply_vector(4'b0010, 2'b01);
    apply_vector(4'b0011, 2'b10);
    apply_vector(4'b1111, 2'b00);
    ...
end
```

Test vectors require synchronous interfaces.

The main problem with input and output test vectors (other than the fact that they are very difficult to specify, maintain and debug), is that they require perfectly synchronous interfaces. If the design under verification contains interfaces in different clock domains, each interface requires its own test vector stream. If any interface contains asynchronous signals, the signals have to be either externally synchronized before vectors are applied, or treated as synchronous signals, therefore under-constraining the verification.

Golden Vectors

A set of reference simulation results can be used.

The next step toward automation of the output verification is the use of golden vectors. It is a simple extension of the manufacturing test process where devices are physically subjected to a series of qualifying test vectors. A set of reference output results, determined to be correct, are kept in a file or database. The simulation outputs are captured in a similar format during a simulation. They are then compared against the reference results. Golden vectors have an advantage over input/output vectors because the expected output values need not be specified in advance.

Text files can be compared using *diff*.

If the simulation results are kept in ASCII files, the simplest comparison process involves using the UNIX *diff* utility. The *diff* output for the simulation results shown in Sample 5-29 is shown in Sample 5-38. You can appreciate how difficult the subsequent task of diagnosing the functional error will be.

Sample 5-38. *diff* output of comparing ASCII view of simulation results

```
14c2
>0505 001010
>0600 001110
--------
<0505 001001
<0600 001110
. . .
```

Waveforms can be compared by a specialized tool.

Waveform comparators can be used also. They are tools similar to waveform viewers and are usually built into one. Waveform comparators compare two sets of waveforms then highlight the differences on a graphical display. The display of a waveform comparator might look something like the results illustrated in Figure 5-14. Identifying the problem is easier since you have access to the entire history of the simulation in a single view.

Figure 5-14.
Waveform
differences in
simulation
results

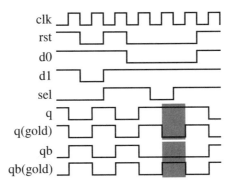

<table>
<tr><td>clk</td></tr>
<tr><td>rst</td></tr>
<tr><td>d0</td></tr>
<tr><td>d1</td></tr>
<tr><td>sel</td></tr>
<tr><td>q</td></tr>
<tr><td>q(gold)</td></tr>
<tr><td>qb</td></tr>
<tr><td>qb(gold)</td></tr>
</table>

Golden vectors
must still be
inspected visually.

The main problem with golden simulation results is that they need to be inspected visually to be determined as valid. This self-checking technique only reduces the number of times a set of simulation responses must be verified visually, not the need for visual inspection. The result from each testbench must *still* be manually confirmed as good.

Golden vectors do
not adapt to
changes.

Another problem: Reference simulation results do not adapt to modifications in the design under verification that may only affect the timing of the result, without affecting its functional correctness. For example, an extra register may be added in the datapath of a design to help meet timing constraints. All that was added was a pipeline delay. The functionality was not modified. Only the latency was increased. If that latency is irrelevant to the functional correctness of the overall system, the reference vectors must be updated to reflect that change.

Golden vectors
require a signifi-
cant maintenance
effort.

Reference simulation results must be inspected visually for every testcase, and modified or regenerated whenever a change is made to the design, each time requiring visual inspection. Using reference vectors is a high-maintenance, low-efficiency self-checking strategy. Verification vectors should be used *only* when a design must be 100% backward compatible with an existing device, signal for signal, clock cycle for clock cycle. In those circumstances, the reference vectors never change and never require visual inspection as they are golden by definition.

Separate the refer-
ence vectors along
clock domains. Reference simulation results also work best with synchronous interfaces. If you have multiple interfaces in separate clock domains, it
is necessary to generate reference results for each domain in a separate file. If a single file is used, the asynchronous relationship
between the clock domains may result in the samples from different
domains being written in a different order. The ordering difference
is not functionally relevant, but would be flagged as an error by the
comparison tool.

Self-Checking Operations

For simple operations on simple devices, it may be possible to verify the output on an operation-by-operation basis. For example, the
task shown in Sample 5-26 can include the verification that the flip-
flop was reset properly as shown in Sample 5-39. Similarly, the task
used to apply the stimulus to load data from a data input shown in
Sample 5-27 can be modified to include the verification of the output, as shown in Sample 5-40. The testcase shown in Sample 5-28
now becomes entirely self-checking.

Sample 5-39.
Verifying the
sync reset
operation

```
task sync_reset;
begin
   rst <= 1'b1;
   d0  <= 1'b1;
   d1  <= 1'b1;
   sel <= $random;
   @ (posedge clk);
   #(Thold);
   if (q !== 1'b0 || qb !== 1'b1) ...
   {rst, d0, d1, sel} <= 4'bxxxx;
   #(cycle - Thold - Tsetup);
end
endtask
```

Make sure the out-
put is verified
properly. The problem with output verification is that you can't identify a
functional discrepancy if you are not looking at it. Using an *if* statement to verify the output in the middle of a stimulus process only
looks at the output value for a brief instant. That may be acceptable,
but this technique does not say anything about the *stability* of that
output. For example, the tasks in Sample 5-39 and Sample 5-40
only check the value of the output at a single point in time.
Figure 5-15 shows the complete specification for the flip-flop. The
verification sampling point is shown as well.

Sample 5-40.
Verifying the
load operation

```
task load;
   input data;
   int [2*8:1] which;
begin
   rst <= 1'b0;
   if (which == "d0") begin
      . . .
   end
   @ (posedge clk);
   #(Thold);
   if (q !== data || qb !== ~data) ...
   {rst, d0, d1, sel} <= 4'bxxxx;
   #(cycle - Thold - Tsetup);
end
endtask
```

Figure 5-15.
Timing
specification
for the flip-
flop

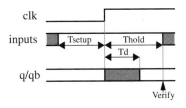

Make sure you
verify the output
over the entire sig-
nificant time
period.

To verify the functionality of the design properly and completely, it is necessary to verify that the output is stable, except for the short period after the rising edge of the clock. That could be verified easily using a static timing analysis tool and a set of suitable constraints to verify against. If you want to perform the verification as part of a functional simulation, the stability of the output cannot be verified easily in the same subprogram that applies the input. The input follows a deterministic data and timing sequence, whereas monitoring stability requires that the testbench code be ready to react to any unexpected changes.

Instead, it is better to use a separate monitor process, executing in parallel with the stimulus. The stimulus subprogram can still check the value. The stability monitor, as shown in Sample 5-41, simply verifies that the output remains stable, whatever its value. The stability of the output signal can be verified in the stimulus procedure, but it requires prior knowledge of the clock period to perform the timing check.

Sample 5-41.
Verifying the stability of flip-flop outputs

```verilog
initial
begin
    // wait for the first clock edge
    @ (posedge clk);
    forever begin
        // Ignore changes for Td after clock edge
        #(Td);
        // Watch for a change before the next clk
        fork: stability_mon
            @ (q or qb) $write("...");
            @ (posedge clk) disable stability_mon;
        join
    end
end
```

Sample 5-42.
Verifying the load operation and output stability

```vhdl
procedure load(data : std_logic;
               which: string) is
begin
    rst <= '0';
    if which = "d0" then
        ...
    end if;
    wait until clk = '1';
    assert q'stable(cycle - Td) and
           qb'stable(cycle - Td);
    wait for Thold;
    rst <= 'X';
    d0  <= 'X';
    d1  <= 'X';
    sel <= 'X';
    assert q = data and qb = not data;
    wait for cycle - Thold - Tsetup;
end load;
```

COMPLEX STIMULUS

This section introduces more complex stimulus generation scenarios through the use of bus-functional models. I start with reactive stimulus, where the stimulus or its timing depends on answers from the device under verification. I also show how to avoid wasting precious simulation cycles by getting caught in deadlock conditions.

Generating inputs may require cooperating with the design.

Applying stimulus to a clock or reset input or applying cycle-by-cycle test vectors is straightforward. You are under complete control of the timing of the input signal. However, if the interface being driven contains handshaking or flow-control signals, the generation

Writing Testbenches: Functional Verification of HDL Models

of the stimulus requires cooperation with the design under verification.

Feedback Between Stimulus and Design

Without feedback, verification can be under-constrained.

Figure 5-16 shows the specification for a simple bus arbiter. If you were to verify the design of the arbiter using test vectors applied at every clock cycle, as described in "Input and Output Vectors" on page 257, you would have to assume a specific delay between the assertion of the *req* signal and the assertion of the *grt* signal. Any delay value between one and five clock cycles would be functionally correct, but the only reliable choice is a delay of five cycles. Similarly, a delay of three clock cycles would have to be made for the release portion of the verification. These choices, however, severely under-constrain the verification. If you want to stress the arbiter by issuing requests as fast as possible, you would want to know when the request was granted and released, so it could be reapplied as quickly as possible.

Figure 5-16.
Specification
for a simple
arbiter

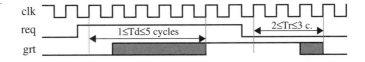

Stimulus generation can wait for feedback before proceeding.

If, instead of using input and output test vectors, you are using encapsulated operations to verify the design, you can modify the operation to wait for feedback from the design under verification before proceeding. You should also include any timing and functional verification in the feedback monitoring to ensure that the design responds in an appropriate manner. Sample 5-43 shows the *bus_request* operation procedure. The procedure samples the *grt* signal at every clock cycle, and immediately returns once it detects that the bus was granted. With a similarly implemented *bus_release* procedure, a testcase that stresses the arbiter under maximum load can be written easily, as shown in Sample 5-44.

Sample 5-43.
Verifying the
bus request
operation

```
procedure bus_request is
    variable cycle_count: integer := 0;
begin
    req <= '1';
    wait until clk = '1';
    while grt = '0' loop
        wait until clk = '1';
        cycle_count := cycle_count + 1;
    end loop;
    assert 1 <= cycle_count and cycle_count <= 5;
end bus_request;
```

Sample 5-44.
Stressing the
bus arbiter

```
test_sequence: process
    procedure bus_request ...
    procedure bus_release ...
begin
    for I in 1 to 10 loop
        bus_request;
        bus_release;
    end loop;
    assert false severity failure;
end process test_sequence;
```

Recovering from Deadlocks

A deadlock may
prevent the
testcase from run-
ning to comple-
tion.

There is a risk inherent to using feedback in generating stimulus: The stimulus now depends on the proper operation of the design under verification to complete. If the design does not provide the feedback as expected, the stimulus generation may be halted, waiting for a condition that will never occur. For example, consider the *bus_request* procedure in Sample 5-43. What happens if the *grt* signal is never asserted? The procedure remains stuck in the *while* loop and never returns.

A deadlocked sim-
ulation appears to
be running cor-
rectly.

If this were to occur, the simulation would still be running, merrily going around and around the *while* loop. The simulation time would advance at each tick of the clock. The CPU usage of your workstation would show near 100% usage. The only symptom that something is wrong would be that no messages are produced on the simulation's output log and the simulation runs for much longer than usual. If you are watching the simulation run and expect regular messages to be produced during its execution, you would quickly recognize that something is wrong and manually interrupt it.

But what if there is no one watching the simulation, such as during
a regression run? Regressions are large scale simulation runs where
all available testcases are executed. They are used to verify that the
functionality of the design under verification is still correct after
modifications. Because of the large number of testcases involved in
a regression, the process is automated to run unattended, usually
overnight and on many computers. If a design modification creates
a deadlock situation, all testcases scheduled to execute subsequently will never run, as the deadlocked testcase never terminates.
The opportunity of detecting other problems in the regression run is
wasted. It will be necessary to wait for another 24-hour period
before knowing if the new version of the design meets its functional
specification.

Eliminate the pos-
sibility of dead-
lock conditions.

When generating stimulus, you must make sure that there is no possibility of a deadlock condition. You must assume that the feedback
condition you are waiting for may never occur. If the feedback condition fails to happen, you must then take appropriate action. It
could include terminating the testcase, or jumping to the next portion of the testcase that does not depend on the current operation, or
retrying the operation after some delay. Sample 5-43 was modified
as shown in Sample 5-45 to avoid the deadlock condition created if
the arbiter failed and the *grt* signal was never asserted.

```
procedure bus_request is
   variable cycle_count: integer := 0;
begin
   req <= '1';
   wait until clk = '1';
   while grt = '0' loop
      wait until clk = '1';
      cycle_count := cycle_count + 1;
      assert cycle_count < 500
         report "Arbiter is not working"
         severity failure;
   end loop;
   assert 1 <= cycle_count and cycle_count <= 5;
end bus_request;
```

If a failure of the feedback condition is detected, terminating the
simulation on the spot, as shown in Sample 5-45, is easy to implement in each operation subprogram. If you want more flexibility in
handling a non-fatal error, you might want to let the testcase handle
the error recovery, instead of handling it inside the operation sub-

program. The subprogram must provide an indication of the status of the operation's completion back to the testcase. Sample 5-46 shows the *bus_request* procedure that includes an *OK* status flag indicating whether the bus was granted. The testcase is then free to retry the bus request operation until it succeeds, as shown in Sample 5-47. Notice how the testcase takes care of avoiding its own deadlock condition if the bus request operation never succeeds.

Sample 5-46.
Returning status in the bus request operation

```
procedure bus_request(ok: out boolean) is
    variable cycle_count: integer := 0;
begin
    ok := true;
    req <= '1';
    wait until clk = '1';
    while grt = '0' loop
        wait until clk = '1';
        cycle_count := cycle_count + 1;
        if cycle_count > 500 then
            ok := false;
            return;
        end if;
    end loop;
    assert 1 <= cycle_count and cycle_count <= 5;
end bus_request;
```

Sample 5-47.
Handling failures in the *bus_request* procedure

```
testcase: process
    variable granted : boolean;
    variable attempts: integer := 0;
begin
    ...
    attempts := 0;
    loop
        bus_request(granted);
        exit when granted;
        attempts := attempts + 1;
        assert attempts < 5
            report "Bus was never granted"
            severity failure;
    end loop;
    ...
end process testcase;
```

Asynchronous Interfaces

Test vectors under-constrain asynchronous interfaces.

Creating synchronous input data and verifying synchronous output values is simple. The inputs are all applied at the same time. The outputs are all verified at the same time. And this process is repeated at regular intervals. In every design, there is some reference signal that can be used to synchronize generation and sampling operations. But many interfaces, although implemented using synchronous finite state machines and edge-triggered flip-flops, are *specified* in an asynchronous fashion. The implementer has arbitrarily chosen a clock to streamline the physical implementation of the interface. If that clock is not part of the specification, it should not be part of the verification. For example, Figure 5-17 shows an asynchronous specification for a bus arbiter. Given a suitable clock frequency, the synchronous specification shown in Figure 5-16 would be functionally equivalent.

Figure 5-17.
Asynchronous
specification
for a simple
arbiter

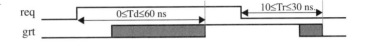

Verify the synchronous implementation against the asynchronous specification.

Even though a clock may be present in the implementation, if it is not part of the specification, you cannot use it to generate stimulus nor to verify the response. You would be verifying against a particular implementation, not the specification. For example, a VME bus is asynchronous. The verification of a VME interface cannot make use of the clock, if a clock is used to implement that interface. If a clock is present, and the timing constraints make reference to clock edges, then you *must* use it to generate stimulus and verify response. For example, a PCI bus is synchronous. A verification of a PCI interface must use the PCI system clock to verify any implementation.

Behavioral code does not require a clock like RTL code.

Testbenches are written using behavioral code. Behavioral models do not require a clock. A clock is an artifice of the implementation methodology and is required only for RTL code. The bus request phase of the asynchronous interface specified in Figure 5-17 can be verified asynchronously with the *bus_request* procedure shown in Sample 5-48 or Sample 5-49. Notice how neither model of the bus request operation uses a clock for timing control. Also, notice how the Verilog version, in Sample 5-49, uses the definitely non-synthesizeable *fork/join* statement to wait for the rising edge of *grt* for a maximum of 60 time units.

Sample 5-48.
Verifying the asynchronous bus request operation in VHDL

```
procedure bus_request(good: out boolean) is
begin
   req <= '1';
   wait until grt = '1' for 60 ns;
   good := grt = '1';
end bus_request;
```

Sample 5-49.
Verifying the asynchronous bus request operation in Verilog

```
task bus_request;
   output good;
begin
   req = 1'b1;
   fork: wait_for_grt
      #60                 disable wait_for_grt;
      @ (posedge grt) disable wait_for_grt;
   join
   good = (grt == 1'b1);
end
endtask
```

Consider all possible failure modes.

There is one problem with the models of the bus request operation in Sample 5-48 and Sample 5-49. What if the arbiter was functionally incorrect and left the *grt* signal always asserted? Both models would never see a rising edge on the *grt* signal. They would eventually exhaust their maximum waiting period then detect *grt* as asserted, indicating a successful completion. To detect this possible failure mode, the bus request operation must verify that the *grt* signal is not asserted prior to asserting the *req* signal, as shown in Sample 5-50.

Were you paying attention?

Pop quiz: The first *disable* statement in Sample 5-50 aborts the *bus_request* task and returns control to the calling block of the statement. Why does it disable the *begin/end* block inside the task and not the task itself?[6] And what is missing from all those task implementations?[7]

---

6. For the answer see "Output Arguments on Disabled Tasks" on page 221.

Sample 5-50.
Verifying all
failure modes
in the asyn-
chronous bus
request opera-
tion

```
task bus_request;
    output good;
begin: bus_request_task
    if (grt == 1'b1) begin
        good = 1'b0;
        disable bus_request_task;
    end
    req = 1'b1;
    fork: wait_for_grt
        #60                     disable wait_for_grt;
        @ (posedge grt) disable wait_for_grt;
    join
    good = (grt == 1'b1);
end
endtask
```

BUS-FUNCTIONAL MODELS

Operations are
abstracted through
bus-functional
models.

Although I have avoided using the term *bus-functional model*, all of
the subprograms abstracting operations on the design shown earlier
are bus-functional models, albeit very simple ones. Operations, also
known as *transactions*, encapsulated using *tasks* or *procedures* can
be very complex. The examples shown earlier were very simple and
dealt with only a few signals. Real-life interfaces are more complex. But they can be encapsulated just as easily. These transactions
may even return values to be verified against expected values or
modify the stimulus sequence. As shown in Figure 4-2, a bus-functional model abstracts transactions on a physical-level interface into
a procedural interface. Bus-functional models can be used to generate stimulus as well as monitor the response of a design. Very often,
a single bus-functional model performs both operations.

CPU Transactions

CPU interfaces are
popular bus func-
tional models.

The first image that probably came to your mind when you read the
term bus-functional model was an interface to a processor.
Abstracting processor bus transactions are the most popular and
common bus-functional models. Figure 5-18 shows the specifica-

---

7. They all include timing control statements. They should have a semaphore to detect concurrent activation. See "Non-Re-Entrant Tasks" on
page 222.

tion for the write cycle for an Intel 386SX processor bus. Sample 5-51 shows the corresponding bus-functional model procedure.

Figure 5-18.
Specification
for the write
cycle of a
386sx
processor

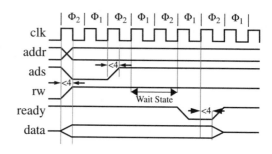

Sample 5-51.
Model for the
write cycle
operation

```
procedure write_cycle (
            wadd : in   std_logic_vector(0 to 23);
            wdat : in   std_logic_vector(0 to 31);
    signal clk  : in   std_logic;
    signal phi  : in   one_or_two;
    signal addr : out  std_logic_vector(0 to 23);
    signal ads  : out  std_logic;
    signal rw   : out  std_logic;
    signal ready: in   std_logic;
    signal data : out  std_logic_vector(0 to 31);
  is
begin
    wait on clk until clk = '1' and phi = 2;
    addr <= wadd after 4 ns;
    ads  <= '0'  after 4 ns;
    rw   <= '1'  after 4 ns;
    data <= wdat after 4 ns;
    wait until clk = '1';
    wait until clk = '1';
    ads  <= '1'  after 4 ns;
    wait on clk until clk = '1' and phi = 2 and
                      ready = '0';
    data <= (others => 'Z') after 4 ns;
end write_cycle;
```

Bus models can
adapt to a differ-
ent number of wait
states.

To generate stimulus for this interface using synchronous test vectors, you would have to assume a specific number of wait cycles to complete the write operation at the right time. With behavioral models of the transaction, you need not enforce a particular number of wait cycles and adapt to any valid bus timing. In Sample 5-51, the wait cycles are introduced by the fourth *wait* statement.

Bus-functional
procedures can
return values.

All of the abstracted transactions shown so far were unidirectional. Data always flowed from the testbench through the bus-functional procedure where the data was applied to the design and outputs were checked for correctness. What if determining the correctness of the output required visibility over multiple operations? What if only the relevant output values for this testcase were known and the others were to be ignored? Bus-functional procedures can just as easily sample output and return it instead of comparing the output against supplied expected values. The sampled value can then be processed by the testbench where the value can be dealt with according to the needs of the testcase. For example, Sample 5-52 shows the *read* operation of the 386SX interface. Notice how the value read is not compared against an expected value. The value read is instead returned through an *out* argument.

Sample 5-52.
Model for the
read cycle
operation

```
task read_cycle (
    input   [23:0] radd;
    output  [31:0] rdat;
begin
    while (phi != 2) @ (posedge clk);
    addr <= #4 radd;
    ads  <= #4 1'b0;
    rw   <= #4 1'b0;
    repeat (2) @ (posedge clk);
    ads  <= #4 '1';
    while (phi != 2 || ready != 1'b0)
        @ (posedge clk);
    rdat = data;
end
endtask
```

You can perform
read-modify-write
operations.

It now becomes easy to perform read-modify-write operations. With abstracted transactions and the full power of a high-level language, you can perform a *read* operation that returns whatever value was read at the specified address, manipulate the read value, then use the modified value in a subsequent write transaction. Sample 5-53 shows a portion of a testcase where a *read_cycle* procedure, similar to the *write_cycle* procedure shown in Sample 5-51, is used to perform a read-modify-write operation.

Sample 5-53.
Performing a
read-modify-
write opera-
tion

```
test_procedure: process
    constant cfg_reg: std_logic_vector(0 to 23)
        := "000000000000000001100010110";
    variable tmp: std_logic_vector(31 downto 0);
begin
    ...
    i386sx_pkg.read_cycle(cfg_reg, tmp, ...);
    tmp(13 downto 9) := "01101";
    i386sx_pkg.write_cycle(cfg_reg, tmp, ...);
    ...
end process test_procedure;
```

From Bus-Functional Procedures to Bus-Functional Model

Bus-functional
procedures are
packaged into bus-
functional models.

A complete bus-functional model is composed of many bus-functional procedures. Each transaction supported by a particular physical interface is implemented using a different procedure. Collected together using the encapsulation mechanism described in "Encapsulating Bus-Functional Models" on page 137, they create a complete bus-functional model for a specific interface.

Procedures are re-
entrant, but bus-
functional models
are not.

In "Non-Re-Entrant Tasks" on page 222, I discussed the problem caused by non-re-entrant procedures. Now that Verilog, like all of the other languages, offers re-entrant subprograms, you'd think that the problem would be solved, right? Wrong. Sample 5-54 shows the *e* bus-functional procedures for the i386SX packed into a single *unit*. Although the *read* and *write* methods are fully re-entrant, what happens if two separate threads concurrently invoke the same method? The local data space of the method is preserved, but not the value of the (global and static) interface signals. The two concurrent transactions will interfere with each other trying to execute at the same time on the same physical interface. The same problem will occur even if two different bus-functional procedures in the same bus-functional model are concurrently invoked.

Put a semaphore
on the bus-func-
tional model.

If two or more threads must read from (or write to) the design, the operations must be coordinated. To pipeline concurrent operations, it is necessary to put a semaphore around the *entire* bus-functional model. Much like a semaphore was used to detect concurrent invocation of a non-re-entrant task in Verilog, it will be used to detect concurrent invocation of transactions in a non-re-entrant bus-functional model. Sample 5-55 shows how an OpenVera bus-functional model can be protected against concurrent transactions using a semaphore. It is up to you to decide, should the semaphore detect a

```
unit i386_sx {
    clk : string;
    addr: string;
    ...
    read(radd: uint (bits: 24)): uint
    @pclk is {
        '(me.addr)' = raddr;
        ...
    };
    write(wadd: uint (bits: 24),
          wdat: uint)
    @pclk is {
        '(me.addr)' = raddr;
        ...
    };
};
```

concurrent transaction, whether to wait for the bus-functional to become available or to terminate with an error.

```
class i386_sx {
    local i386_sx_port sigs;
    local integer sem;
    ...
    task new(i386_sx_port sigs) {
        this.sigs = sigs;
        this.sem = alloc(SEMPAHORE, 0, 1, 1);
    }
    function bit [31:0] read(bit [32:0] radd) {
        void = semaphore_get(WAIT, this.sem, 1);
        @0 this.sigs.$addr = raddr;
        ...
        semaphore_put(this.sem, 1);
    }
    task write(bit [23:0] wadd,
               bit [31:0] wdat) {
        void = semaphore_get(WAIT, this.sem, 1);
        @0 this.sigs.$addr = raddr;
        ...
        semaphore_put(this.sem, 1);
    }
}
```

HVLs are not significantly better for physical-level bus-functional models.	With all of my enthusiasm for HVLs expressed in earlier chapters, why are almost all of the examples in the previous sections in Verilog or VHDL? Because HVLs are not significantly better than HDLs in implementing physical-level bus-functional models. Low-level bus-functional models simply translate an abstracted representation of a transaction into 1's and 0's applied or sampled at individual clock cycles. Regardless of the language used, signal assignments, signal sampling and waiting for the next clock edge require almost identical statements because it is the same level of abstraction.
HVLs are significantly better above the physical level.	HVLs become clearly superior once we stop dealing with the physical interface because of their support for high-level data types, object-orientedness and randomization. HVLs allow for a simpler transaction-layer interface. The transaction descriptors will be easier to model and manipulate using object-oriented methods and transaction descriptors will be easier to generate using constrainable randomization.
HVLs can call HDL subprograms.	If you have existing bus-functional model subprograms, there is no need to re-implement them in OpenVera or *e*. Both languages have a mechanism for invoking Verilog *tasks* and VHDL *procedures*. If you have to implement a new bus-functional model, you probably are better off implementing it in the HVL you are using. It will let you design a more flexible and higher-level transaction-layer interface.
You must understand the HVL interface model.	HVLs are interfaced to the HDL simulation through the simulator's external program interface. When that interface was implemented, the HVL designers created specific mechanisms for driving and sampling signals in Verilog or VHDL. It is important to understand that interfacing mechanism to create functionally correct bus-functional models and minimize the performance impact of crossing the HDL/HVL boundary.

OpenVera's Interface Model

Function and timing are separated.	In OpenVera, the function of a physical-level bus-functional model is described using sequential statements in a *task* or a *function*. The timing of the physical signals—what is the clock signal, what is the active edge, when they are sampled with respect to the active edge, how much hold time before a new value is driven—is defined separately in the *interface* or in the *signal_connect()* task. Interface sig-

nals (potentially from different *interface* declarations) are then collected into a single *virtual port binding*. Through the virtual port binding, the functional description can drive and sample interface signals without needing to be aware of their detailed timing.

Interface declarations are synchronous. Figure 5-19 shows a functional and timing diagram of OpenVera's interface model. The active edge of the clock is specified on a per-signal and per-direction basis (using the P or N prefix). The sample (Ts) and hold (Th) times are also specified for each signal and each direction. Whenever a signal *drive* or *sample* statement is executed, if the simulation is not aligned currently with the active edge of the signal, the simulation waits for the next active edge before performing the drive or sample[8]. Each *drive* or *sample* statement specifies a number of cycle delays using the @ notation. An @1 specification always will wait until the next active edge. But an @0 specification *may* wait until the next active edge if the simulation is not currently aligned with it, as described previously. Note how referring to interface signals within an expression completely bypasses the interface synchronization (but not delay) mechanism.

Figure 5-19.
OpenVera
interface
model

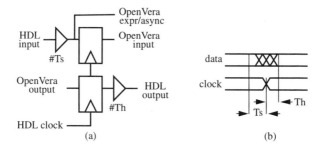

(a)   (b)

That was a good and a bad idea. Separating timing from function was a good idea. The separation lets the user of a bus-functional model shift the sampling or driving point with respect to the clock without having to modify the code. It lets that same user modify the clock signal from an internal to an external source with similar ease. When writing the bus-functional model, you can concentrate on the functionality without worrying about whether you are on the correct phase of the clock or remembered to insert that #4 delay before all nonblocking assignments to the data signal. This design decision has its trade-off. If the inter-

---

8. This is similar to the behavior of the *sync* statement in *e*.

---

face signals are not sampled using an internally consistent clocking scheme, the bus-functional model is simply not going to work as intended. Unexpected and unwanted delays will be inserted, stretching a single-cycle operation over multiple cycles.

Prefer *static* inter-face signals.

Static signal bindings i.e., using *interface* and *bind* declarations, are more efficient than dynamic bindings i.e., using *signal_connect()* tasks. With static binding, the OpenVera-to-HDL interface is able to perform additional optimizations that are not available if dynamic binding is used. With dynamic binding, Vera must keep all potentially useful information available at all times because new signal connections can be created at any time. Executing a single *signal_connect* makes it impossible to perform the static binding optimization. Note that the previous sentence said "executing". You can have calls to the *signal_connect* task in your code without having to use the dynamic binding mode if it is never executed.

Bus-Functional Models in OpenVera

Any method can sample, drive or wait for time.

In OpenVera, any method can sample or drive a signal, even a constructor. This means that time can advance in any method because of the implicit synchronization that occurs when sampling or driving signals. This flexibility is fine when writing a bus-functional model. But from a user's perspective, it becomes impossible to know whether time will advance when I invoke a method. It may be important that I perform an operation in zero-time. Therefore, it is valuable to know if a method can suspend the current execution thread or not.

Use a naming convention to identify methods that include timing control statements. Personally (and in this book), I use the "_t" suffix to indicate such methods. You should *never* cause the execution thread to be suspended in a constructor (i.e., the *new* task). Would you expect time to advance between the two declarations in Sample 5-56? When initializing output signals in a constructor, use the non-blocking assignment to avoid suspending the execution thread.

Don't wait for clock edges.

OpenVera's syntax for physical-level operations is borrowed heavily from Verilog. It is therefore natural to use a Verilog coding style when coding using OpenVera. But this style can create problems. Consider the synchronous ROM bus-functional model in Sample 5-57. This is obviously a model for a synchronous interface

Sample 5-56.
Constructor
with timing
control state-
ments

```
class i386_sx {
   task new() {
      . . .
      @0 ads = 1'b1;
      @0 data = 'bz;
   }
}
. . .
{
   i386_sx cpu = new;
   eth_mii eth = new;
   . . .
}
```

active on the rising edge of the clock, right? Wrong. Timing synchronization is specified in the *interface* declaration, not by the
sequential statements. Should the port signals be bound to the *interface* declaration shown in Sample 5-58, everything will operate on
the falling edge of the clock. This style may introduce unexpected
delays in performing a read transaction. Worst, it may return the
wrong results under certain pathological signal timing conditions.
A better style, shown in Sample 5-59, goes along with the Open-
Vera flow and lets the *interface* synchronization mechanism define
the timing of the transaction. Another advantage of this second
style is the use of the windowed *expect* to detect the case where the
ROM does not respond and avoid the deadlock condition found in
if *rdy* is never asserted.

Sample 5-57.
Verilog cod-
ing style in
OpenVera

```
port rom_port {
   clk, addr, cs, rdy, data
}
class sync_rom_bfm {
   rom_port sigs;
   . . .
   function bit [31:0] read(bit [31:0] radd) {
      @ (posedge sigs.$clk);
      sigs.$addr = radd;
      sigs.$cs   = 1'b1;
      while (!sigs.$rdy) {
         @ (posedge sigs.$clk);
      }
      read = sigs.$data;
      sigs.$ads = 1'b0;
   }
}
```

Sample 5-58.
Interface declaration active on the negative edge

```
interface rom_if {
    input           clk  CLOCK;
    output [31:0] addr NHOLD   #1;
    output          cs   NHOLD   #1;
    input           rdy  NSAMPLE #-1;
    input  [31:0] data NSAMPLE #-1;
}
```

Sample 5-59.
Proper coding style in Open-Vera

```
class sync_rom_bfm {
    rom_port sigs;
    ...
    function bit [31:0] read(bit [31:0] radd) {
        @1 sigs.$addr = radd;
        @0 sigs.$cs   = 1'b1;
        @,100 sigs.$rdy == 1'b1;
        read = sigs.$data;
        @0 sigs.$ads = 1'b0;
    }
}
```

Data edges involve sampling.

Another Verilog coding style that likely will yield unexpected results is shown in Sample 5-60. Waiting for an edge on an input signal still involves its sampling by the interface clock. The edge will be detected only when two consecutive samples have a different value, as illustrated in Figure 5-20. Refer to "Asynchronous Signals in OpenVera" on page 281 for details on how to sample and wait for input signals asynchronously.

Sample 5-60.
Waiting for synchronous data signal

```
interface arb_if {
    input clk  CLOCK          hdl_node "top.clk";
    input req  NSAMPLE #-2 hdl_node "top.req";
    input ack  NHOLD   #1  hdl_node "top.ack";
}
...
@ (posedge rom_if.req);
...
```

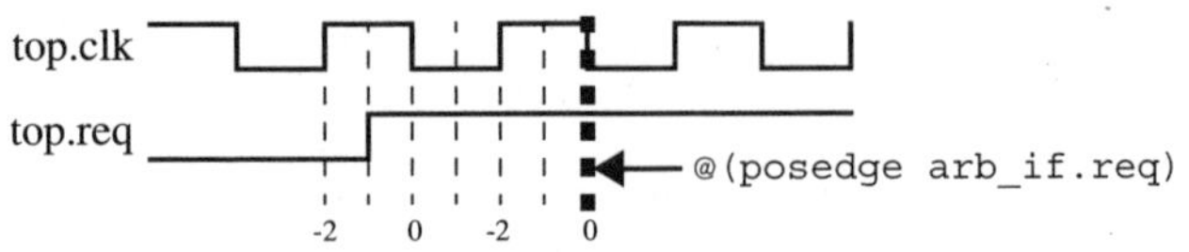

Figure 5-20.
Synchronous signal timing in OpenVera

Always drive
using nonblocking
assignment.

A common technique to avoid race conditions between clock and data signals is to sample and drive the signals on different clock edges. OpenVera's sample and hold interface delays eliminates this requirement when both the source and destination of a signal use the same clock signal. But you can always count on the ingrained habits of engineers. As the writer of a bus-functional model, you have no control over the binding of the virtual port signals. A user may very well decide to drive the signals on the other edge of the clock. Furthermore, certain protocols where the clock and data signals are generated together require this 180 degree phase between sampling and driving.

Using the *interface* declaration in Sample 5-61, the code in Sample 5-62 will not cause the time to advance after the assertion of the *rdy* signal. But should the interface be modified to use different clock edges to sample and drive signals, as shown in Sample 5-63, the same code section will cause time to advance by a full clock cycle. To avoid this dependency on *interface* declaration, either perform all of your sampling before any driving, or use the nonblocking assignment, as shown in Sample 5-64

Sample 5-61.
Single edge
interface dec-
laration

```
interface rom_if {
    input           clk   CLOCK;
    output  [31:0]  addr  PHOLD    #1;
    output          cs    PHOLD    #1;
    input           rdy   PSAMPLE  #-1;
    input   [31:0]  data  PSAMPLE  #-1;
}
```

Sample 5-62.
Potential time
advance

```
...
@,100 sigs.rdy == 1'b1;
sigs.cs = 1'b0;
read = sigs.data;
...
```

Sample 5-63.
Dual edge
interface dec-
laration

```
interface rom_if {
    input           clk   CLOCK;
    output  [31:0]  addr  NHOLD    #1;
    output          cs    NHOLD    #1;
    input           rdy   PSAMPLE  #-1;
    input   [31:0]  data  PSAMPLE  #-1;
}
```

Sample 5-64.
Avoiding
unexpected
time advances

```
. . .
@,100 sigs.rdy == 1'b1;
sigs.cs <= 1'b0;
read = sigs.data;
. . .
```

Do not mix clock domains.

The potential time advance in Sample 5-62 can be made even worse if signals from different clock domains are bound to the same virtual port. For example, the binding shown in Sample 5-65 will result in unpredictable behavior that will depend on the frequency and phase relationship of the clocks in each interface. You should avoid mixing signals from different clock domains in the same port binding.

Sample 5-65.
Multiple clock
domains
bound to sin-
gle port

```
interface rom_in_if {
    input           clk  CLOCK;
    output [31:0] addr PHOLD    #1;
    output          cs   PHOLD    #1;
}
interface rom_out_if {
    input           clk  CLOCK;
    input           rdy  PSAMPLE #-1;
    input  [31:0] data PSAMPLE #-1;
}
bind rom_port rom_port_0 {
    addr  rom_in_if.addr;
    cs    rom_in_if.cs;
    rdy   rom_out_if.rdy;
    data  rom_out_if.data;
};
```

Provide an interface declaration template.

Functional correctness of OpenVera bus-functional models depends on the correctness of the interface signals they are bound to. Despite having the best intentions, users of your bus-functional model *will get it wrong*. You will then be blamed for writing a bus-functional model that does not work. To help avoid this situation and facilitate the integration of your bus-functional model in someone else's testbench, always provide a template for a suitable *interface* declaration and port binding. Identify which parameters can be modified, and which ones should not.

Asynchronous Signals in OpenVera

All interface signals are synchronous.

There is no way of declaring an interface signal as asynchronous. Every *interface* declaration contains a reference clock signal. Even if you leave out the declaration for the *CLOCK* signal (which is a valid syntax), a default clock is used by the *interface*. This default clock is the signal named *SystemClock* in the generated shell file. If you let an *interface* declaration default to the default clock and do not specify a waveform on the default clock signal, the simulation is going to hang on the first attempt to drive or sample a signal in that *interface* declaration. Why? Because Vera will attempt to synchronize the operation with an active edge that will never occur. A poor solution would be to declare all input signals as *CLOCK* signals, but that would require one *interface* declaration per input signal. Furthermore, it would not be possible to declare multi-bit input signals as clock signals are restricted to single-bit signals.

Drive and sample using *async*.

Recall that referring to interface signals in expressions is asynchronous. But any sampling (and that includes @ *posedge* statements) or driving operation requires going through the synchronous interface. For asynchronous operations, sampling and driving operations can be qualified with the *async* suffix to bypass the alignment with the active clock edge. Any assignment operation will execute immediately, without waiting for the next active edge (assuming @0 is specified of course).

Specify #0 sample and hold delays.

There is one catch: The *async* qualifier only bypasses the synchronization mechanism. Like the reference in expressions, this qualifier does not bypass the sample or hold delay mechanisms. If you specified a hold delay of #2 on an output interface signal, driving it with the *async* qualifier will produce the assigned value on the HDL signal two timescale units later. Conversely, sampling an input signal with a sample delay of #2 will sample the value it had two timescale units prior to the current simulation time. The effect of the *async* qualifier and of the sample and hold delays is shown in Sample 5-66 and illustrated in Figure 5-21. When declaring an asynchronous signal, locate in any *interface* declaration, but specify #0 for sample and hold delays.

Sample 5-66.
Asynchronous
signals in
OpenVera

```
interface sigs {
    input   req0  PSAMPLE  #0   hdl_node  "top.req";
    input   req2  PSAMPLE  #-2  hdl_node  "top.req";
    input   rw0   PSAMPLE  #0   hdl_node  "top.rw";
    input   rw2   PSAMPLE  #-2  hdl_node  "top.rw";
    output  ack0  PHOLD    #0   hdl_node  "top.ack";
    output  ack2  PHOLD    #2   hdl_node  "top.ack";
}

program asynchronous {
    @(posedge sigs.req0 async);
    sigs.ack2 = 1'b0 async;
    sigs.ack0 = 1'b0 async;
}
```

Figure 5-21.
Asynchronous
signal timing
in OpenVera

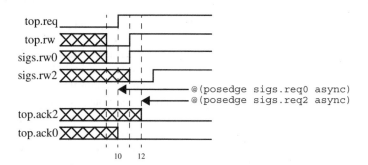

Synchronous Bus-Functional Models in *e*

Bus functional
procedures are
time-consuming
methods.

Unlike OpenVera, *e* makes a clear distinction between methods that
may cause time to advance, and those that do not. Thus, it is very
easy to enforce (or rely on) a zero-delay execution sequence
because methods (which consume no time) are not allowed to
invoke time-consuming methods (TCMs), nor are methods allowed
to execute time consuming actions, such as *wait*. The latter are different from the former in that TCMs have, in their declaration, an
event identified as a default sampling event. Sample 5-67 and Sample 5-68 show two method declarations in *e*. The first one is a regular method. The second is a time-consuming method. For a bus-
functional procedure, the second declaration style must be used.

Sample 5-67.
Declaration of
non-TCM

```
unit eth_mii {
    event clk is rise('(clk)') @sim;

    send(frame: mac_frame) is {
        ...
        wait cycle @clk;  // Illegal
        ...
    };
};
```

Sample 5-68.
Declaration of
TCM in *e*

```
unit eth_mii {
    event clk is rise('(clk)') @sim;

    send(frame: mac_frame) @clk is {
        ...
        wait cycle;  // OK ("@clk" implicit)
        ...
    };
};
```

TCMs imply syn-
chronous opera-
tions with
asynchronous sig-
nals.

The event specified in the time-consuming method declaration will be used as the default sampling event in all *wait* and *sync* actions[9] inside the method. For example, in Sample 5-68, no explicit sampling event was specified in the *wait* action. The sampling event used is the TCM default sampling event. Furthermore, there is an implicit "sync cycle" action at the beginning of the TCM. This forces the execution of the TCM to be aligned with its default sampling event at all times. This consistent reference to a single event makes the execution of a TCM inherently synchronous to that event (which may or may not be related to a clock signal).

Even though *e*'s interface to HDL signals is asynchronous, TCMs naturally map to synchronous bus-functional transactions. Because all of the events you'll be waiting for will be sampled (by default) using the TCM sampling event, the signal sampling and driving points will always be aligned with that event. If that default sampling event happens to be a clock, then the TCM will be synchronous to that clock. Sample 5-69 shows a bus-functional procedure to transmit an ATM cell on a Utopia Level 1 interface illustrated in Figure 5-22.

---

9. In *e*, sequential statements are called "actions".

Sample 5-69.
Utopia Level 1
transmit bus-
functional pro-
cedure

```
unit phy_utopia1 {
    RxClk: string;
    event rx_clk is rise('(RxClk)') @sim;
    ...
    tx(cell: atm_cell) @rx_clk is {
        var bytes: list of byte;
        ...
        for each in bytes {
            sync true('(RxEnb)' != 1'b0));
            '(RxEmpty)' = 1'b1;
            '(RxData)'  = it;
            wait cycle;
        };
        '(RxEmpty)' = 1'b0;
    };
};
```

Figure 5-22.
Utopia Level 1
PHY-to-ATM
cell
transmission
with octect-
level
handshake

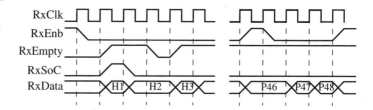

There is no race
condition between
the HDL and *e*
simulator.

The mention that signal sampling and driving points were aligned should have raised race-condition alarm bells in your mind. Fortunately, race conditions do not actually occur because of the co-simulation cycle used (see "The Co-Simulation Cycle" on page 196). Specman Elite is invoked after all possible delta cycles in the HDL simulator have been processed. Thus, all signal values in the simulation will have been simulated as far as possible and will have reached their final stable state. Similarly, all of Specman Elite's delta cycles are executed before returning to the HDL simulator, ensuring stable and final output values. Nevertheless, if it is necessary to shift the timing of the output signals from the sampling of the input signals, there are two mechanisms that can be used: output delays and separate events.

Output delays can
be specified.

Using the *verilog variable* or *vhdl driver* statement, you can specify a delay to be introduced in any output signal. This delay will be physically implemented in the *stub file* and modeled using HDL constructs. For example, Sample 5-70 shows how to introduce a two timescale unit delay in a VHDL output signal. The Specman Elite documentation does not specify how the output delay is implemented in Verilog. Whereas the *vhdl driver* statement can specify *inertial* or *transport* delay model, there is no such option in the *verilog variable* statement. It is not specified whether the delay is implemented using continuous assignments, blocking assignments or nonblocking assignments. If the Verilog delay model used is important (and it is *very* important on asynchronous signals), refer to the generated code in the *specman.v* file. If a different Verilog delay model should be used, you may have to generate your own interface code and internal output variable using the *verilog code* statement.

Sample 5-70.
Specifying
output delay in
e.

```
unit phy_utopia1 {
    RxClk : string;
    RxData: string;
    event rx_clk is rise('(RxClk)') @sim;
    ...
    vhdl driver '(RxData)' using
        delay=2, mode=INERTIAL;
};
```

HDL statements may not be subject to aspect-oriented extensions.

If the functional correctness of a bus-functional model depends on the presence or the value of output delays, it would be nice if you could guarantee that they were specified at all times. The only way to do so would be to include the *verilog variable* or *vhdl driver* statement in the bus-functional model code. The problem with *verilog variable* statements is that, once in there, they cannot be modified without modifying the code itself. To modify a delay specified using a *verilog variable* statement requires that a different source file be loaded.

On the other hand, *vhdl driver* statements are unit members. They can thus be located in a *when* extension. Different *when* extensions can be used to specify different delays for different instances of the bus-functional model. As a rule, *verilog variable* and *vhdl driver* statements should be added only to the user of a bus-functional model, never its author. This addition enables the user to adapt the timing of the bus-functional model to the particular requirement of the design under verification. As an author, your bus-functional models should never depend on the presence of output delays.

Have separate
sampling and driv-
ing events.

If the functional correctness of a bus-functional model depends on
the timing of the outputs being separated from the timing of the
inputs, you can use two separate events. One event is used to synchronize the sampling of the inputs; the other event is used to synchronize the driving of the outputs. Which event is used to
synchronize the TCM depends on which activity must be done first.
If the first operation is to sample signal, then use the input synchronization event. Otherwise, use the output synchronization event.

Sample 5-71 shows the bus-functional model from Sample 5-69
using separate sample and drive events. This techniques makes it
very easy to sample inputs on one edge and drive the outputs on the
other. This technique can be coupled with *verilog* or *vhdl* statements to introduce additional delays on individual outputs after the
output synchronization event. Using this method, the output synchronization can be modified using aspect-oriented extensions
without modifying the original code by redefining the drive event
using *is only*. Sample 5-72 shows an example of modifying the output synchronization event of the bus-functional model in Sample 5-
71. Used in concert with *when* extensions, the output synchronization event can be modified on a per-instance basis.

Sample 5-71.
Using sepa-
rate sample
and drive
events

```
unit phy_utopial {
    RxClk: string;
    event sample is rise('(RxClk)') @sim;
    event drive  is fall('(RxClk)') @sim;
    ...
    tx(cell: atm_cell) @sample is {
        var bytes: list of byte;
        ...
        for each in bytes {
            sync true('(RxEnb)' != 1'b0) @sample;
            wait cycle @drive;
            '(RxEmpty)' = 1'b1;
            '(RxData)'  = it;
            wait cycle @sample;
        };
        '(RxEmpty)' = 1'b0;
    };
};
```

<table>
<tr>
<td>

Sample 5-72.
Modifying the
drive event
</td>
<td>

```
extend phy_utopia1 {
    event drive is only {rise('(RxClk)');
                          delay(2)} @sim;
};
```
</td>
</tr>
</table>

Asynchronous Bus-Functional Models in *e*

TCM synchroniza-
tion event is only a
default.

If a TCM requires a synchronization event in its declaration, and
that synchronization event is used in all of *wait* or *sync* actions in
the TCM, how can you model asynchronous behavior using *e*? The
TCM synchronization event is provided only as a convenience
when describing an synchronous bus-functional model. It becomes
the default sampling event for temporal expressions in the *wait* and
sync statements instead of the *sys.any* event. But the TCM synchronization event is only a default. As shown in Sample 5-71, you can
specify a different temporal sampling event explicitly.

Use @*sim* as tem-
poral sampling
events.

When detecting edges on asynchronous signals, simply use the
@*sim* pre-defined temporal sampling event, as shown in Sample 5-
73. If you define an explicit event for the relevant asynchronous
edges, be careful to specify that event as the sampling event when
waiting for that edge, as in Sample 5-74. If you do not, as in Sample
5-75, the sampling event will default back to the TCM synchronization event.

<table>
<tr>
<td>

Sample 5-73.
Asynchronous
TCM in *e*
</td>
<td>

```
unit cpu_if {
    ...
    write(wadd: uint (bits: 24),
          wdat: uint (bits: 16)) @sys.any is {
    '(addr)' = wadd;
    '(data)' = wdat;
    '(wr)'   = 1'b1;
    wait rise('(rdy)') @sim;
    '(wr)'   = 0;
    '(data)' = 16'hZZZZ;
    wait fall('(rdy)') @sim;
};
```
</td>
</tr>
</table>

Use *sys.any* for the
TCM event.

Using explicit temporal sampling events for *wait* and *sync* actions
solves the synchronization problem once inside the TCM. But what
about the initial implicit *sync* action at the beginning of the TCM?
When invoking a TCM, if its synchronization event has not been
emitted in the current simulation cycle, it will wait until that event

Sample 5-74.
Waiting for
asynchronous
event

```
unit cpu_if {
    . . .
    event is_rdy is rise('(rdy)') @sim;
    event is_ovr is fall('(rdy)') @sim;
    write(wadd: uint (bits: 24),
          wdat: uint (bits: 16)) @sys.any is {
        . . .
        wait cycle @is_rdy;
        . . .
        wait cycle @is_ovr;
    };
```

Sample 5-75.
Improperly
waiting for
asynchronous
event

```
unit cpu_if {
    . . .
    event is_rdy is rise('(rdy)') @sim;
    event is_ovr is fall('(rdy)') @sim;
    write(wadd: uint (bits: 24),
          wdat: uint (bits: 16)) @sys.any is {
        . . .
        wait @is_rdy;
        . . .
        wait @is_ovr;
    };
```

is emitted. This waiting is not desirable for asynchronous bus-functional models where there is no such synchronization event.

To prevent a TCM from waiting at all, use the *sys.any* event as the TCM synchronization event, as shown in Sample 5-73. Because the *sys.any* event is always emitted in all simulation cycles, the TCM will never wait. This technique goes against a guideline (that is almost guaranteed to exist in your *e* coding guidelines) that says never to use the *sys.any* event. This guideline is designed to prevent inefficient temporal sampling, such as the one found in Sample 5-75. By not specifying an explicit temporal sampling event, the temporal expression will default to *sys.any*, which is the highest frequency event in *e*. Sampling a temporal expression more often than required quickly decreases simulation performance. But guidelines are only guidelines. If you know and understand what you are doing, they can be ignored safely.

In an asynchronous bus-functional model, always specify an explicit temporal sampling event for all *wait* and *sync* actions. The same inefficient model can be rendered very efficient by specifying the lowest frequency signal possible as the temporal sampling event. In Sample 5-74, the very edge we were waiting for is used as the temporal sampling event, which will occur exactly once. It's hard to find a lower frequency than that.

Configurable Bus-Functional Models

Protocols can have configurable elements.

A protocol specification may contain configuration options. For example, the assertion level for a particular control signal may be configurable to either high or low. Each option has a small impact on the operation of the interface. Taken individually, you could create a different task or procedure for each configuration. The problem would be relegated to the testcase in deciding which flavor of the operation to invoke. You would also have to maintain several nearly identical models.

Simple configurable elements become complex when grouped.

Taken together, the number of possible configurations explodes factorially.[10] It would be impractical to provide a different procedure or task for each possible configuration. It is much easier to include configurability in the bus-functional model. An RS-232 interface, shown in Figure 5-23, is the perfect example of a highly configurable, yet simple interface. Not only is the polarity of the parity bit configurable, but also its presence, as well as the number of data bits transmitted. And to top it all, because the interface is asynchronous, the duration of each pulse is also configurable. Assuming eight possible baud rates, five possible parities, seven or eight data bits, and one or two stop bits, there are 160 possible combinations of these four configurable parameters.

Figure 5-23. Specification for the RS-232 interface

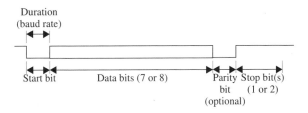

10. Exponential growth follows a K^n curve. Factorial growth follows a n! curve, where n! = 1 x 2 x 3 x 4 x ... x (n-2) x (n-1) x n.

Write a config-
urable bus-func-
tional model.

Instead of writing 160 flavors of the same transaction, it is much easier to model the configurability itself, as shown in Sample 5-76. Configuration parameters tend to remain static during an entire simulation (unless the corresponding design can be re-configured on-the-fly, and this on-the-fly reconfiguration is the objective of the test). They must also be consistent across different bus-functional procedures within the same bus-functional model. Rather than passing "constant" information through the interface of each bus-functional procedure, it is better located in the bus-functional model encapsulating structure (see "Encapsulating Bus-Functional Models" on page 137) alongside the bus-functional procedures where it can be accessed directly.

For Verilog, configuration parameters would be implemented as variables in the encapsulating *module*. For *e*, they would be implemented as fields in the encapsulating *unit*. For OpenVera, they would be implemented as properties in the encapsulating *class*. The only exception is VHDL: Because there can be only one instance of a package in a simulation, a shared variable would not support different configurations for different instances of the interface. The configuration parameters must be passed through the interface of each procedure using a record type argument. What important safety measure is missing from Sample 5-76?[11]

RESPONSE MONITORS

Response transac-
tions can be encap-
sulated.

Earlier in this chapter, we encapsulated input transactions to abstract the stimulus generation from individual signals and waveforms to generating sequences of operations. A similar abstraction can be used for verifying the response. The repetitiveness of output signals within a transaction can be taken care of and verified inside the bus-functional model. Then the testbench only needs to worry about the correctness of the data carried by the transaction. Sample 5-77 is an example of a bus-functional procedure for an RS-232 monitor.

---

11. The time consuming method should be protected using a semaphore to prevent concurrent access to the interface signals. See "From Bus-Functional Procedures to Bus-Functional Model" on page 272.

Sample 5-76.
Model for a
configurable
bus-functional
model

```
unit rs_232 is {

    baud_rate: uint [1200, 2400, ..., 115960];
    parity   : [NONE, ODD, EVEN, MARK, SPACE];
    n_bits   : uint [7..8];
    n_stops  : uint [1..2];

    send(data: byte) @sys.any is {
        var duration: uint = 1000000 / me.baud_rate;

        `(tx)' = 1'b0;
        wait delay(duration);
        for i from 0 to me.n_bits-1 {
            `(tx)' = data[i];
            wait delay(duration);
        };
        if (me.parity != NONE) {
            if (n_bits == 7) data[7] = 1'b0;
            case (me.parity) is {
                ODD  : `(tx)' = ~^data;
                EVEN : `(tx)' = ^data;
                MARK : `(tx)' = 1'b1;
                SPACE: `(tx)' = 1'b0;
            };
            wait delay(duration);
        };
        `(tx)' = 1'b1;
        wait [me.n_stops] * delay(duration);
    };
    ...
};
```

Verifying the data
in the response
monitor is too
restrictive.

The response verification operation, as encapsulated in Sample 5-77, has a very limited application. It can be used only to verify that the output value matches a pre-defined expected value, with no parity error. Can you imagine other possible uses? What if the output value can be any value within a predetermined set or range? What if the output value is to be ignored until a specific sequence of output values is seen? What if the output value, once verified, needs to be fed back to the stimulus generation? What if the parity value is expected to be incorrect? What if data values were to be ignored if the parity bit is invalid? The usage possibilities are endless. It is not possible, a priori, to determine all of them nor to provide a single interface that satisfies all of their needs.

Sample 5-77.
RS-232 serial
receive bus-
functional
checker

```
class rs_232 {
...
task recv(bit [7:4] expect) {
    integer    period = 1000000 / this.baud_rate;
    bit [7:0] data;
    integer    i      = 0;

    @ (negedge this.sigs.rx async); // Wait 4 start
    delay(period / 2);      // Sample mid-pulse
    data[7] = 1'b0;         // Handle 7 data bits
    for (i = this.n_bits-1; i >= 0; i--) {
        delay(period);
        data[i] = rx;        // 7-8 data bits
    }
    if (data != expect) ...
    if (this.parity != this.NONE) then  // Parity?
        delay(period);
        this.sigs.rx == calc_parity(...) async;
    }
    delay(period);          // Stop bit
    rx == '1' async;
}
```

Separate monitor-
ing from value
verification.

The most flexible implementation for an response transaction bus-functional procedure is simply to return to the caller whatever output value was just received. It will be up to a "higher authority" to determine if this value is correct or not. The RS-232 receiver was modified in Sample 5-78 to return the byte received without verifying its correctness. The correctness of the parity is also returned, this time using a parameter passed by reference (equivalent to an *out* parameter in VHDL or Verilog).

Consider all possi-
ble failure modes.

The bus-functional procedure shown in Sample 5-78 has some potential problems and limitations. What if the output signal being monitored is dead and the start bit is never received? This procedure will wait forever. It may be a good idea to provide a maximum delay to wait for the start bit via an additional argument, as shown in Sample 5-79, or to compute a sensible maximum delay based on the baud rate. Notice how a default argument value is used in the procedure definition to avoid forcing the user to specify a value when it is not relevant, as shown in Sample 5-79, or to avoid modifying existing code that was written before the additional argument was added.

```
function bit [7:0] recv(var bit ok) {
    integer period = 1000000 / this.baud_rate;
    integer i = 0;

    @ (negedge this.sigs.rx async); // Wait 4 start
    delay(period / 2);      // Sample mid-pulse
    recv[7] = 1'b0;         // Handle 7 data bits
    for (i = this.n_bits-1; i >= 0; i--) {
        delay(period);
        recv[i] = rx;       // 7-8 data bits
    }
    ok = 1;
    if (this.parity != this.NONE) then  // Parity?
        delay(period);
        ok = (this.sigs.rx === calc_parity(...));
    }
    delay(period);          // Stop bit
    rx == '1' async;
}
```

```
function bit[7:0] recv(var bit ok,
                       integer timeout = 0) {
    ...
    if (timeout > 0) {
        fork
        {
            delay(timeout);
            ok   = 0;
            recv = 8'hXX;
            return;
        }
        join none
    }
    @ (negedge this.sigs.rx async);
    terminate;
    ...
end recv;
```

Do not arbitrarily
constrain the trans-
action.

The width of pulses is not verified in the implementation of the RS-232 receive operation in Sample 5-78. Should it? If you assume that the procedure is used in a controlled 100% digital environment, then verifying the pulse width might make sense. This procedure also could be used in system-level verification, where the serial signal was digitized from a noisy analog transmission line as illustrated in Figure 5-24. In that environment, the shape of the pulse, although unambiguously carrying valid data, most likely does not meet the rigid requirements of a clean waveform for a specific baud rate. Just as in real life, where modems fail to communicate properly if their baud rates are not compatible, an improper waveform shape is detected as invalid data being transmitted.

Figure 5-24.
Modification
to the serial
signal in a real
system

Generation and monitoring pertains to the ability to initiate a transaction.

We have already seen that input transactions sometimes have to monitor some output signals from the design under verification. The same is true for a response monitor. Sometimes, the monitor has to provide data back as an answer to an "output" transaction. This reporting blurs the line between stimulus and response. Isn't a stimulus bus-functional procedure that verifies the control or feedback signals from the design also doing response checking? Isn't a monitor procedure that replies with control flow signals back to the design also doing stimulus generation? The terms *generator* and *monitor* become meaningless if they are attached to the direction of the signals being generated or monitored. They regain their meaning if you attach them to the initiation of transactions. If a procedure *initiates* the transaction under full control of the testbench, it is a *stimulus generator*. If the procedure sits there and waits for a transaction *to be initiated by the design*, then it is a *response monitor*.

Monitors must always be monitoring.

Bus-functional model procedures must be invoked by the testbench. Invoking a bus-functional procedure either initiates a stimulus transaction or initiates the expectation of a response transaction. What if the design happens to initiate a response transaction but the testbench had not called the appropriate bus-functional model response procedure? At best, the design will detect that the testbench is not ready to receive data because of some control flow signals were left at an appropriate level at the completion of the previous response transactions. But this would result in back-pressure building up inside the design and would not verify the design under maximum throughput. Typically however, output data would "spill" and a gap would be created in the output data stream. At worst, the design will fail to operate correctly because the output transaction protocol will be violated due to missing feedback signals. What if the testbench invokes the response procedure just a few cycles too late, after the design has already initiated a response transaction? A transaction protocol violation is likely to be reported. To avoid the false errors introduced by the misalignment of the response transaction in the testbench and the design, response monitors should always be active and monitoring the design output interface.

Autonomous Monitors

Decouple timing of transaction and timing of response checking.

Since the testbench is not responsible for the initiation of the response transaction, why give it the responsibility for the initiation of the response monitoring procedure? Testbenches are not usually interested in the timing of the output transaction; testbenches are interested only in verifying that the output data is correct. Therefore, we can decouple the monitoring of the physical interface signals from the retrieval of the output data. As illustrated in Figure 5-25, an independent thread continuously monitors the output transactions. Output data is extracted from each transaction and put into a FIFO. The testbench retrieves the next output data that was received from the front of the FIFO.

Figure 5-25.
Structure of an autonomous monitor

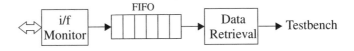

Add a concurrent thread in the bus-functional model.

The bus-functional response procedure is invoked within a concurrent thread running inside the bus-functional model, as shown in Sample 5-80. Because *module*, *unit* and *class* constructs in Verilog, *e* and OpenVera can contain execution threads (*always* block, time-consuming method *start*ed in the *run()* method and method called in a *fork/join none* statement in the constructor, respectively), it is very easy to implement within the bus-functional model encapsulation. The only problem is VHDL: Procedures are encapsulated using *packages*, but *packages* cannot contain *processes*. *Processes* can be encapsulated in *entities*, but *entities* cannot contain externally-callable procedures[12]. There is a (convoluted) way out that is described in "VHDL Test Harness" on page 325.

12. AAAAAAAAAAaaaaaaaaaaaaaaarrrrrrrrrrrggggggggggghhhhhhhhhh!

Sample 5-80.
Autonomous
RS-232
response monitor

```
unit rs_232 {
    !data_bfr: list of byte;
    rx(ok: *bool): byte @sys.any is ...;

    rx_daemon() @sys.any is {
        var data: byte;
        var ok   : bool;

        while TRUE {
            data = me.rx(ok);
            if (ok) {
                me.data_bfr.push(data);
            };
        };
    };

    run() is also {
        start me.rx_daemon();
    };
};
```

Output data is
buffered in a list.

As output data is extracted from the response transactions, it is added to a FIFO to be retrieved later by the testbench. For simple interfaces, such as RS-232, the buffered data is a simple byte. For more complex interfaces, such as SONET/SDH, the buffered data would be an entire frame. It is not possible to predict how many such data items will need to be buffered before we get the testbench's attention. Therefore, it must be accumulated in a list (see "Lists" on page 157) that will grow as data is collected from the output and shrink as data is retrieved by the testbench.

Data collection
could be optional.

What if, for a particular testcase, a testbench does not need to examine the output data from an interface? Data would accumulate in the FIFO, consuming an ever increasing amount of memory. Hopefully, the simulation would terminate before running out of memory—but that is not likely. An alternative is not to activate the monitoring process to avoid data from being accumulated in the first place. But what if the testcase in question causes a protocol violation on the output interface that we thought was not relevant? Because there is no active monitor on that interface, the error will go unnoticed. The better alternative is to have the monitor extract data from output transaction, verifying adherence to the protocol and providing necessary feedback signals but turn off data accumulation. As shown in Sample 5-81, protocol correctness will be monitored without data being accumulated.

```
class phy_utopia1 {
    local bit is_sink = 0;
    local integer cell_bfr;
    local function atm_cell rx();

    task rx_daemon() {
        while (1) {
            atm_cell cell = this.rx();
            if (cell != null && !this.is_sink) {
                mailbox_put(this.cell_bfr, cell);
            }
        }
    }

    task new(bit is_sink = 0) {
        this.is_sink = is_sink;
        this.cell_bfr = alloc(MAILBOX, 0, 1);
    }
}
```

Can provide
blocking or non-
blocking model.

To remove the testbench from knowing the detailed implementation of the data buffering mechanism, provide a procedure to retrieve the next data element that was received. This raises one question: What do we do when there is no data for the testbench? One solution is to wait for data to become available suddenly. But what if the testbench needs to turn its attention elsewhere while the retrieval procedure is stuck waiting for data? A better solution is to give the choice to the testbench whether to wait if there is no output data. If the testbench specifies a nonblocking retrieval mode and there is no available data, a suitable indication that no data was available must be returned. In Sample 5-82, a *null* reference is returned.

Decoupling
implies that trans-
action timing is
not relevant.

Decoupling the monitoring of the physical signals from the verification of the data output requires that the timing of the data not be relevant to determine functional correctness. As long as the data comes out, the design works. This decoupling is actually one of the great benefits of using behavioral testbenches: As the design is modified and pipeline stages are added or removed, the testbench need not be modified. But what if it is functionally important that data comes out after a specific (or range of) number of clock cycles? If it is important, it must be stated in the design specification document. If it is in the design specification document, it must be verified. If it must be verified, the testbench must be able to verify the timing of the output data.

```
unit rs_232 {
    event new_cell;
    ...
    rx_deamon() @sys.any is {
        ...
        while TRUE {
            cell = me.rx(ok);
            if (ok and not is_sink) {
                me.cell_bfr.push(cell);
                emit me.new_cell;
            };
        };
    };

    get_cell(blocking: bool): atm_cell @sys.any is
    {
        while (blocking and cell_bfr.size() == 0) {
            wait cycle @new_cell;
        };
        if (cell_bfr.size() == 0) {
            return null;
        };
        return cell_bfr.pop0();
    };
    ...
};
```

Transaction timing can be verified through time-stamping.

Verifying timing does not mean that it must be verified where and when the transaction started (or completed). Timing also can be verified by comparing timestamps. To satisfy the need of both types of testbenches, one where transaction timing is relevant, the other where it is not, timing information should be added to the extracted data. The testbench is then free to compare or ignore the timing information. There is one problem though: *Where* is the timestamp information stored? The output data structure is unlikely to have additional fields to store that information. Modifying the original data structure to add the necessary field may be a bad idea. If the original data structure potentially is reusable in other project or testbenches, you are adding project- or testbench-specific information to a common object. If everyone did the same, it quickly would grow into an unmaintainable mess that could not be trusted to be functionally correct. Instead, extend the original object to add your required information, leaving the original data structure intact. Sample 5-83 shows an extension and timestamping example.

Sample 5-83.
Timestamping
output data

```
class stamped_atm_cell extends atm_cell {
    bit [63:0] started;
    bit [63:0] completed;
}

class phy_utopia1 {
    ...
    local function stamped_atm_cell rx() {
        @1 this.sigs.$TxEnb == void;
        if (this.sigs.$TxEnb == 1'b1) {
            @ (negedge this.sigs.$TxEnb);
        }
        rx = new;
        rx.started = {get_time(HI), get_time(LO)};
        ...
        rx.completed = {get_time(HI), get_time(LO)};
    }
    ...
}
```

Slave Generators

Response monitors
may need to reply
with "input" data.

How do you verify an interface used by the design to fetch data? Typical examples include the instruction fetch interface on a processor or a EPROM interface on a design. The transactions are initiated by the design, not the testbench. Therefore, it falls under the category of response monitor. But the transactions do not produce any output data. Instead, they require and consume input data. It is the responsibility of the testbench to supply the data to complete the transaction in a timely manner. Of course, in those cases, the correctness of the data is implied. It will have to be verified elsewhere when it (or its descendent) shows up at another interface.

Slave generators
must "callback"
the testbench.

Because of the time-sensitive nature of the transaction, it is not possible to decouple the monitoring of the output interface and the generation of the reply data. The testbench must be ready to supply data at all times. The difficulty is how to get the testbench's attention when required. In software engineering terms, this is called a callback. VHDL and Verilog do not have any built-in callback mechanism. In *e*, the bus-functional model would provide a callback method designed to be extended as required by the testbench. In OpenVera, callback methods are implemented using *virtual* methods.

Emulate callbacks by calling a procedure in HDLs. In VHDL and Verilog, a callback mechanism can be emulated by having the testbench constantly call a "standby" procedure. The callback occurs when the procedure returns. The testbench can execute testcase-specific code to generate the input data that is immediately supplied by calling the standby procedure again. Sample 5-84 shows a bus-functional model monitoring the instruction fetch interface of a CPU. Rather than pre-generating the code before the simulation and statically loading it into a memory model, this bus-functional model lets the testbench dynamically generate instructions on-the-fly. Notice the "fetched" procedure. It is the standby procedure used by the testbench to provide the instruction opcode fetched from the specified address. Sample 5-85 shows how a testbench could use this standby procedure to implement a random instruction generator stream.

Sample 5-84.
Bus-functional model with standby procedure

```
module code_mem(addr, as, data, rdy);
...
event fetch, ready;

always
begin
    ...
    @ (negedge as);
    fetched.address = addr;
    -> fetch;
    @ (ready);
    data = fetched.opcode;
    rdy = 1'b0;
    ...
end

task fetched;
    output [31:0] address;
    input  [31:0] opcode;
begin
    -> ready;
    @ (fetch);
end
endtask
endmodule
```

Provide empty callback method in *e*. With aspect-oriented programming in *e*, you can extend a method within the bus-functional model to include testbench-specific code. The only detail is that the method must exist to begin with. A testbench can then extend this callback method at its leisure. The initial implementation of a callback method should provide an innocuous

Sample 5-85.
Using the
standby proce-
dure

```
module testbench()

code_mem pmem(...);

always
begin
    reg [31:0] addr, opcode;
    pmem.fetched(addr, opcode);
    opcode = generate_opcode(addr);
end
...
endmodule
```

input value so that it is not necessary for all testbenches to extend
the callback method if it is not relevant. The callback method is
called at the appropriate point in the bus-functional model, executing either the default or an extended implementation. In *e*, callback
methods should not use arguments to pass information back and
forth between the bus-functional model and the testbench.

Instead, "public" fields should be used. Because a method must be
extended by reproducing the entire method declaration *exactly*, any
modification in the interface to the callback method (such as adding, modifying or removing an argument) will require that all testbenches extending that method be modified to match. If all
arguments are passed via globally-visible fields, there is no argument to modify creating a more stable interface.

Using fields over arguments has another advantage: They can be
constrained. More details on constraint strategy will be discussed in
"Random Stimulus" on page 354. Sample 5-86 shows a similar
instruction fetch bus-functional model as in Sample 5-84. The callback method is extended in a testbench as shown in Sample 5-87.

Use *virtual* call-
back method in
OpenVera.
In OpenVera, callback methods are implemented using *virtual*
methods. The default implementation of a callback method should
be to return an innocuous value to eliminate the requirement that all
testbenches overload the callback method to render a bus-functional model functionally correct. Testbench-specific code can
replace the default implementation by providing an overloaded definition in a derived class. Information can be passed between the
bus-functional model and the callback method through arguments
or through public or *protected* properties. Sample 5-88 shows a

Sample 5-86.
Bus-func-
tional model
callback
method in *e*

```
unit code_mem is {

    !address: uint (bits: 32);
    !opcode : uint (bits: 32);
    ...
    monitor_daemon() @sys.any is {
        ...
        wait fall('(adstb)') @sim;
        me.address = '(addr)';
        me.fetch();
        data = me.opcode;
        '(rdy)' = 1'b0;
        ...
    }

    fetched() is {
        me.opcode = NOP.opcode();
    };
};
```

Sample 5-87.
Extending the
callback
method

```
import code_mem;
extend code_mem {
    !instr: instruction;
    fetched() is also {
        gen me.instr;
        me.opcode = me.instr.opcode()
    };
};
```

similar instruction fetch bus-functional model as in Sample 5-84. The callback method is extended in a testbench as shown in Sample 5-89.

Callback methods
create a reusable
slave generator.

Why bother with this callback method extensions and standby procedure calls? Why not simply go in the bus-functional model, add the code we need directly in there and be done with it? That would be the simple way out, but one that will create maintenance challenges later on. This approach makes one big assumption: that you have access to the source code to being with. If you wrote the bus-functional model yourself, you do. But it also could be a very large bus-functional model purchased from a third party who will only supply compiled or encrypted code to protect their interests. What if different testbenches need different extensions to the bus-functional model? Are you going to create a different copy for each? What about reusing that bus-functional model in the next revision

Sample 5-88.
Bus-func-
tional model
callback
method in
OpenVera

```
class code_mem {
    ...
    task monitor_daemon() {
        bit [31:0] opcode;
        ...
        @ (negedge this.sigs.$as async);
        opcode = this.fetched(this.sigs.addr);
        this.sigs.$data = opcode async;
        this.sigs.$rdy  = 1'b0   async;
        ...
    }

    virtual function
        bit [31:0] fetched(bit [31:0] addr) {
        fetched = NOP.opcode();
    }
}
```

Sample 5-89.
Overloading
the callback
method

```
#include "code_mem.vrh"
class my_code_mem extends code_mem {
    instruction instr;
    virtual function
        bit [31:0] fetched(bit [31:0] addr) {
        void = this.instruction.randomize();
        fetched = this.instr.opcode();
    }
}
```

of the project or in a different project altogether? By specializing a bus-functional model to the specific needs of your testbench(es), you have made reusing it and incorporating upgrades and bug fixes more difficult. You saved a little initially, but lost a lot more in the long run

Time may be
allowed to
advance in call-
back methods.

There was one question I conveniently ignored in the callback method discussion. Must they all execute in zero-time or is time allowed to advance in a callback method? The answer is: It depends. If the transaction protocol includes handshaking and flow-control indicators, it is possible to have time advance in the execution of a callback method. This technique would introduce delays in the response back to the design. Other transaction protocols may suffer a total breakdown if any delay is introduced, in which case the callback method must execute in zero-time. Even in such cases, it may be possible to allow *some* time to be advanced within a callback method as the response is often not required until the beginning of the next clock cycle.

e has a built-in mechanism for enforcing time restrictions: Use a TCM for callback methods that are allowed to advance time and use a regular method for callback methods that must execute in zero-time. With HDLs and OpenVera, you have to rely on the testbench to respond appropriately. To detect misuse of zero-delay or time-limited callback methods, provide a timer to detect callback method invocations that do not return within the appropriate time window, as shown in Sample 5-90.

Sample 5-90.
Timing duration of callback methods

```
class code_mem {
    . . .
    task monitor_deamon() {
        bit [31:0] opcode;
        . . .
        @ (negedge this.sigs.$as);
        fork
            opcode = this.fetched(this.sigs.addr);
            {
                delay(...);
                printf("Too much time in cllbck\n");
                exit(-1);
            }
        join any
        terminate;
        this.sigs.$data = opcode;
        . . .
    }
}
```

Multiple Possible Transactions

The next transaction on an output interface may not be predictable.

You may be in a situation where more than one type of transaction can happen on an output interface. Each would be valid and you cannot predict which specific operation will come next. An example would be a processor that executes instructions out of order. You cannot predict (without detailed knowledge of the processor architecture) whether a read or a write cycle will appear next on the data memory interface. The functional validity is determined by the proper access sequence to related data locations.

Sample 5-91.
Processor test
program

```
load A,   R0
load B,   R1
add  R0,  R1,  R2
sto  R2,  X
load C,   R3
add  R0,  R3,  R4
sto  R4,  Y
```

Verify the
sequence of
related operations.

For example, consider the testcase composed of the instructions in Sample 5-91. It has many possible execution orders. From the perspective of the data memory, the execution is valid if the conditions listed below are true.

1. Location A is read before location X and Y are written.

2. Location B is read before location X is written.

3. Location C is read before location Y is written.

4. Location X must be written with the value A+B.

5. Location Y must be written with the value A+C.

These are the sufficient and necessary conditions for a proper execution of the test program. Verifying for a particular order of the individual cycles over-constrains the testcase, unless the specification demands in-order instruction execution.

Use a transaction
descriptor.

How do you write a response monitor when you do not know what kind of transaction comes next? You must write a bus-functional procedure that identifies the next transaction after it has started. It verifies the preamble to all transactions on the output interface until it becomes unique to a specific transaction. It then builds a transaction descriptor containing any information collected so far to identify, to the testbench via the callback procedure, which transaction is currently underway. It is then up to the testbench to supply the necessary (and correct) information to complete the verification of the transaction.

Sample 5-92 shows a response monitor bus-functional procedure that identifies whether the next transaction from the design is a read or a write cycle. Since the address already has been sampled by the time the decision of the type of cycle was made, the address is returned along with the current cycle type. These two values make up the transaction descriptor. Small transaction descriptors can be implemented using discrete values. More complex transaction descriptors should be implemented using records (see "Records" on page 146). Sample 5-93 shows how this transaction descriptor is used by the testbench to determine the next course of action.

Sample 5-92.
Monitoring many possible output transactions

```
module ram_if(addr, cs, ale, rw, data, rdy);
  . . .
  always
  begin
      while (cs == 1'b1) @ (negedge ale);
      mem_cycle.address  = addr;
      mem_cycle.is_write = (rw == 1'b1);
      mem_cycle.data     = data;
      -> do_cycle;
      @ (cycle_done);
  end

  task mem_cycle;
      output [31:0] address;
      output        is_write;
      inout  [31:0] data;
  begin
      @ (do_cycle);
      ->cycle_done;
  end
  endtask
endmodule
```

Sample 5-93.
Handling many possible output operations

```
initial
begin: test_procedure
    reg [31:0] addr;
    reg [31:0] data;
    reg        is_write;

    mem_cycle(addr, is_write, data);
    case (is_write)
    0: data = read_cycle(addr);
    1: write_cycle(addr, data);
    endcase
    . . .
end
```

TRANSACTION-LEVEL INTERFACE

Testbenches are removed from the physical level.

As illustrated in Figure 5-26, the purpose of bus-functional models is to remove the testbench from the repetitive physical-level details. The bus-functional model lets the testbench concentrate on the data to be supplied, on the data that was produced and how it is supposed to have been transformed. Once your have a reliable set of bus-functional models, it makes writing testbenches faster and easier. In this section, I will describe how to design a transaction-layer interface. The next chapter will describe how to structure a testbench on top of a transaction layer.

Figure 5-26. Transaction layer testbench

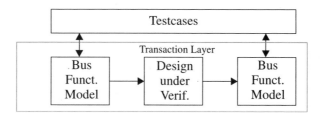

The transaction interface must be designed.

Throughout this chapter, the procedural interface of the bus-functional models evolved from the particular transaction being discussed. It was optimized according to the particular point I was trying to make while presenting advantages and disadvantages of various alternatives. Each procedure was written and evolved independently of each other. The purpose of this chapter was the generation and monitoring of physical-level signals, not the design of a transaction interface. Using this process to write a complete bus-functional model will likely result in an awkward and clumsy transaction interface dictated by the physical-level details, not the requirements of the testbenches that must use it. The transaction interface of a bus-functional model must be designed and planned, just like the testcases or the design.

Declare.

When designing a bus-functional model, write the transaction interface first. This is akin to writing the *h* file in C. Code the signal, variable, procedure, task, function and method declarations that make up the entire transaction interface of your bus-functional model. Leave the body or the implementation of each procedure empty. This style will let you focus on the transaction interface across all of the transactions. A good starting point is one procedure per transaction. But as you start thinking about the needs of the testcases, their number may grow or shrink. This is the stage where you address questions like "Should this method be blocking?" and "How much work can I do without the testbench's attention?" You have to strike a balance between abstraction and controllability. You have to consider the requirements of each testcase, the self-checking mechanism and functional coverage measurement.

Document.

Next, write the user documentation for the bus-functional model. Yes, documentation. It is the only way to ensure that the documentation will reflect the content of the bus-functional model accurately and that it will exist at all. It also presents an opportunity to think about the purpose of each element of the transaction interface. The documentation will have to describe the functionality and interaction of each element, often highlighting inconsistencies or difficulties that were not considered when coding the interface. Review and iterate over the declaration and the documentation until your have specified a bus-functional model that will meet all of your requirements.

Implement.

With the declaration and documentation of each procedure completed, the implementation of the bus-functional model becomes a simple coding exercise. During the implementation, you will discover misunderstanding in the specification of the physical interface specification. You will encounter functionality that cannot be implemented as intended. You will find inconsistencies in the bus-functional model specification. Update the transaction interface *and* the documentation as required.

Variable-Length Transactions

Transactions can transfer a variable number of bytes.

Most of the transactions used so far were simple fixed-size transactions. The amount of data transmitted to or from the design was identical in each occurrence of the transactions. For example, an RS-232 interface always transmits a single byte. Most physical interfaces nowadays deal with variable-length data. For example, all ethernet interfaces deal with MAC frames between 64 and 1,518 bytes long. In a PCI interface, the maximum number of bytes that can be transferred in a read or write cycle is not even specified. A variable-length protocol can be built on top of a fixed-length physical interface. For example, a PPP transaction over an RS-232 link will transmit and receive variable-length packets, one byte at a time. How do you design a transaction interface that can handle those differences? The easiest solution is to provide enough memory for the largest amount of data and a "length" specification indicating how many data elements are actually valid. But always specifying the maximum number of data is inefficient and wastes memory.

Use a list or an array.

In OpenVera or *e*, the response is simple: Use a list or an array. Sample 4-35 shows the implementation of a MAC frame, using an array for the variable data. Sample 5-67 shows that same variable-length MAC frame used as an argument of a procedure, making it possible to send MAC frames of any (even invalid) lengths efficiently. The high-level capabilities of these languages make designing a high-level transaction interface a breeze.

Similarly, for VHDL and Verilog, you can use the list emulation techniques shown in "Lists" on page 157. In VHDL, you may be tempted to use unconstrained arrays, but those only work for input data. In VHDL, the size of an unconstrained array argument is determined by the caller and cannot be changed. For output data, where the amount of data is unknown, a priori, it would not be possible to increase or decrease the size of the array from within the procedure to match the amount of data received. The only remaining possibility is using an *access* to an unconstrained array type. But that also requires knowing, a priori, the amount of data that will be received so the correct-size array can be allocated. Some protocols transmit the data size in the early portion of the transaction, but many simply transmit data for as long as they wish and you have to live with the amount of data received.

In Verilog, use a buffer with length indication.

In Verilog, since list emulation requires the declaration and allocation of the maximum-size list, you might as well save yourself the trouble and use an maximum-size array with a length indication. Unfortunately, arrays cannot be passed through interfaces. Instead, locate the array as a static variable inside the task implementing the bus-functional procedure. Make sure this procedure it not automatic to make the buffer static and globally visible. See Sample 5-94 for an example. The testbench can access the variable-length data through a hierarchical reference to the data buffer, as shown in Sample 5-95.

Sample 5-94.
Variable-
length transac-
tion interface
in Verilog

```
module eth_mii(...);
  ...
task rx;
    reg [7:0]    frame [0:1517];
    reg integer frame_length;
begin
    ...
end
endtask

endmodule
```

Sample 5-95.
Using a vari-
able-length
transaction
interface in
Verilog

```
module testbench;
  ...
eth_mii mii(...);

always
begin
   mii.rx;
   if (mii.frame_length !== expected_length) ...
   for (i = 0; i < mii.frame_length; i = i + 1) {
       if (mii.frame[i] !== expected[i]) ...
   }
end

endmodule
```

Split Transactions

Transactions may
be composed of
subtransactions.

Many high-performance bus protocols have split transactions. For example, a "read" transaction could be composed of a separate address and data tenures. The bus master can perform the address tenure. While the target device performs the work and buffers data, the master can perform other bus transactions to other devices. The master either polls or is interrupted by the first device when it is ready to complete the read transaction. The master then performs the second tenure, transferring data from the device, completing the read transaction. The same target device may be able to handle several split and non-split transactions concurrently. Split transactions may also include out-of-order completion.

Provide tenure
procedures.

Whatever transaction-layer interface you decide on, you will have to perform those tenures. They should be implemented as separate subprograms or methods. As long as they exist, you should make them public to enable a testbench to have detailed control over the atomic tenures on the physical interface. For example, a testbench may need to create a specific sequence of tenures and transactions to exercise a particular corner case. If the testbench is too far removed from the physical interface, it may not be possible to create. The low-level tenure procedures should be used only by the very few testcases responsible for verifying the physical interface logic.

Provide complete transaction procedure.

For any verification project, the bulk of the testcases are concerned with higher-level functionality, not physical interface logic. They do not depend on the split nature of the transaction and do not require detailed control of the physical interface. Having to deal with the tenures would make writing those testbenches tedious and cumbersome. You should provide an interface to the split transaction that makes it look like an atomic operation. Internally, it would execute the tenures as required. But from the testbench's perspective, it would appear like an ordinary transaction. The procedure would wait until the split transaction completes and returns with the completed data and status.

This is not sufficient to support random transaction generation.

Providing a transaction-layer interface that includes direct access to the tenures and a completed split transaction is fine for directed testcases. What if you wanted to verify the interface operation by performing random sequences of ordinary and split transactions, performing as many overlapping transactions concurrently as possible? A random transaction generator would decide which procedure to call. To be able to perform random transactions in the middle of a split transaction, it would be necessary to use the low-level tenure procedures. This would require that the random generator, after having generated the start of a split transaction, generates any necessary completion polling tenure and eventually the split transaction termination tenure. All of that intermixed with other split and ordinary transactions. It is possible to write such a random transaction generator. But it would likely be impossible to constrain (see "Random Stimulus" on page 354 for more details on writing constrainable generators).

Provide nonblocking complete transaction procedure.

To support a random transaction methodology, you should provide a complete nonblocking split transaction procedure. I use the term nonblocking because it may not necessarily execute in zero-time. The procedure may very well wait while the initial tenure of the split transaction is applied. But it is nonblocking in that it would not wait for the completion of the split transaction. It would return instead with an indication of the status of the split transaction—whether it was accepted by the target device—and a mechanism for alerting the testbench that the transaction has been completed. Sample 5-96 shows an implementation in *e* of an interface to a split read transaction. The response of the nonblocking procedure is a reference to a transaction completion descriptor object instance that contains a status indication, a list that will later contain the data returned and an event that will be emitted when the transaction is completed. Internally, it would manage all of the details of completing the transaction, even with the presence of other transactions. Sample 5-97 shows how such a procedure would be used. It is easy to fork separate threads that will wait for the completion of the split transaction and verify their correctness.

Sample 5-96.
Nonblocking
split transac-
tion procedure

```
struct split_read_response {
    status: [DECLINED, PENDING, RETRY, ABORT, OK];
    data  : list of byte;
    event completed;
};

unit bfm {
    ...
    split_read(addr: uint (bits: 24);
               len : uint): split_read_response
    is {
        result = new;
        if (!me.setup_split_read(addr, len)) {
            result.status = DECLINED;
            return;
        };
        result.status = PENDING;
        me.register_split_read(addr, len, result);
    };
};
```

Retries and Completion Status

Transactions may
be retried.

In many protocols, transactions may fail not because they are invalid but because one of the parties involved in the transaction is busy or is out-of-sync. The transaction should be retried at a later time. Only after a certain number of retries is a transaction considered failed. Should the bus-functional model do the retrying or should you let the testbench worry about it?

Sample 5-97.
Using a non-blocking trans-action proce-dure

```
var resp: split_read_response;

resp = master0.split_read(0x000FFF, 32);
if (resp.status == PENDING) {
    start check_split_response(0x000FFF, 32, resp);
};
...
check_split_response(addr: uint (bits: 24);
                     len : uint;
                     resp: split_read_response)
@sys.any is {
    sync cycle @resp.completed;
    if (data.size() != len) {
        dut_error(...);
        return;
    }
    ...
}
```

Let the transaction procedure do the retry.

The bus-functional model should make its best possible effort to complete a transaction. That includes retrying transactions that did not complete initially Testbenches should not be burdened with the repetitive retry operations. On the other hand, a testbench may need to have control over the number of retries, or whether to even allow a transaction to be retried. The transaction-layer interface of trans-actions that can be retried should have a parameter specifying the maximum number of attempts. Once the transaction has been tried for the specified number of attempts, it is considered failed. A test-bench that does not wish a transaction to be retried would simply specify a single attempt. If the language supports it, provide a default value for the number of attempts (usually specified in the protocol) that will not need to be specified for each invocation—only when a different value is required. Sample 5-98 shows an example of an MII ethernet transaction-layer interface with control over the number of transmission attempts.

Symbol-Level Control

Transactions are composed of phys-ical symbols.

For most interfaces, the transaction data is not transmitted in paral-lel, in a single cycle. Instead, the transaction data is divided into "symbols" transmitted sequentially over multiple cycles. For exam-ple, a byte transmitted over an RS-232 physical interface is trans-lated into one-bit symbols. A 1 KB PCI memory read is translated into 256 32-bit symbols. Some symbols may be added to the trans-

Sample 5-98.
Retried trans-
action

```
class mii {
    ...
    function bit send(mac_frame frame,
                      integer   attempts = 10) {
        while (attempts-- > 0) {
            ...
            send = 1;
            return;
            ...
        }
        send = 0;
    }
}
```

action data by the protocol for framing, synchronization, or error protection. For example, a MAC frame is prefixed with an 8-byte preamble. A USB packet is prefixed with an 8-bit synchronization pattern.

Symbol-level
parameters must
be controllable.

For each symbol, an interface protocol usually has several controllable parameters such as symbol-level flow control and status indication. But a transaction-layer interface deals with information for the entire transaction, not individual symbols. This is fine when verifying functionality that resides behind the interface. But when verifying the implementation of the interface itself, it is necessary to have detailed control over all relevant symbol-level parameters. One way would be to provide a symbol-layer interface that could be used instead of the transaction-layer interface. This technique would require writing completely different testbenches to verify the physical interface from the ones used to verify higher-level functions because the testbenches would use the symbol-layer interface. This style is fine in a directed methodology, but in a random-based methodology you will not be able to leverage the same generation and self-checking environment for both classes of testcases.

Provide a symbol-
level callback
method.

A better approach is to extend the callback method approach from the transaction level ("Slave Generators" on page 299) to the symbol level. Before transmitting or receiving a symbol, a callback method should be called with the necessary arguments or global variables to let a testbench modify the default symbol-level behavior. This callback method can let time advance if the protocol supports symbol-level control flow. Otherwise, the callback method must complete in zero-time. For example, Sample 5-99 shows the symbol-level callback method for a slave PCI memory read interface. It controls symbol-level flow control by introducing time advances that will delay the assertion of the *target-ready* signal. It also controls byte-enable indications, has the possibility of aborting the transaction or signaling a parity error for this symbol.

Sample 5-99.
Symbol-level
callback
method

```
struct pci_symbol {
    data : uint;
    be   : uint (bits: 4);
    abort: bool;
    perr : bool;
};
unit pci_slave {
    ...
    pre_symbol_tx(symbol: pci_symbol) @sys.any is
        empty;
    ...
    mem_read_tx() @pci_clk is {
        ...
        while (...) {
            ...
            '(trdy)' = 1'b1;
            me.pre_symbol_tx(sym);
            if (sym.abort) {
                '(abrt)' = 1'b0;
                return;
            };
            '(trdy)'    = 1'b0;
            '(data)'    = symbol.data;
            '(cmd_be)'  = symbol.be;
            ...
        };
        ...
    };
};
```

Use the symbol-level callback to inject errors.

From the transaction-layer, it is simple to inject transaction-level errors such as a bad CRC or a packet that is too short. But how can you inject errors at the symbol level, such as corrupting a symbol, violating the handshake protocol or unexpectedly terminating a transaction? As long as you have control over the symbol-level parameters in a callback method, why not take this opportunity to use the symbol-level callback method to inject symbol-level errors?

Simply add parameters for the errors that can be injected. By default, they are set to not inject errors. If they are not modified in the callback method, no errors will be injected. Sample 5-100 shows the symbol-level callback method for a MII ethernet interface. From that callback method, it is possible to corrupt the symbol, cause the TX_EN signal to be deasserted, cause the TX_ER signal to be asserted or abort the frame altogether. When implementing a bus-functional model, it is necessary to provide every mechanism for breaking the protocol that the design should be able to sustain.

Sample 5-100.
Symbol-level
callback
method with
error injection

```
class mii_symbol {
    bit [3:0] data;
    bit        tx_en;
    bit        tx_er;
    bit        abort;
};
class eth_mii {
    ...
    virtual task pre_symbol_tx(mii_symbol symbol);

    task tx(mac_frame frame) {
        ...
        while (...) {
            ...
            this.pre_symbol_tx(sym);
            if (sym.abort) {
                this.sigs.$tx_en = 1'b0;
                return;
            }
            this.sigs.$txd   = sym.data;
            this.sigs.$tx_en = sym.tx_en;
            this.sigs.$tx_er = sym.tx_er;
            ...
        }
        ...
    }
}
```

SUMMARY

Model your clock signals in VHDL or Verilog. Be careful about time resolution issues, delta cycle alignment and implicit synchronization of asynchronous signals.

Encapsulate repetitive physical-level operations into bus-functional procedures that provide an effective transaction-level interface.

Understand the interface model of your HVL and use it in the way it was intended to be used. Provide mechanisms for users to modify the timing of input and output operations with the lowest possibility of introducing errors.

Provide callback methods in response monitors to request reply data. Provide callback methods in bus-functional procedures to enable access to symbol-level protocol parameters and inject symbol-level errors.

Collect all of the bus-functional procedures for a physical interface into a bus-functional model. Detect concurrent activation of bus-functional procedures within the same bus-functional model using a semaphore.

CHAPTER 6   ARCHITECTING TESTBENCHES

A testbench need not be a monolithic block. Although Figure 1-1 shows the testbench as a large thing that surrounds the design under verification, it need not be implemented that way. The design is also shown in a single block, and it is surely not implemented as a single unit. Why should the testbench by any different? Figure 6-1 depicts the architecture of a generic testbench. In this chapter, I will describe how to implement each component.

Figure 6-1.
Typical
testbench
architecture

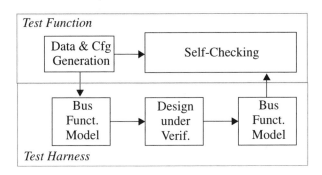

The previous chapter was about low-level testbench components.

In Chapter 5, we focused on the generation and monitoring of the low-level signals going into and coming out of the device under verification. I showed how to abstract them into transactions using bus-functional models. The emphasis was on the stimulus and response of interfaces and the need for managing separate execution threads underneath a useful procedural interface. If you prefer a bottom-up approach to writing testbenches, I suggest you start with the previous chapter.

This chapter focuses on the structure of the testbench.

This chapter concentrates on implementing the many testcases and filling the functional coverage models that were identified in your verification plan. I show how best to structure the bus-functional models into a transaction-level test harness. This test harness will create the platform on top of which the self-checking structure and stimulus sources will be built. I also describe how to create random generators that can be constrained easily, with a minimum of modifications, from testbench to testbench.

A hypothetical design will be used to illustrate the concepts.

Figure 6-2 shows the interfaces around a hypothetical ATM switch node design. A management interface allows a processor to read and write internal registers to configure the switch node. This design will be used throughout this chapter to illustrate important concepts.

Figure 6-2.
4x4 ATM
switch design

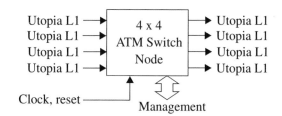

TEST HARNESS

Create a transaction layer test harness.

All of the testbenches have to interface through an instantiation to the same design under verification. It is safe to assume that they all require the use of the same bus-functional models to generate stimulus and to monitor response. Instead of a monolithic block, the testbenches should be designed with a layer of transaction-level bus-functional models. This transaction-level layer, common to all testbenches for the design under verification, is called the *test harness*. The *test functions* required to implement the testcases identified in the verification plan are built on top of the test harness, as illustrated in Figure 6-3. The *test function* and the *harness* together form a *testbench*.

Encapsulate the test harness.

The encapsulation of a transaction-level test harness in Verilog, OpenVera or *e* is relatively simple. The bus-functional model

Figure 6-3.
Structure of a
transaction-
level testbench

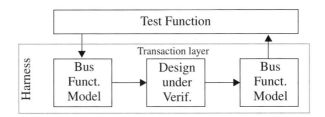

instances and their connectivity to the design under verification are
encapsulated in a *module, class* or *unit,* respectively. The various
test functions then instantiate the encapsulated test harness. The
high-level interface of each bus-functional model is accessed using
a hierarchical name. Sample 6-1 and Sample 6-2 show the public
interface of the bus-functional models required to interface to the
ATM switch node. They were designed and implemented using the
techniques described in the previous chapter. The transaction-level
test harnesses, encapsulating these bus-functional models as illustrated in Figure 6-4, are shown in Sample 6-3, Sample 6-4 and
Sample 6-5.

Sample 6-1.
Utopia Level 1
bus-functional
model in
OpenVera

```
class atm_cell {
    ...
}
port utopia_L1_port {
    clk;
    data;
    soc;
    enb;
    clav;
}
class utopia_L1 {
    task new(utopia_L1_port tx_sigs,
             utopia_L1_port rx_sigs);
    task send(atm_cell cell);
    function atm_cell receive();
}
```

The test harness
includes every-
thing needed to
operate the design.

The test harness should be self-contained and provide all signals
necessary to operate the design under verification properly. In addition to all the low-level bus-functional models, the test harness
should include the clock generators and reset procedure. Notice
how the OpenVera and *e* test harnesses do not include the clock
generator. As mentioned in "Random Generation of Reference Sig-

Sample 6-2.
Management
interface bus-
functional
model in *e*

```
unit mgmt_if {
    addr: string;
    data: string;
    rd  : string;
    wr  : string;
    rdy : string;

    write(wadd: uint (bits: 16),
          wdat: uint (bits: 16))
        @sys.any is empty;

    read(radd: uint (bits: 16)): uint (bits: 16)
        @sys.any is empty;
}
```

Figure 6-4.
Test harness
for 4x4 ATM
switch design

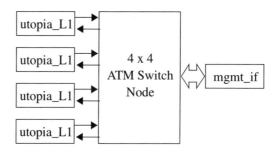

nal Parameters" on page 239, they should be generated in the HDL simulation. An HVL test harness is always tightly coupled with an HDL test harness. The HDL test harness instantiates the design, provides signals to be monitored and driven by the HVL and supplies clock and constant signals to the design. The HDL test harness for the *e* test harness shown in Sample 6-5 is shown in Sample 6-6.

Test functions use
procedures in the
test harness.

A complete test harness provides a transaction-layer abstraction of the design to be verified. It provides a foundation on which the data generation mechanism, the self-checking structure and the functional coverage measurements are built. Test functions are implemented by using the high-level interface elements in the bus-functional models and test harness itself. These interface elements are accessed using their hierarchical names in a single instance of the test harness. Sample 6-7 shows a partial test function that configures the device then injects an ATM cell in one of the ports.

Sample 6-3.
Verilog test
harness for
ATM switch
node

```
module harness;
wire TxClk_0,  TxClk_1,  TxClk_2,  TxClk_3;
...
wire RxClav_0, RxClav_1, RxClav_2, RxClav_3;
wire [15:0] addr, data;
wire        rd, wr, rdy;
reg  reset, clk;

switch_node dut(TxClk_0, ..., clk, reset);

utopia_L1 atm_port_0(TxClk_0, ..., RxClav_0);
utopia_L1 atm_port_1(TxClk_1, ..., RxClav_1);
utopia_L1 atm_port_1(TxClk_2, ..., RxClav_2);
utopia_L1 atm_port_2(TxClk_3, ..., RxClav_3);

mgmt_if   cpu(addr, data, rd, wr, rdy);

task reset;
   ...
endtask

always
begin
   #5 clk = 0;
   #5 clk = 1;
end

endmodule
```

Sample 6-4.
OpenVera test
harness for
ATM switch
node

```
#include "utopia_L1.vrh"
#include "mgmt_if.vrh"
#include "switch.if.vri"
class harness {
    utopia_L1 atm_port[4];
    mgmt_if   cpu;
    task new() {
        this.atm_port[0] = new(tx_0, rx_0);
        this.atm_port[1] = new(tx_1, rx_1);
        this.atm_port[2] = new(tx_2, rx_2);
        this.atm_port[3] = new(tx_3, rx_3);
        this.cpu = new(mgmt_0);
    }
    task reset();
}
```

Sample 6-5.
e test harness
for ATM
switch node

```
import utopia_L1;
import mgmt_if;
unit harness {
    atm_port[4]: list of utopia_L1 is instance;
    cpu        : mgmt_if is instance;
    keep me.hdl_path() == "~/top";
    keep for each in atm_port {
        it.TxClk == appendf("TxClk_%d", index);
        ...
        it.RxClav == appendf("RxClav_%d", index);
    };
    keep cpu.addr == "addr";
    ...
    keep cpu.rdy  == "rdy";

    reset() @sys.any is empty;
};
```

Sample 6-6.
Verilog har-
ness for the *e*
test harness

```
module top;
reg  TxClk_0,  TxClk_1,  TxClk_2,  TxClk_3;
...
wire RxClav_0, RxClav_1, RxClav_2, RxClav_3;
reg  [15:0] addr;
wire [15:0] data;
reg         rd, wr;
wire        rdy;
reg         reset, clk;

switch_node dut(TxClk_0, ..., clk, reset);

always
begin
    #5 clk = 0;
    #5 clk = 1;
end

endmodule
```

<table>
<tr>
<td>

Sample 6-7.
Test function
using a test
harness

</td>
<td>

```
module my_test;

harness th();

initial
begin: test_function
    reg 'ATM_CELL_TYP cell;
    th.write(16'h0001, 16'h0010);
    ...
    th.atm_port_0.send(cell);
    ...
    $finish;
end

endmodule
```

</td>
</tr>
</table>

VHDL TEST HARNESS

Bus-functional procedures are encapsulated in a package.

As described in "Encapsulating Bus-Functional Models" on page 137, VHDL bus-functional procedures are encapsulated in procedures with *signal-class* arguments. These procedures can be used outside of the *package* by each testbench that requires them. Sample 6-8 shows the package declaration of bus-functional model procedures for the Intel 386SX processor. Notice how all the signals for the processor bus are required as *signal*-class arguments in each procedure.

Bus-functional model procedures are cumbersome to use.

Sample 6-9 shows a process using the procedures declared in the package shown in Sample 6-8. They are very cumbersome to use as all the signals involved in the transaction must be passed to the bus-functional procedure. Furthermore, there would still be a lot of duplication across multiple testbenches. Each would have to declare all interface signals, instantiate the component for the design under verification and properly connect the ports of the component to the interface signals. With today's ASIC and FPGA packages, the number of interface signals that need to be declared, then mapped, can easily number in the hundreds. If the interface of the design were to change, even minimally, all testbenches would need to be modified.

<table>
<tr><td></td><td>

```
package i386sx is

subtype add_typ is std_logic_vector(23 downto 0);
subtype dat_typ is std_logic_vector(15 downto 0);

procedure read(         raddr: in     add_typ;
                        rdata: out    dat_typ;
                 signal clk  : in     std_logic;
                 signal addr : out    add_type;
                 signal ads  : out    std_logic;
                 signal rw   : out    std_logic;
                 signal ready: in     std_logic;
                 signal data : inout dat_typ);

procedure write(        waddr: in     add_typ;
                        wdata: in     dat_typ;
                 signal clk  : in     std_logic;
                 signal addr : out    add_type;
                 signal ads  : out    std_logic;
                 signal rw   : out    std_logic;
                 signal ready: in     std_logic;
                 signal data : inout dat_typ);
end i386sx;
```

</td></tr>
</table>

<table>
<tr><td></td><td>

```
use work.i386sx.all;
architecture test of bench is
    signal clk  : std_logic;
    signal addr : add_type;
    signal ads  : std_logic;
    signal rw   : std_logic;
    signal ready: std_logic;
    signal data : dat_typ;
begin

    duv: design port map (..., clk, addr, ads,
                          rw, ready, data, ...);

    testcase: process
        variable data: dat_typ;
    begin
        ...
        read(some_address, data,
             clk, addr, ads, rw, ready, data);
        ...
        write(some_other_address, some_data,
             clk, addr, ads, rw, ready, data);
        ...
    end process testcase;
end test;
```

</td></tr>
</table>

A package is the wrong encapsulation mechanism.

It is not possible to use the same test harnessing strategy in VHDL as with the other languages. Packages cannot be instantiated in an architecture because they are not a structural element. They must somehow first be encapsulated into an entity. Furthermore, VHDL's strict scoping rules make it impossible to invoke a procedure in an entity using a hierarchical name.

Bus-Functional Entity

An entity is a better encapsulation mechanism.

Finding a way to encapsulate bus-functional procedures into an entity would have several advantages:

- It would be possible to have multiple instances of the bus-functional model, each with its own internal state information.

- It would be possible to include processes, allowing the inclusion of autonomous behavior and nonblocking procedural interfaces, as described in "Autonomous Monitors" on page 295.

- It would eliminate having to specify all of the interface signals in each procedure call. They would be mapped once, when the entity is instantiated in the test harness.

Use transaction-level signals to control the bus-functional entity.

Because VHDL does not allow hierarchical access into entities, it will be necessary to use signals on the port of the entity to control the bus-functional procedures embedded in it. Figure 6-5 depicts the control structure of a bus-functional entity in VHDL. The physical interface signals flow through the entity's port and are connected to the design under verification. An additional pair of signals is used to exchange transaction-level information between the test function and the bus-functional procedures. Whether multiple processes or a single process is used to implement the functionality of the bus-functional model depends on the required features of the transaction-level interface.

This creates a client/server relationship.

The relationship between the test function and the bus-functional entity is a *client/server* relationship. Control signals are used to exchange transaction-level information between the client and the server. Typically, the test function (the client) initiates a transaction by causing an event on the control signal to the bus-functional entity (the server). The test function then waits for the bus-functional entity to indicate the completion and status of the transaction through an event on the other control signal. A single control signal could be used, but the signal must be resolved.

Figure 6-5.
Control
structure of a
VHDL bus-
functional
entity

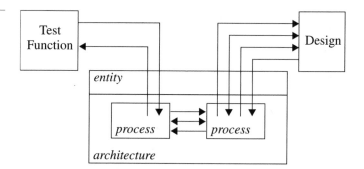

Control signals
should be of
record type.

The control signals must be able to transport all of the information that completely describes a transaction to be executed and the result of the transaction once it has been completed. This is a perfect application for *records*. They can collect data of different types under a single structure. Even if a transaction can be described using a single scalar value, it is a good idea to use a record anyway: A record will allow the addition of other transaction description information without requiring changes to the existing test functions or test harness. Sample 6-17 shows the package declaring the control record types for the 386SX bus-functional entity.

Sample 6-10.
Client/server
control signal
types

```
package i386sx_pkg is

type kind is (read, write);

type to_srv_ctl is record
   do   : kind;
   addr: integer range 0 to 16#FFFF#;
   data: integer range 0 to 16#FFFF#;
   go   : boolean;
end record;

type frm_srv_ctl is record
   data: integer range 0 to 16#FFFF#;
   go   : boolean;
end record;

end package i386sx;
```

Use a toggling
Boolean to force
an event.

Notice how each control record in Sample 6-10 contains a Boolean field named *go*. This field is toggled by the client or the server to force the occurrence of an event on the signal. This way, the process waiting on the other side of the control signal will be triggered after every assignment to the control signal, even if the transaction description is identical to the previous one. Had the toggling Boolean field not been present, the second of two consecutive identical operations, as shown in Sample 6-11, would be missed.

Sample 6-11.
Performing
two identical
operations
back-to-back

```
to_srv <= (do   => write,
           addr => (others => '1'),
           data => (others => '0'));
wait on frm_srv;
to_srv <= (do   => write,
           addr => (others => '1'),
           data => (others => '0'));
```

The bus-functional entity contains a server process that waits for events on the input control record. Such an event indicates that a client requires a transaction to be performed. The process performs the transaction, then returns any result information back to the client by creating an event on the output control signal. Sample 6-12 shows a simple server process for the i386SX bus-functional entity.

Abstracting the Client/Server Protocol

The client must operate the control signals to the server process properly.

Sample 6-13 shows a client process accessing the services provided by the i386SX bus-functional entity shown in Sample 6-12. Notice how the client process waits for an event on the return signal to detect the end of the operation.

Encapsulate the client/server operations in procedures.

Defining a communication protocol on signals between the client and the server processes does not seem to accomplish anything. Instead of having to deal with a physical interface documented in the design specification, we have to deal with an arbitrary protocol with no specification. Just as the operation on the physical interface can be encapsulated, the operations between the client and server also can be encapsulated in procedures. This encapsulation removes the client process from knowing the details of the protocol with the server. The protocol can be modified without affecting the testcases using it through the procedures encapsulating the operations. The server access procedures should be located in the package containing the type definition and signal declarations. Their implementation is tied closely to these control signals and should be located

Sample 6-12.
Server process in bus-functional entity

```
use work.i386sx_pkg.all;
entity i386sx is
    port(clk    : in     std_logic;
         addr   : out    std_logic_vector;
         ads    : out    std_logic;
         rw     : out    std_logic;
         ready  : in     std_logic;
         data   : inout  std_logic_vector;
         to_srv : in     to_srv_ctl;
         frm_srv: out    frm_srv_ctl);
end entity i386sx;

architecture server of i386sx is
begin
    process
        variable rdat   : integer;
        variable toggle: boolean = TRUE;
    begin
        wait on to_srv;
        case (to_srv.do) is
        when read =>
            read(to_srv.addr, rdat,
                 clk, addr, ads, rw, ready, data);
        when write =>
            write(to_srv.addr, to_srv.data,
                  clk, addr, ads, rw, ready, data);
        end case;
        frm_srv <= (data => rdat, go => toggle);
        toggle := not toggle;
    end process;
end architecture server;
```

with them. Sample 6-14 shows how the *read* and *write* access procedures would be added to the package previously shown in Sample 6-10.

Client processes use the server access procedures.

The client processes are now free from knowing the details of the protocol between the client and the server. To perform an operation, they simply need to use the appropriate access procedure. The pair of control signals to and from the server must be passed to the access procedure to be driven and monitored properly. Sample 6-15 shows how the client process, originally shown in Sample 6-13, is now oblivious to the client/server protocol.

The testcase must still pass control signals to and from the bus-functional access procedures. So, what has been gained from the starting point shown in Sample 6-9? The answer is: a lot.

Sample 6-13.
Client process
controlling the
bus-functional
server

```
use work.i386sx_pkg.all;
architecture test of bench is
begin
    i386sx_client: process
        variable data  : integer;
        variable toggle: boolean := TRUE;
    begin
        . . .
        -- Perform a read
        to_srv <= (do   => read,
                   addr => ...;
                   go   => toggle);
        toggle := not toggle;
        wait on frm_srv;
        data := frm_srv.data;
        . . .
        -- Perform a write
        to_srv <= (do   => write;
                   addr => ...;
                   data => ...;
                   go   => toggle);
        toggle := not toggle;
        wait on frm_srv;
        . . .
    end process i386sx_client;
end test;
```

Sample 6-14.
Client/server
access pack-
age

```
package i386sx_pkg is
. . .
procedure read(       addr  : in  natural;
                      data  : out natural;
                signal to_srv : out to_srv_ctl;
                signal frm_srv: in  frm_srv_ctl);

procedure write(      addr  : in  natural;
                      data  : in  natural;
                signal to_srv : out to_srv_ctl;
                signal frm_srv: in  frm_srv_ctl);
end package i386sx_pkg;
```

Testbenches are
now removed from
the physical
details.

No matter how many signals are involved in the physical interface, you need only pass two signals to the bus-functional access procedures. The testcases are completely removed from the physical interface of the design under verification. Pins can be added or removed and polarities can be modified without affecting the existing testcases. It also becomes possible to create a test harness as shown in Figure 6-4.

Sample 6-15.
Client process using server access procedures

```
use work.i386sx_pkg.all;
architecture test of bench is
begin
    i386sx_client: process
        variable data: integer;
    begin
        ...
        -- Perform a read
        read(..., data, to_srv, frm_srv);
        ...
        -- Perform a write
        write(..., ..., to_srv, frm_srv);
        ...
    end process i386sx_client;
end architecture test;
```

Test Harness

The test harness contains declarations and functionality common to all testbenches.

The test harness contains elements common to all testcases. These common elements for a single design are:

- Declaration of the interface signals
- Instantiation of the design under verification
- Instantiation of the bus-functional entities
- Mapping of interface signals to the ports of the design
- Mapping of interface signals to the ports of bus-functional entities

Sample 6-16 shows the outline of a test harness using the i386SX bus-functional entity connected to a i386SX slave device.

Use global control signals.

The control signals have to be visible to both the client process in the top-level testcase architecture and the bus-functional instance in the test harness. This visibility can be accomplished in two ways:

- Passing them as ports on the test harness entity
- Making them global signals in a package

Using global signals eliminates the need for any signal to flow through the test harness entity. It makes instantiating the test harness in each test function extremely simple as there are no interface signals to declare nor map. Should the control signals be large records, it will be more efficient to use a single shared variable to

```
use work.i386sx_pkg.all;
architecture test of harness is
    signal clk    : std_logic := '0';
    signal addr   : std_logic_vector;
    ...
    signal to_srv : to_srv_ctl;
    signal frm_srv: frm_srv_ctl;

begin
    dut: entity xyz    port map (clk, ...);
    cpu: entity i386sx port map (clk, ...,
                                 to_srv, frm_srv);

    process
    begin
        wait for 5 ns;
        clk <= not clk;
    end process;
end architecture server;
```

transfer the transaction data in both directions and use a pair of simple Boolean signals to handshake between the client and the server.

```
use work.i386sx_pkg.all;
package th is

signal to_srv : to_srv_typ;
signal frm_srv: frm_srv_typ;

end package th;
```

In a test harness for a real design, there may be a dozen different bus-functional entities, each with different control signals and access procedures. A real-life client test function, creating a complex testcase, uses all of them. It may be difficult to trace the source of each identifier and ensure that all identifiers are unique across all access packages. In fact, making identifier uniqueness a requirement would place an undue burden on the authoring and reusability of these packages.

The identifier tracing and collision problems can be eliminated by using qualified names when using control record types and access procedures. Sample 6-18 shows a test function using qualified names to access the *read* procedure out of the *i386sx_pkg* package. Notice how the *use* statement for the package does not specify *.all* to make all of the identifiers it contains visible.

<table>
<tr>
<td>

Sample 6-18.
Client tests
function using
qualified iden-
tifiers

</td>
<td>

```
use work.i386sx_pkg;
use work.th;
architecture test of bench is
begin
    i386sx_client: process
        variable data: integer;
    begin
        . . .
        -- Perform a read
        i386sx_pkg.read(..., data,
                        th.to_srv, th.frm_srv);
        . . .
    end i386sx_client;
end test;
```

</td>
</tr>
</table>

Multiple Server Instances

Provide an array of
control signals for
multiple instances
of the same server
processes.

Designs often have multiple instances of identical interfaces. For example, the ATM switch node example design has four Utopia Level 1 input and output ports, all using the same physical protocol. Each can be stimulated or monitored using separate instances of the bus-functional entity. The client test functions need to have a way to identify which instance they want to operate on to perform operations on the proper port on the design.

Using an array of control signals, one pair for each server, meets this requirement. Sample 6-19 shows the test harness package containing a global array of control signals, while Sample 6-20 shows the instances of the bus-functional entity, each driven using a different pair of control signals.

<table>
<tr>
<td>

Sample 6-19.
Array of cli-
ent/server con-
trol signals for
multiple bus-
functional
entity
instances

</td>
<td>

```
use utopia_L1_pkg;
package th is
. . .

type to_uL1_ctl_ary  is array(integer range <>)
    of utopia_L1_pkg.to_srv_ctl;
type frm_uL1_ctl_ary is array(integer range <>)
    of utopia_L1_pkg.frm_srv_ctl;

signal to_uL1  : to_uL1_ctl_ary (0 to 3);
signal frm_uL1 : frm_uL1_ctl_ary(0 to 3);

end package th;
```

</td>
</tr>
</table>

Sample 6-20.
VHDL test
harness for
ATM switch
node

```
use work.th;
architecture test of harness is
   ...
begin
   dut: entity switch_node port map (clk, ...);

   U0: entity utopia_L1
         port map (..., to_srv(0), frm_srv(0));
   U1: entity utopia_L1
         port map (..., to_srv(1), frm_srv(1));
   U2: entity utopia_L1
         port map (..., to_srv(2), frm_srv(2));
   U3: entity utopia_L1
         port map (..., to_srv(3), frm_srv(3));
   ...
end architecture test;
```

You may be able to
use the *for-gener-*
ate statement.

If the physical signals for the multiple instances of a port are declared properly using arrays, a *for-generate* statement can be used to replicate the instances of the bus-functional entity automatically. Sample 6-21 illustrates this replication.

Sample 6-21.
Generating
multiple
instances of a
server process

```
use work.rs232.all;
architecture test of bench is
   signal TxClk: std_logic_vector(0 to 3);
   ...
begin

   dut: entity switch_node
         port map(TxClk_0 => TxClk(0), ...);

   U: for I in TxClk'range generate
      L1: entity utopia_L1
            port map (..., to_srv(I), frm_srv(I));

   end generate U;
end test;
```

Testbench genera-
tion tools can help
in creating the test
harness and access
packages.

If you are discouraged by the amount of work required to implement a VHDL test harness and access packages, remember that it will be the most leveraged verification code. It will be used by all testbenches so investing in implementing a test harness that is easy to use returns the additional effort many times. Testbench generation tools, such as Quickbench by Forte Design Systems, can automate the generation of the test harness from a graphical specification of the timing diagrams describing the interface operations.

DESIGN CONFIGURATION

Most designs require configuration.

Unless you are verifying a very simple design, or an implementation unit of a much larger design, it will be necessary to perform certain configuration operations before it will be possible to apply data to and observe data from the design. Configuration may be as simple as enabling some data path, or it may be as complicated as generating, then downloading, firmware code. It may involve writing to internal registers, writing to an embedded memory or setting external pins to particular levels.

Avoid using one or two configurations.

Because of the often complex nature of a device configuration, it is not unusual for verification to proceed with only one or two device configurations. Device configurations are maintained as a series of register write operations that "magically" produce a configuration. The testbenches are then written according to the configuration usually loaded. Unfortunately, this will likely prevent some bugs from being exercised. Some unexpected correlation may exist between different configuration parameters. If these parameters are not exercised, the correlation will not be highlighted.

Abstracting Design Configuration

Model the configuration.

Instead of relying on an implicit knowledge of the current design configuration, why not create a formal model of the configurable elements of the design? That model could then be used by the self-checking structure (see "Self-Checking Testbenches" on page 341) to determine the correctness of the response. For example, a design could have an input pin that can be used to select between two different management interfaces. As illustrated in Sample 6-22, you can use an enumerated type to model the currently selected interface. That enumerated type can then be passed to the test harness to instantiate the proper bus-functional model based on the interface configuration. The test harness would also use the value to determine the polarity used to drive the interface selection pin.

Do not model the implementation of the configuration.

Even though the configuration of the device is expressed in terms of 1's and 0's in various bit fields in various registers, it is not necessary to use the same approach when modeling a device configura-

```
type mgmt_if_mode [INTEL, MOTOROLA];

struct device_cfg {
   mgmt_if: mgmt_if_mode;
};

unit harness {
    cfg: device_cfg;
    cpu: mgmt_if is instance;
    keep cpu.mode == me.cfg.mgmt_if;

    when INTEL harness {
       run() is also {
          'int_mot' = 1'b1;
       };
    };

    when MOTOROLA harness {
       run() is also {
          'int_mot' = 1'b0;
       };
    };
};
```

tion. Instead of maintaining an image of the register values, model
the purpose and function of the configuration. A high-level description of the device configuration will be much easier to use in the
self-checking structure and won't necessitate the interpretation of
low-level bit fields.

For example, the configuration of the switch table in the example
ATM switch node design could be implemented as a series of bits in
a register. If bit x in register y is set, then any cell with a VPI value
equal to y is forwarded to output port x. As shown in Sample 6-23,
the same information can be modeled in a more abstract fashion by
using an array of a list of integers. Cells with a VPI value of y are
forwarded to all ports whose number is found in the list at index y
of the array.

Sample 6-23.
Modeling con-
figuration
function, not
implementa-
tion

```
class to_ports {
   bit [1:0] number[];
}
class device_cfg {
   to_ports switch_table[256];
}
```

Collect all device configuration information in a single record.

As shown in Sample 6-22 and Sample 6-23, it is good practice to collect all device configuration information under a single descriptor object. This technique makes it easier to pass it to the test harness and the self-checking structure. Collecting device configuration information will also make it possible to create constraints and relationships between various configuration items and include methods to ensure internal consistency.

Configuring the Design

Compile the configuration description into bit fields.

Once the device configuration is captured in an instance of the configuration descriptor, it will be necessary to configure the design to match. This step will necessitate the translation of the various configuration items into the appropriate bit field values in the appropriate registers. This step may seem like a daunting task—and it usually is—but it is simply formally coding the process you would have to perform intellectually otherwise. This translation will provide for a better documentation of the device configuration process. This translation will also ensure that configuring the design to match is repeatable, should the location of various bits fields be reorganized or their encoding modified. Sample 6-24 shows how the switch table in Sample 6-23 could be compiled into the corresponding bit field values in the corresponding registers.

Sample 6-24.
Translating a high-level configuration descriptor in OpenVera

```
class device_cfg {
    to_ports switch_table[4];

    task apply(cpu: mgmt_if) {
        integer i, j, ok;
        bit [15:0] entry;

        for (i = 0; i < 256; i++) {
            for (ok = assoc_index(FIRST,
                        switch_table[i].number, j);
                 ok;
                 ok = assoc_index(NEXT,
                        switch_table[i].number, j))
            {
                entry[switch_table[i].number[j]] = 1;
            }

            cpu.write(16'h0800 + i, entry);
        }
    }
}
```

Design Configuration

Grow the configu-
ration capability.

It is not necessary to implement the translation process of the entire configuration descriptor from day one. The first simulations will likely be performed using a simple device configuration, leaving the bulk of the configuration parameters in their default state. Therefore, translate only that part of the configuration descriptor that is relevant for these first simulations. As more and more configuration parameters are being verified or are supported by the self-checking structure, they should be added to the translation process similarly. Eventually, you will end up with the entire configuration descriptor appropriately translated and programmed into the device.

Assert that unsup-
ported configura-
tions are not used.

While the configuration translation process does not support certain configuration parameters, you must ensure that they are not accidentally used in a simulation. The translation procedure must check that all unsupported configuration parameters are at their default values. Sample 6-25 shows the translation process for the management interface configuration signal. Because one of the modes is not currently available (because the bus-functional model may not be ready yet), it will report an error if the unsupported configuration is attempted.

Sample 6-25.
Detecting
unsupported
configurations

```
type mgmt_if_mode [INTEL, MOTOROLA];

struct device_cfg {
    mgmt_if: mgmt_if_mode
};

unit harness {
    cfg: device_cfg;
    cpu: mgmt_if is instance;
    keep cpu.mode == me.cfg.mgmt_if;

    when INTEL harness {
        run() is also {
            dut_error("Intel mgmt i/f not avail!");
        };
    };

    when MOTOROLA harness {
        run() is also {
            'int_mot' = 1'b0;
        };
    };
};
```

Random Design Configuration

Randomize the configuration.

Once you have a descriptor capable of coherently describing any possible configuration of your design, why bother specifying it manually? You'd probably always be specifying the same configuration anyway, which is exactly the problem we were trying to avoid. If you can generate random instructions, packets or data items, why not generate a random device configuration as well? By using a different randomly generated device configuration in each simulation run, you quickly will cover many more combinations. If an unintended correlation between parameters exists, it is likely to be exposed.

Add constraints to match limitations.

But what about the limitations of your configuration translation procedure? If an unsupported configuration is generated, it will cause an error to be reported. You should maintain a set of constraints that match the current limitation in the device configuration support. As more and more of the configuration parameters are supported, constraints are removed. Sample 6-26 shows the constraints that would be added to all simulation runs temporarily to prevent a configuration, unsupported by the test harness shown in Sample 6-25, from being generated.

Sample 6-26. Support limitation constraints

```
extend device_cfg {
    keep mgmt_if != INTEL;
};
```

Add constraints to generate simple debug configurations.

Likewise, a completely random configuration is not likely to be useful for the first simulations. In the early stage of a project, a design will contain many functional bugs. They will be easier to identify and debug if a simple configuration is used. The first simulations should be executed with constraints on the configuration descriptor to generate a simple configuration. Once the design simulates successfully, these constraints are removed to increase immediately the number of configuration combinations that can be verified.

Use functional coverage to identify configurations that were verified.

If the device configuration is generated randomly, how do you know which configurations you have verified? Simple: A device configuration is treated just like a feature of the design. All interesting and relevant configurations should be identified in the verification plan. They should be included in the functional coverage model of the design. Functional coverage measurements will identify which configurations were indeed verified and the ones that remain to be verified.

Randomize system configuration.

If your design can be used in different system configurations, why limit the testbench structure to just one of the possibilities? The configuration of the system under verification can be generated randomly as well. For example, a USB hub design could be surrounded by a randomly selected set of devices. Some devices would be low speed, others full speed. Some could have asynchronous endpoints, others have multiple interfaces with alternate settings. Some devices actually could be another instance of the USB hub with further devices connected to it. Based on the randomly generated system configuration, the necessary instances of the design under verification are created and connected to the necessary instances of bus-functional models.

SELF-CHECKING TESTBENCHES

Testbenches must be self-checking.

As discussed in "Verifying the Response" on page 95, visually inspecting simulation results to determine functional correctness is not an acceptable long-term strategy. Whatever intellectual process you would go through to identify an error visually in the simulation result must be coded in your testbench. This technique will let the testbench detect errors and declare success or failure on its own. Coding error detection into your testbenches will free you to work on other tasks while the design is autonomously subjected to hundreds of simulations.

Define what to check.

The problem with verification is that you cannot find an error where you are not looking. It is therefore necessary, during the verification planning stage, to identify all of the failure modes that must be checked for and how they can be detected. Typical correctness criteria include data transformation, data ordering, protocol correctness, data losses and design state. The requirement of the self-checking mechanism must be specified and reviewed to ensure that a potential failure will not go undetected.

It will be the most complex portion of your testbench.

After the completion of a project, you will find that the largest, most complex component of the testbenches is the self-checking structure. It will have been the portion that required the most authoring and maintenance effort. The self-checking structure is also the most critical portion as it is responsible for declaring the functional correctness of the design. It will embody a duplication of the specified functionality of the design under verification.

Why is this section so short?

If the self-checking structure is the most complex and largest portion of a verification project, why is it such a small portion of this book? That's because the bulk of the functionality in the self-checking structure will be to model the expected functionality of the design under verification. That is unique to every design and cannot be described in a generic fashion in a book. Each class of function requires different approaches and different mechanisms for identifying failures. Each class of design could be the topic of its own book.

General implementation techniques are presented.

This section presents various techniques for implementing the self-checking structure. Which one to use depends on the class of design under verification. The techniques can also be used in combination. Some techniques depend on the availability of reference models. Others rely on the availability of unmodified data payloads. Once you have specified the requirements of the self-checking structure, use the necessary techniques to implement them.

Hard Coded Response

Some techniques require hard-coded stimulus and configuration.

The self-checking strategy used to verify the muxed flip-flop in "Self-Checking Testbenches" on page 256 relied on hard coded response checking. The function and configuration of the device under verification was very simple. The response could be checked for each individual input value. To hardcode a response in a testbench requires a known configuration and a known input stream. It is therefore only applicable to directed testcases. Sample 6-27 shows the pseudo-code for a directed testcase with a hardcoded response on the ATM switch node design. The objective of this testcase is to verify that cells from every input port can be switched to every output port.

It must be replicated for each testcase.

Each testcase is supposed to verify a different feature of the design. This requirement needs a different configuration or a different input data stream. This setup will yield a different response. If a hard-coded response strategy is used, it will be necessary to replicate the response checking in each testbench.

Sample 6-27.
Pseudo-code
for hardcoded
response

```
Program configuration:
   for x in 0..3:
      vpi x -> output port #x
for out_port in 0..3:
   fork
   {
      for in_port in 0..3:
         generate atm cell with:
            vpi == out_port;
            vci == in_port;
            send cell on port(in_port);
   }
   {
      for in_port in 0..3:
         wait for cell on port(out_port);
         assert cell.vpi == out_port;
         assert cell.vci == in_port;
   }
   join
```

Errors can slip
through easily.

Because the response being checked is crafted to the testcase, it tends to ignore other potential problems. It is assumed that the other functions operate correctly and that any problem would be caught by the testcase targeting those functions. Should an unexpected correlation or corner case exist, it will likely go undetected if it is accidentally created in a testcase that focuses on different features.

Data Tagging

Packets have
untouched data
fields.

Many designs use some of the input information for processing, sometimes transforming it, but leave other portions of the input untouched and forward it, intact, all the way through the design to an output. Examples abound in the datacom industry. They include ethernet hubs, IP routers, ATM switches and SONET framers.

Use the untouched
fields to encode
the expected tran-
formation.

The portion of the data input that passes untouched through the design under verification can be put to good use. It is often called *payload* and the term *packet* or *frame* often is used to describe the unit of data processed by the design. You must first determine, through a proper check, that the payload information is indeed not modified by the design. Subsequently, the payload information can be used to describe the expected destination, position and transformation for this packet. For each packet received, the output monitor uses the information in the payload to determine if the packet was processed appropriately.

This simplifies the testbench control structure.

This self-checking strategy usually lends itself to the simplest self-checking structures. All of the intelligence is located in independent output monitors. The control of this type of testbench is simple because all the processing (stimulus and specification of the expected response) is performed in a single location: the stimulus generator. Some minor orchestration between the generators may be required in some testcases when it is necessary to synchronize traffic patterns to create interesting scenarios. Figure 6-6 shows the structure of a testbench using data tagging to verify the example switch node design example.

Figure 6-6.
Testbench structure for the switch node design

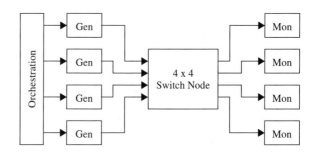

Include all necessary information in the payload to determine functional correctness.

The payload must contain all necessary information to determine if a particular packet came out of the appropriate output, in the proper sequence and with the appropriate transformation of its control information. For example, assume the success criteria is that the packets for a given input stream be received in the proper order by the proper output port. The payload should contain a unique stream identifier, a sequence number and an output port identifier, as shown in Figure 6-7.

Figure 6-7.
Example packet payload structure

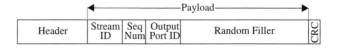

The output monitor needs to verify that the output identifier matches its own identifier. It also needs to verify that the sequence number is equal to the previously received sequence number in that stream plus one, as outlined in Sample 6-28. A CRC value is used to verify that the payload was indeed not modified by the design.

Sample 6-28.
Implementation using payload information to determine functional correctness

```
while (1) {
    atm_cell cell;

    cell = th.uL1_0.receive();
    // Cell was corrupted?
    if (cell.payload[47] !== cell.payload_crc()) {
        ...
        continue;
    }
    // Cell is for this port?
    if (cell.payload[0] !== my_id) ...;
    // Packet in correct sequence?
    if (last_seq[cell.payload[1]] + 1 !=
        {cell.payload[2], cell.payload[3]}) ...;
    // Reset sequence number
    last_seq[cell.payload[1]] =
        {cell.payload[2], cell.payload[3]};
}
```

Use data tagging in collaboration with scoreboarding.

Should it be possible for a packet to have a payload too short to contain all of the tag information, another self-checking strategy— such as scoreboarding—must be used in concert with data tagging. When present in the payload, the tag information is used by the output monitor to quickly search the scoreboard to confirm correctness of the received object. When not available, the scoreboard is searched normally, to ensure that the received object is indeed expected. For more details on scoreboarding, see "Scoreboarding" on page 348.

Reference Models

You can use a reference model.

As illustrated in Figure 6-8, the reference model and the design under verification are subjected to the same stimulus and their output is monitored and compared for discrepancies constantly.

Figure 6-8.
Using a reference model to predict output

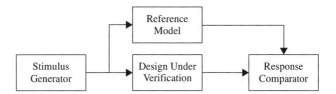

Reference models rarely exist.

The problem with this strategy is that reference models rarely exist. Reference models are available only during a re-design exercise where backward compatibility is required and when they form an integral part of the specification. Pure backward compatible re-designs are rare as the re-design is often used as an opportunity to increase performance, add to the number of ports or add new features. That only leaves reference models that exist as part of the specification.

It is a popular strategy for processors.

Some classes of designs are not fully specified on paper. Rather, they are specified using an executable model that was used to explore architectural and performance trade-offs. Because the model is the specification, it is golden by definition. It is the typical approach used for general-purpose and digital signal processors.

The model need not run concurrently.

Often, the difficulty of integrating the reference model with the design simulation prevents it from being simulated concurrently with the design. The output is thus compared in a post-processing step, as illustrated in Figure 6-9. The input can be generated externally similarly when the reference model includes a suitable data generator, as depicted in Figure 6-10.

Figure 6-9. External output comparison

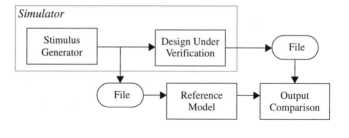

Figure 6-10. External input generation

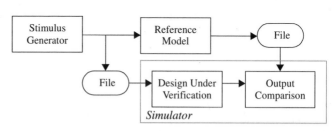

It is a force behind C-based verification.

The fact that these reference models are usually implemented in C or C++ is a force behind using C or C++ as a simulation language for the design and verification. It is thought that by using a common language the design and verification can proceed smoothly from system-level and architectural-specification down to detailed implementation. As I write these lines in the middle of an economic downturn it appears that economic conditions[1] more than methodology make this solution attractive. Time will tell whether it proves to be the solution everyone claims it is.

Transfer Function

Model the data transformation.

A transfer function is used to reproduce any data transformation performed by the design to determine which output value to expect, as illustrated in Figure 6-11. The transfer function is often used in concert with a scoreboard (see "Scoreboarding" on page 348). The transfer function uses the design configuration descriptor to perform the same transformation. Data transformation is not limited to computation and modification of fields and values inside each data item. Data transformation also includes the generation of new data items (for example IP segments from a TCP packet), the identification of ordering and destination for the data item and computation of the next state of the design when executing an instruction.

Figure 6-11. Reproducing the transformation to predict output

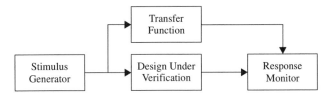

It's not the same as a reference model.

Isn't a transfer function the same thing as a reference model? Figure 6-11 sure looks like Figure 6-8![2] The answer is no, for several reasons. First, a transfer function does not exist a priori. Second, it is not golden by definition. As simulations will be run and errors reported, there will be as many errors in the transfer function as in the design itself. Third, a transfer function does not have low-level interfaces. When a packet or an instruction is sent from the stimulus generator to the transfer function, it does not use a physical-level protocol and low-level signals. Instead, each data item is passed as an atomic object using a procedural interface. Similarly,

1. Most C simulation tools are free.
2. In fact, one was cut and pasted from the other!

the transformed data is stored into the scoreboard or forwarded to the response monitor using a high-level procedural interface.

Scoreboarding

A scoreboard is a data structure.

The definition of scoreboard is definitely not standardized across the industry. For some, it is the entire self-checking structure, including the transfer function or reference model, the expected data storage mechanism and the output comparison function. In this book, the definition of scoreboard is limited to the data structure used to hold the expected data for ease of comparison against the monitored output values.

A scoreboard holds expected data.

A scoreboard is a data structure that holds data expected to be received by the output monitor. As illustrated in Figure 6-12, the transfer function adds data to the scoreboard. Any data received by the output monitor is compared against the data in the scoreboard. If an identical data item is found, the design produced the expected response. If an identical data item is not found in the scoreboard, an error is reported. At the end of the simulation, any data items left in the scoreboard were lost in the design—which may or may not be an error.

Figure 6-12. Scoreboarding

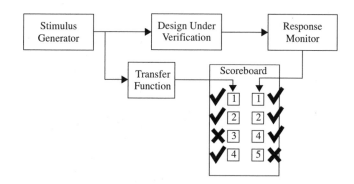

The data structure depends on the self-checking requirements.

Just as there is no single definition of scoreboard, there isn't a single scoreboard kind or structure. Each scoreboard is designed to meet the needs of the self-checking requirements. Some scoreboards are simply scalar variables holding just one data item at a time. Some scoreboards are lists of lists of data items. A scoreboard may be centralized into a single data structure or it may be distributed in the comparison functions attached to each output monitor.

The self-checking structure of a testbench may be composed of a single scoreboard or of a series of scoreboards daisy-chained one into one another.

Use a list if ordering is important. If the design is supposed to maintain the original order of the input data stream, the scoreboard is usually implemented using a list. The output produced by a functionally correct design would be found, in order, in the list. If multiple data streams are multiplexed onto a single data stream with no ordering relationship between them, use one list per input stream. Any data item received from the design must be found at the head of one of the lists. The scoreboard for the example ATM switch node design would be composed of four sets of four lists: one set per output port, one list per input port. An ATM cell would be added to the appropriate list based on the input port where it is injected and the expected destination ports.

Optimize the look-up function. When a new data item is received from the design by an output monitor, it must be compared against a data item in the scoreboard. For simple designs, it may be necessary only to look at the data item at the head of a very specific list. For more complex designs where data losses are possible or ordering is difficult to predict, output data may come in an apparent random order. It is therefore important to make the look-up operation as efficient as possible to identify the output data as valid or not quickly. If you have to search through all of the data items in the entire scoreboard, simulations will take forever to run. In *e*, use *keyed lists*. In OpenVera, use associative arrays.

Refer to a data structure book. Designing a scoreboard is about designing a suitable data structure that will meet the self-checking requirements of the design. It has to be efficient, both in terms of runtime and storage. A scoreboard that is expected to hold thousands of very large packets must be given a lot of careful attention. You have to watch out for memory leaks as objects are discarded after being compared. It would be pointless for me to describe in this book what has been the object of several other books. Lists, hash functions, circular buffers, lookup tables, queues, indexing strategies and the like have already been described better than I ever could. I recommend you look up the computer science section of your local technical bookstore for a textbook on data structures.

Integration with the Transaction Layer

The self-checking structure must be visible globally.

The self-checking structure must be accessible from almost every component of the testbench. The configuration generator needs to pass the device configuration descriptor to it. The stimulus generators need to forward generated input data to it. The bus-functional models need to inform the self-checking structure of any unexpected events that occurred during the data transmission. Output monitors must indicate that new output data has been received to verify its correctness.

Encapsulate the self-checking structure.

Whatever strategy is used to make the self-checking structure visible to all components of the testbench, it will be easier if it is encapsulated as a single object. Provide a transaction-level interface that will be used to notify the self-checking structure of new data being injected or received. Sample 6-29 shows the definition of a possible self-checking structure for the example ATM switch node.

Sample 6-29. Definition of the self-checking structure for the ATM switch node

```
unit self_check {
    cfg: device_cfg;

    sent (cell    : atm_cell,
          on_port: uint (bits: 2)) is empty;
    received(cell    : atm_cell,
             on_port: uint (bits: 2)) is empty;
};
```

Make the self-checking structure a global instance.

One way to make the self-checking structure visible to all components of the testbench is to make its instance global. In *e*, include it in the *sys* struct. In OpenVera, make it a program variable.

Pass the self-checking structure instance to all bus-functional models.

Another way to make the self-checking structure visible to all components is to extend each component with a reference to the scoreboard instance. The test harness is then built of these extended bus-functional models, all sharing a reference to the same scoreboard structure instance. Sample 6-30 shows how the OpenVera bus-functional model and harness defined in Sample 6-4 are extended and Sample 6-31 shows how those in *e* from Sample 6-5 are extended. For OpenVera, the instance of the self-checking structure is passed via an additional constructor argument whereas a constraint is used to set the reference value in *e*.

Sample 6-30.
Extending an
OpenVera bus-
functional
model and test
harness to
include self-
checking

```
#include "self_checking.vrh"
#include "harness.vrh"
class sc_utopia_L1 extends utopia_L1 {
    self_checking sc;
    task new(self_checking sc, ...) {
        super.new(...);
        this.sc = sc;
    }
}

class sc_harness extends harness {
    self_checking sc;
    task new() {
        sc_utopia_L1 uL1;
        super.new();
        this.sc = new;
        my_uL1 = new(this.sc, tx_0, rx_0);
        this.atm_port[0] = my_uL1;
        ...
    }
}
```

Sample 6-31.
Extending an *e*
bus-functional
model and
harness to
include self-
checking

```
import self_checking;
import harness;
extend utopia_L1 {
    sc: self_checking is instance;
};

extend harness {
    sc: self_checking is instance;
    keep for each in me.atm_port {
        it.sc == me.sc;
    };
};
```

For each transac-
tion on the design,
notify the score-
board.

Once the self-checking structure is visible by all components of the testbench, it is simple to extend the callback methods in the bus-functional models to invoke the proper transaction-level procedures in the self-checking structure at the proper time. Sample 6-32 shows how to complete the integration of the self-checking structure with the transaction-layer test harness.

Sample 6-32.
Calling the
transaction-
level proce-
dures in call-
back methods

```
import self_checking;
import harness;
extend utopia_L1 {
    port_num: uint;
    sc: self_checking is instance;
    keep for each in me.atm_port {
        it.port_num == index;
    };
    post_cell_tx(cell: atm_cell) is also {
        me.sc.sent(cell, me.port_num);
    };
    post_cell_rx(cell: atm_cell) is also {
        me.sc.received(cell, me.port_num);
    };
};
```

DIRECTED STIMULUS

Stimulus is hard-
coded.

Directed stimulus is specified in the verification plan and hard-coded for each testcase. Executing a testcase requires simulating the testbench that includes the directed stimulus for that testcase. It is used to implement each directed testcase specified using the approach defined in "Directed Testbenches Approach" on page 104. Sample 6-33 shows a directed testcase used to debug a CPU interface: It "generates" a write cycle followed by a read cycle at the same address and verifies that the readback value is correct.

Sample 6-33.
Directed
debug stimu-
lus

```
program simple
{
    harness th = new;
    bit [15:0] actual;

    th.cpu.write(16'h00FF, 16'hABCD);
    actual = th.cpu.read(16'h00FF);
    if (actual !== 16'hABCD) ...
}
```

Can include ran-
dom filling.

Directed stimulus need not be specified 100%. The part that is coded explicitly usually only pertains to the testcase being implemented. The data that is deemed irrelevant for this testcase is usually filled with random—but valid—values. For example, the content of an ethernet frame could be filled with random values, except for the VLAN tag that is the objective of the testcase. The sequence of VLAN tag values would be hardcoded in the testcase while the remaining data fields would be generated randomly. Sample 6-34 shows an example of directed stimulus for an instruction stream. The content of the operands is randomized while the actual sequence of opcodes is hardcoded.

Sample 6-34.
Directed instruction sequence stimulus

```
void = instr.randomize() with {
    opcode == CMP;
};
repeat (3) {
    void = opcode.randomize() with {
        opcode == NOP;
    };
};
void = instr.randomize() with {
    opcode == BLT;
};
```

Other streams can be generated randomly.

Directed testcases often concentrate on a single data stream when creating a stimulus sequence. The other streams are left idle or can be generated randomly. In the ATM switch node example, directed stimulus can be specified for the cell stream on port #0 while random traffic is injected in the other ports, as shown in Sample 6-35.

Sample 6-35.
Random background stimulus

```
unit testcase {
    th: harness is instance;

    directed_stim() @sys.any is {
        ...
        me.th.atm_port[0].send(cell);
        ...
    };

    bg_noise(on_port: uint) @sys.any is {
        var cell: atm_cell;
        while (1) {
            gen cell;
            me.th.atm_port[on_port].send(cell);
        };
    };

    run() is also {
        start me.directed_stim();
        start me.bg_noise(1);
        start me.bg_noise(2);
        start me.bg_noise(3);
    };
};
```

Directed stimulus implements testcases.	Even though some random stimulus was used, the nature of directed stimulus is always tied to a specific testcase. Each directed testbench can be tied to a specific testcase. It was written specifically to implement that testcase and no other. If there are one hundred directed testcases to write, there will be one hundred sets of directed stimulus. Any random stimulus included in directed testcases is usually ignored in the response checking.

RANDOM STIMULUS

Generators create the stimulus.	Random stimulus is created by a random generator that can be constrained according to the requirement of the verification plan. Executing a testcase requires simulating the random testbench with a seed that will hit the functional coverage point that corresponds to the testcase. They are used to create automated verification environments specified using the approach defined in "Coverage-Driven Random-Based Approach" on page 109.
Random generation is more than calling $random.	Stimulus generation today has evolved far beyond random generation of individual scalar values using the $random system task. The generation components of a verification environment are designed to generate two subsets of all possible stimuli autonomously. The first is that subset that is legal, i.e., the possible inputs. The second is that subset of the first that is defined by the functional coverage models of the verification plan.

Atomic Generation

Generating a stream of random data is easy.	Sample 6-36 shows a simple random ATM cell generator. It should be encapsulated in an object, like a bus-functional model. A callback method is used to enable its integration within a testbench, as shown in Sample 6-37. The same code could be used to generate CPU instructions, bus cycles, digitized signal samples or any other input data stream required by the design under verification.
Define termination mechanisms.	The simple random generator will always generate 100 data items then terminate. This number is completely arbitrary and is unlikely to satisfy the needs of all testbenches. During initial design debug stage, generating just a single data item is often required. And a random testbench must run for much longer to increase the likelihood that functional coverage points will be hit. Generators should have several termination mechanisms that can be armed at the start

Sample 6-36.
Simple ran-
dom generator

```
class atm_gen {
    integer id;

    virtual task post_cell_gen_t(atm_cell);

    task new(integer id) {
        this.id = id;
    }

    task main_t() {
        atm_cell cell;

        repeat (100) {
            cell = new;
            void = cell.randomize();
            post_cell_gen_t(cell);
        }
    }
}
```

Sample 6-37.
Integration of
a generator
object

```
extend atm_gen {
    bfm: utopia_L1 is instance;
    post_cell_gen(cell: atm_cell) @sys.any
    is also {
        me.bfm.send(cell);
    };
};

unit random_tb {
    th     : harness is instance;
    gen[4]: list of atm_gen is instance;
    keep for each in gen {
        it.id  == index;
        it.bfm == value(th.atm_port[index]);
    };
};
```

of the simulation (such as the number of objects to generate) or externally triggered by the testbench. Sample 6-38 shows a generator that, by default, will generate the maximum number of objects. It will also not start immediately, leaving time for the testbench to configure the device. The generator also can be suspended at any-time by turning the *run* event OFF. Sample 6-39 shows how a debug testcase can constrain the generator to generate a single cell on a randomly selected port and no cells on the others.

<table>
<tr>
<td>

Sample 6-38.
Random generator with termination mechanisms

</td>
<td>

```
class atm_gen {
   integer id;
   event run;
   integer cell_count;

   virtual task post_cell_gen_t(atm_cell);

   task new(integer id) {
      this.id = id;
      this.cell_count = -1;
      trigger(OFF, this.run);
   }

   task main_t() {
      atm_cell cell;

      while (this.cell_count != 0) {
         sync(ALL, this.run);
         cell = new;
         void = cell.randomize();
         post_cell_gen_t(cell);
         this.cell_count--;
      }
   }
}
```

</td>
</tr>
<tr>
<td>

Sample 6-39.
Debug testcase injecting a single cell on random port

</td>
<td>

```
extend sys {
   the_one: uint (bits: 2);
};

extend atm_gen {
   keep this.id == sys.the_one => cell_count == 1;
   keep this.id != sys.the_one => cell_count == 1;
};
```

</td>
</tr>
</table>

Adding constraints is hard.

What if a testcase requires that a stream of cells with the same VPI value be injected? Or that only write cycles within a narrow address range be generated? Or samples with negative values? Or no branch instructions? Adding constraints to the simple generator shown in Sample 6-38 requires that the entire generation method be replaced, as shown in Sample 6-40. This results in a lot of duplicated code and a methodology similar to using directed stimulus. As illustrated in Figure 6-13, a different random generator would be created for each testcase. The idea behind the HVL productivity cycle depicted in Figure 2-17 is to write just one generator that can be steered toward the uncovered functional coverage points by adding constraints with as little modifications as possible, as illustrated in Figure 6-14.

Sample 6-40.
Replacing random generation method

```
extend atm_gen {
    main() @sys.any is only {
        var cell: atm_cell;

        while (me.cell_count != 0) {
            if (!me.run) {
                wait cycle @go;
            };
            gen cell keeping {
                it.vpi == 8'h00;
            };
            me.post_cell_gen(cell);
            me.cell_count--;
        };
    };
};
```

Figure 6-13.
Different random generators

Figure 6-14.
Constraining a single random generator

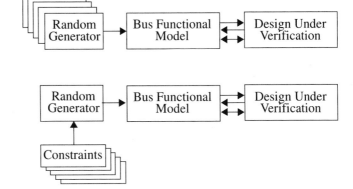

Adding Constraints in *e*

You can add constraints to the generated type.

The simplest mechanism for adding constraints is to add them to the type being generated. For example, the constraint forcing the generation of a stream of ATM cells with a VPI value equal to 0 can be added to the ATM cell type as shown in Sample 6-41. The problem with this approach is that the constraint will apply to every instance of the object in the entire simulation. If the objective was to inject a specific condition on one specific stream, or use a different constraint on a different stream, this approach will not work.

Sample 6-41.
Adding constraints to the generated type

```
extend atm_cell {
    keep vpi == 8'h00;
};
```

Always generate fields.

The problem with the simple generator is that it uses a variable to generate data items. You can only add constraints to fields, not variables. By generating a field instead of a variable, as shown in Sample 6-42, you can add constraints limited to the generated instance, not all instances of the object. Sample 6-43 shows how the constraint on the VPI value can be added to this new version of the generator.

Sample 6-42.
Generating a field instead of a variable

```
unit atm_gen {
    ...
    !cell: atm_cell;

    main() @sys.any is {
        while (me.cell_count != 0) {
            ...
            gen me.cell;
            me.post_cell_gen(me.cell);
            ...;
        };
    };
};
```

Sample 6-43.
Adding constraints to the generated field

```
extend atm_gen {
    keep me.cell.vpi == 8'h00;
};
```

Use unique identifier to enable stream-specific constraints.

The constraint in Sample 6-43 still applies to all instances of the generator. To specify a constraint for a single instance, use a conditional expression that involves a unique identifier for that instance. As shown in Sample 6-44, the field *id* can be used to express a stream-specific constraint. As long as this field is set appropriately, as in Sample 6-37, each generator will use a different set of constraints.

Do not add constraints to the harness.

You may be tempted to add stream-specific constraints by referring to the appropriate generator instance in the test harness, as shown in Sample 6-45. *This technique will not work!* Constraints have a

Sample 6-44.
Adding
stream-spe-
cific constraint

```
extend atm_gen {
    keep me.id == 0 => me.cell.vpi == 8'h00;
    keep soft me.id == 3 =>
        me.cell.is_bad == select {
            1: TRUE;
            9: FALSE;
        };
};
```

scope. They only apply if the generation is triggered from a level at or above the level they are specified. In the case of the generator, a cell is generated by the *gen* action of a field in the generator itself. That is in a lower scope that the constraints in Sample 6-45, which are at the harness level. Therefore, the constraints will not be applied. The constraints in Sample 6-44 are at the same scope level as the generated field and thus will apply. The constraints in Sample 6-45 will only apply when the harness itself is generated.

Sample 6-45.
Invalid con-
straint scope

```
extend harness {
    keep gen[0].cell.vpi == 8'h00;
    keep soft gen[3].cell.is_bad == select {
        1: TRUE;
        9: FALSE;
    };
};
```

Extend
pre_generate() or
post_generate(), as
needed.

Constraints are powerful, but sometimes cannot express a particular condition that needs to be generated. You can execute procedural code before or after the generation process by extending the pre-defined *pre_generate()* and *post_generate()* methods. You could use *pre_generate()* to initialize some non-generated fields or use *post_generate()* to compute or corrupt CRC values, as illustrated in Sample 6-46. CRC values should always be computed in the *post_generate()* method to ensure that they are computed last, based on the final value of all the covered fields. Using a constraint to set a CRC field can yield the wrong value if the generation order causes the CRC value to be generated early.

Sample 6-46.
Corrupting
CRC value

```
extend atm_cell {
    is_bad: bool;
    post_generate() is also {
        if (is_bad) {
            gen me.crc keeping {
                it != me.crc;
            };
        };
    };
};
```

Adding Constraints in OpenVera

You can add constraints to the randomized type.

The simplest mechanism for adding constraints is to add them to the class being randomized. For example, the constraint forcing the generation of a stream of ATM cells with a VPI value equal to 0 can be added to the ATM cell class as shown in Sample 6-47. The problem with this approach is that the constraint will apply to every instance of the object in all future simulations and models. If the objective was to inject a specific condition for a specific testcase or simulation, this approach will not work.

Sample 6-47.
Adding constraints to the generated class

```
class atm_cell {
    ...
    constraint vpi_is_0 {
        vpi == 8'h00;
    }
}
```

You can turn constraints ON and OFF.

If a constraint is not supposed to apply at all times, it should be turned OFF in the constructor. It can then be turned ON only when required. For example, the constraint added to the ATM cell class Sample 6-47 is turned OFF in Sample 6-48. The constraint will only apply to a randomized instance only if it is explicitly turned ON.

Always randomize a public property.

The problem with the simple generator is that it randomizes a variable to generate data items. The variable is not visible externally to allow the testcase to turn the relevant constraint blocks ON. By randomizing a public property instead of a variable, as shown in Sample 6-49, you can control the constraints in the randomized instance as shown in Sample 6-50. For a constraint to apply to a single

<table>
<tr>
<td>

Sample 6-48.
Adding con-
straints turned
OFF by
default

</td>
<td>

```
class atm_cell {
    ...
    constraint vpi_is_0 {
        vpi == 8'h00;
    }

    task new() {
        this.constraint_mode(OFF, "vpi_is_0");
    }
}
```

</td>
</tr>
</table>

stream, turn the relevant constraint block in the randomized property for the generator instance of that stream.

<table>
<tr>
<td>

Sample 6-49.
Randomizing
a public prop-
erty instead of
a variable

</td>
<td>

```
class atm_gen {
    ...
    atm_cell cell;

    task main_t() {
        while (this.cell_count != 0) {
            ...
            this.cell = new;
            void = this.cell.randomize();
            this.post_cell_gen_t(this.cell);
            ...
        }
    }
}
```

</td>
</tr>
</table>

<table>
<tr>
<td>

Sample 6-50.
Controlling
constraints in
randomized
instance

</td>
<td>

```
program corner_case
{
    random_th th = new;
    ...
    void = th.gen[0].cell.constraint_mode(ON,
                                "vpi_is_0");
    ...
}
```

</td>
</tr>
</table>

Always random-
ize the same
instance.

There is still a problem with the generator in Sample 6-49: It keeps
randomizing a different instance. The randomized instance keeps
changing and must be controlled as shown Sample 6-50 before each
and every randomization. That's too much work. Instead, randomize a single instance whose value is then copied into a new instance.
These new instances will create the stream of generated data items,
as shown in Sample 6-51.

Sample 6-51.
Randomizing
a single
instance

```
class atm_gen {
    . . .
    atm_cell cell;

    task new {
        this.cell = new();
    }

    task main_t() {
        while (this.cell_count != 0) {
            atm_cell new_cell;
            . . .
            void = this.cell.randomize();
            new_cell = this.cell.copy();
            this.post_cell_gen_t(new_cell);
            . . .
        }
    }
}
```

Add constraints to
a derived class.

The constraint in Sample 6-48 was added to the original class modeling the ATM cell. If all of the constraints required to generate all of the input conditions to meet your coverage goals are added to that one class, it will soon become unmanageable. Furthermore, a generic object model, such as an ATM cell, can be reused across projects. It should not be polluted with project or testcase-specific additions. Constraints should be added in a derived class as shown in Sample 6-52. Use a different extension for each testcase.

Sample 6-52.
Adding con-
straints in a
derived class

```
class constrained_atm_cell extends atm_cell {
    constraint vpi_is_0 {
        vpi == 8'h00;
    }

    task new() {
        this.constraint_mode(OFF, "vpi_is_0");
    }
}
```

Replace the ran-
domized instance
with an instance of
the derived class.

This technique does not appear to be helpful, as the generator is still making use of the base class, not the derived class. Therefore the new constraints are not used. One solution would be to change the generator to use an instance of the derived class, but you'll end-up modifying the generator for each constraint set. Remember that a derived class is compatible with its base class and that the *randomize()* method is a virtual method. We can simply sneak an instance of the derived class in lieu of the original randomized instance, and the generator won't even be aware that it is now generating a data stream subject to additional constraints! Sample 6-53 shows how to do so for a specific instance of a generator.

Sample 6-53.
Adding constraints via a derived class.

```
class constrained_atm_cell extends atm_cell {
    constraint vpi_is_0 {
        vpi == 8'h00;
    }
}
program corner_case
{
    random_th th = new;
    . . .
    {
        constrained_atm_cell cell = new;
        th.gen[0].cell = cell;
    }
    . . .
}
```

Extend
pre_randomize()
or
post_randomize(),
as needed.

Constraints are powerful, but sometimes cannot express a particular condition that needs to be generated. You can execute procedural code before or after the randomization process by extending the predefined *pre_randomize()* and *post_randomize()* methods. You could use *pre_randomize()* to initialize some non-randomized fields or use *post_randomize()* to compute or corrupt CRC values, as illustrated in Sample 6-54. CRC values must be computed in the *post_randomize()* method because methods cannot be used in constraint expressions. When overloading the *pre_randomize()* or *post_randomize()* methods, do not forget to invoke their original version in the parent class using *super.pre_randomize()* or *super.post_randomize()*. This approach will ensure that any procedural operations required to randomize the parent class successfully is executed.

Sample 6-54.
Corrupted
CRC value

```
class may_be_bad_atm_cell extends atm_cell {
    bit is_bad;
    task post_randomize() {
        super.post_randomize();
        if (is_bad) {
            bit [7:0] crc = random();
            while (crc == this.crc) {
                crc = random();
            }
        }
    }
}
```

Constraining Sequences

Generating atomic elements is not interesting.

In the previous section, data items were generated randomly independent of each other. This technique is going to create some interesting conditions but is unlikely to generate all of the conditions required to meet your functional coverage goals. The design verification has data and temporal behavior, each of which must be verified. The temporal properties of applied stimuli must be as flexible and diverse as the data properties to verify the temporal behavior easily. You must include a mechanism that will make it possible to express constraints describing a sequence of data items that will exercise the temporal features of the design.

Provide a unique data item identifier.

It is possible to express stream-specific constraints using a unique stream identifier in a conditional expression. The same mechanism can be used to specify constraints applicable to data items at a specific index within a sequence. The random generator from Sample 6-42 has been modified in Sample 6-55 to increment an object index after each random generation. Constraints specific to the position of the object within the sequence can then be specified using that unique object identifier, as shown in Sample 6-56.

Generate a list in *e*.

Using a unique object identifier allows specifying sequence-specific constraints. But they can be specified only as independent values. It is not possible to express constraints that refer to previously generated values. For example, how would you generate a sequence of ATM cells with random VPI values with no two consecutive

Sample 6-55.
Generating
constrainable
sequences

```
unit atm_gen {
    ...
    !cell_idx: uint;
    !cell: atm_cell;

    main() @sys.any is {
        while (me.cell_count != 0) {
            ...
            gen me.cell;
            me.post_cell_gen(me.cell);
            ...;
            me.cell_idx += 1;
        };
    };
};
```

Sample 6-56.
Specifying
sequence-spe-
cific con-
straints

```
extend atm_gen {
    keep vpi == cell_idx % 4;
};
```

identical values? In *e*, the solution is to generate a list instead of a
single object. List constraints can refer to any previous elements in
the list using the *prev* or *index* implicit variables. Sample 6-57
shows the previous generator, modified to generate a list instead of
a single object. Note how a soft constraint is used to generate a single element by default. Sample 6-58 shows how to add constraints
to avoid generating two identical consecutive VPI values.

Sample 6-57.
Generating
sequences
using a list

```
unit atm_gen {
    ...
    !cells: list of atm_cell;
    keep soft cell.size() == 1;

    main() @sys.any is {
        while (me.cell_count != 0) {
            ...
            gen me.cells;
            me.post_cell_gen(me.cells);
            ...;
            me.cell_idx += me.cells.size();
        };
    };
};
```

Sample 6-58.
Specifying
sequence-spe-
cific list con-
straints

```
extend atm_gen {
    keep cells.size().reset_soft();
    keep for each in cells {
        index > 0 => it.vpi != prev.vpi;
    };
};
```

Keep a list of pre-
viously generated
values in Open-
Vera.

OpenVera does not support list constraints. But you can keep a copy of all previously generated objects in the sequence in a list that you are then free to refer to in constraints. Sample 6-59 shows how sequence constraints based on previously generated values can be added to the generator in Sample 6-51.

Sample 6-59.
Referring to
previously
generated val-
ues in
sequence con-
straints

```
class constrained_atm_cell extends atm_cell {
    integer cell_idx = 0;
    atm_cell prev[];

    constraint no_two_same_vpi {
        if (cell_idx > 0) {
            vpi != prev[cell_idx-1].vpi;
        }
    }

    task post_randomize() {
        super.post_randomize();
        this.prev[this.cell_idx++] = this.copy();
    }
}
```

Define a sequence
for debug testcase.

During a consulting engagement, I had spent several days helping a customer write a random-based self-checking environment to verify a CPU interface on an RTL design. After explaining and implementing the concepts shown in this section, the engineer I was working with interjected, "But the first testcase I'll want to run, when the RTL is delivered tomorrow, is a simple *write* followed by a *read*. I don't need this fancy generation and constraint mechanism yet." I replied that his first testcase was simply a very simple scenario: Constrain the sequence of transactions to a length of two, the first transaction to be a *write* cycle and the second transaction to be a *read* cycle at the same address as the previous one. Instead of writing a separate testcase, only a few additional lines creating a simple sequence were sufficient. Once the initial debug of the design was over, these constraints were removed, subjecting the RTL code to a lot of different input sequences with no additional testbench development effort. The sequence constraints can be found in Sample 6-60.

Sample 6-60.
Defining a
first-test
sequence

```
extend transaction_gen {
    keep transactions.size() == 2;
    keep for each in transactions {
        index == 0 => it.kind == WRITE;
        index == 1 => it.kind == READ and
                      it.address == prev.address;
    };
};
```

Need to define
multiple
sequences.

The sequences defined using the constraint mechanism shown in this section would be generated only once per simulation and each simulation would generate only one sequence per generator. It is more efficient to have multiple interesting sequences—or scenarios—be generated randomly, one after another in a single simulation. Different instances of the same generator could generate the same scenario at the same time or generate different scenarios.

Define scenarios
to increase func-
tional coverage.

Left unconstrained, random generators will generate valid but most likely uninteresting input sequences. By defining scenarios, generators will be constrained to generate a series of constrained sequences, focused on interesting cases. One scenario is usually the "random" scenario i.e., no constraints at all. As holes in functional coverage are identified, scenarios are added to the verification environment to steer the generators toward the uncovered areas of the solution space.

Scenarios can
define directed
testcases.

If you have the necessary control variables, it is possible to specify a directed testcase as a series of constraints. By specifying a set of constraints for which there is only one solution, you have created a scenario that implements a directed testcase.

Defining Scenarios in OpenVera

Use the stream
generator.

Scenarios are best described using the stream generator. A stream generator ruleset is defined using the *randseq* sequential statement. A scenario is described as a *production rule*, often making use of other production rules. Thus scenarios can be described in terms of other scenarios. A second type of production rule describes a weighted choice between equivalent scenarios. For someone with an RTL background, the reverse-YACC syntax of the stream generator is really bizarre. But once you realize it is a simple front-end to small subprograms and the *randcase* statement, it becomes easy to understand.

A production rule is like a task definition.

A production rule is similar to defining a small task or function. When invoked, it will "execute" its definition. For example, the ruleset in Sample 6-61 is functionally equivalent to defining the tasks in Sample 6-62. A rule can return values and be passed arguments. Refer to chapter 12 of the OpenVera User's Manual for a complete description of the stream generator.

Sample 6-61. Stream generator ruleset

```
randseq (SCENARIOS) {
    SCENARIOS: (&1) RANDOM_CELL
             | (&1) TEN_WITH_SAME_VPI;

    RANDOM_CELL: {
        atm_cell cell;
        void = this.cell.randomize();
        cell = this.cell.copy();
        this.post_cell_gen_t(cell);
    }

    TEN_WITH_SAME_VPI:
        RANDOM_CELL {
            void = this.cell.rand_mode(OFF, "vpi");
        } <repeat (9) RANDOM_CELL> {
            void = this.cell.rand_mode(ON, "vpi");
        }
    }
}
```

The choice weights can be expressions.

The weights assigned to the different choices in an alternative rule can be expressions, just like the weights in the equivalent *randcase* statement. By making these weights public properties of the generator object, each simulation run can pick and choose which scenarios will be enabled and the probability a particular scenario will be generated. For example, in Sample 6-63, the simulation run will only use the "debug testcase" scenario in the generator shown in Sample 6-64.

Do not create random-length sequences using recursive rules.

Rules can be recursive. For example, Sample 6-65 shows a ruleset that creates a sequence of ATM cells. The number of cells in the sequence is random, determined by the number of times the *CELL_STREAM* rule is selected over the *RANDOM_CELL* rule. This style works just fine except for one thing: The length of the

Sample 6-62.
Equivalent
task defini-
tions

```
task SCENARIOS() {
    randcase {
        1: RANDOM_CELL();
        1: TEN_WITH_SAME_VPI();
    }
}

task RANDOM_CELL() {
    atm_cell cell;
    void = this.cell.randomize();
    cell = this.cell.copy();
    this.post_cell_gen_t(cell);
}

task TEN_WITH_SAME_VPI() {
    RANDOM_CELL();
    void = this.cell.rand_mode(OFF, "vpi");
    repeat (9) {
        RANDOM_CELL();
    };
    void = this.cell.rand_mode(ON, "vpi");
}
```

Sample 6-63.
Selecting sce-
narios

```
program initial_debug {
    random_harness th = new;
    th.trans_gen.null_weight = 0;
    ...
}
```

Sample 6-64.
Variable
weight sce-
nario selection

```
class transaction_gen {
    transaction trans;
    ...
    integer null_weight = 1;
    integer debug_weight = 1;

    task main_t() {
        while (...) {
            randseq (SCENARIOS) {
                SCENARIOS: (&null_weight) NULL
                         | (&debug_weight) DBG_TST;

                ...
            }
        }
    }
}
```

sequence is not distributed evenly. The probability that the length of the sequence is 1 is 10%. The probability that it is 2 is 9% (0.9). The probability that it is 3 is 8.1% (0.1 x 0.9 x 0.9 x 0.1). The probability that it is N is $0.9^N/10$. Furthermore, the length of the sequence cannot be constrained other than by playing with the selection weights.

Sample 6-65.
Recursive
ruleset

```
CELL_STREAM: (&90)  CELL_STREAM RANDOM_CELL
           | (&10)  RANDOM_CELL;

RANDOM_CELL: {
    atm_cell cell;
    void = this.cell.randomize();
    cell = this.cell.copy();
    this.post_cell_gen_t(cell);
}
```

Generate the length first, then generate the sequence.

A better approach is to generate the length of the sequence first. That value can be subjected to constraints and have an even distribution. Once the length of the sequence is decided, you generate the sequence using a *repeat* statement. Notice how Sample 6-66 encapsulates the randomized scalar sequence length value into a class. It simply makes it possible to generate its value using *randomize()* and subject it to constraints. Had *random()* been used, it would not have been possible to constrain the generated value using any of the techniques shown in this section.

Rulesets and rules cannot be virtual.

An important limitation of the stream generator is that production rules and entire rulesets cannot be virtual. It is not possible to extend an existing generator using a stream generator ruleset without modifying its source code. If you want to be able to add new production rules or modify existing rules or entire rulesets, I suggest you use the equivalent subprogram style, as shown in Sample 6-62. If each task or function is defined as *virtual*, it will be simple to extend a ruleset by creating a derived generator class. Virtual tasks and functions can still make use of the stream generator and the *randseq* statement to describe their respective scenarios.

<table>
<tr>
<td>

Sample 6-66.
Generating
variable-
length
sequences

</td>
<td>

```
class seq_length {
    integer value;
    constraint valid {
        value > 0;
    }
    constraint reasonable {
        value < 50;
    }
}

class atm_gen {
    seq_length len;
    ...
    task main_t() {
        ...
        randseq (...) {
            ...
            CELL_STREAM: {
                void = this.len.randomize();
            } <repeat (this.len.value) RANDOM_CELL>;
            ...
        }
    }
}
```

</td>
</tr>
</table>

Defining Scenarios in *e*

Generate a *struct* containing a list of objects and a kind.

To be able to generate multiple scenarios one after the other, it is necessary to generate multiple lists of objects, one after another. Each generated list defines a scenario. The generator in Sample 6-57 is modified in Sample 6-67 to generate a *struct* containing a list of objects instead of a list. Fields unique to the generator instance and the scenario instance are also included to enable the specification of stream-specific or scenario-order-specific constraints.

Define a new scenario using a *when* extension.

The default *random* scenario simply generates a random-length list of random cells. To create new scenarios representing interesting sequences of object instances, add a new enumeral to the enumerated type identifying the scenario then specify the scenario constraints in a *when* extension of the scenario *struct*, as shown in Sample 6-68. The scenario becomes available to all instances of the generator by simply loading the file containing the scenario definition.

Sample 6-67.
Generating
scenarios

```
type atm_scenario_kind: [RANDOM];

struct atm_scenario {
    stream_id  : uint;
    scenario_id: uint;
    kind       : atm_scenario_kind;
    cells      : list of atm_cells;
};

unit atm_gen {
    ...
    scenario: atm_scenario;

    main() @sys.any is {
        while (...) {
            ...
            gen me.scenario keeping {
                it.stream_id == me.id;
                it.scenario_id == me.scenario_count;
            };
            for each in me.scenario.cells {
                me.post_cell_gen(it);
            };
            ...;
            me.scenario_count += 1;
        };
    };
};
```

Sample 6-68.
Defining new
scenarios

```
extend atm_scenario_kind: [SAME_VPI];

extend SAME_VPI atm_scenario {
    keep cells.size() > 1;
    keep for each in cells {
        index > 0 => it.vpi == prev.vpi;
    };
};
```

Testcases can con-
strain the scenario
selection.

Using the unique stream and scenario identifiers, a particular testcase can select specific scenarios to be generated by specific generators at a specific time. For example, Sample 6-69 is a testcase that applies random background traffic to ports 1 through 3 but sequences of 10 identical VPI values on port #0 after 10 random cells.

Sample 6-69.
Selecting sce-
narios

```
extend atm_scenario {
    keep soft kind == RANDOM;
    keep stream == 0 && scenario_id >= 10 =>
        kind == SAME_VPI and cells.size() == 10;
};
```

Use *sequences*
from the eRM
toolkit.

The scenario generation mechanism shown in Sample 6-67 works well for simple scenarios. If hierarchical scenarios are required—for example, to describe code structures where a "loop" scenario is composed of a sub-scenario for the loop body[3]—use the *sequence* mechanism described in Chapter 5 of the *e* Reuse Methodology (eRM) Developer Manual. The scenario generation mechanism automatically creates generator declarations similar to those shown in Sample 6-67. It also includes a mechanism for generating scenarios one object at a time instead of using lists or generating them procedurally instead of using constraints on lists.

---

3. And if that sub-scenario is also a "loop" scenario, you get nested loops.

SUMMARY

Encapsulate VHDL bus-functional procedures in a server entity. Provide equivalent client access procedures in a package.

Create a transaction-layer test harness encapsulating all of the bus-functional model instances and clock generators connected to the design under verification.

Provide a high-level device configuration descriptor. Interpret the descriptor value to configure the design under verification at the beginning of the simulation.

Generate the device configuration randomly. Use functional coverage measurements to determine which configuration (or combination of configurations) has been verified. Use constraints to limit configurations to currently supported or interesting values and combinations.

Make your testbench self-checking. Build the self-checking structure on top of the transaction-layer test harness.

Self-checking can be implemented using a reference model, by tagging data, by using a transfer function with a scoreboard or any combination of the above.

Generate directed stimulus by invoking the proper transaction-level procedures directly.

When writing random generators, provide a constraint mechanism that can described all of the interesting and relevant input sequences.

Provide unique identifiers for each generator instance and data instance to allow stream-specific and order-specific constraints to be expressed.

Write random generators that generate random scenarios. Add scenarios to increase your functional and code coverage.

Directed and initial debug testcases can be described as tightly constrained random scenarios.

## CHAPTER 7

SIMULATION MANAGEMENT

Simulation must
be managed.

In "Revision Control" on page 68, I described how tools can help manage source code. In "Issue Tracking" on page 74, I described how issues and bugs can be tracked to ensure they are resolved. In this chapter, I address the simulation management issues. I describe how to debug your testbenches efficiently using behavioral models. Often overlooked but important topics, such as terminating your simulation, reporting errors and determining success or failure are covered. We also discuss configuration management: How do you know you are simulating what you think you are simulating?

BEHAVIORAL MODELS

This section demonstrates how behavioral models can benefit a design project. These benefits can be realized only if the model is written with the proper perspective. This section also shows how to model exceptions properly and explains how to demonstrate a behavioral model to be equivalent to an RTL model.

Testbenches
need a model to
be debugged.

You have decided which testcases and functional coverage measurements are needed to verify a design functionally. Your best verification engineers are developing the test harness, self-checking structure and random-generators. Hardware design engineers are working furiously on the RTL model, but it will not be available for several weeks. Meanwhile, the test harness and verification environment continue to be written. When all is said and done, the amount of code written for the verification will surpass the amount of RTL code. You are looking at writing thousands of lines of code without being able to debug them. Furthermore, when writing a constrainable random environment, you need a model to exercise your generators to ensure that they offer the constraint capabilities required to generate interesting input scenarios.

Behavioral models are used to debug testbenches.

What if someone walked up to you and offered you a model, available about at the same time as early versions of verification environment can be simulated, that runs one hundred times faster than the RTL model and that looks and feels just like the real thing? You could start debugging your verification environment even before the RTL is ready. Because this model simulates faster, the debug cycles would be shorter. By the time the RTL is available to simulate, you'd probably have most of the scenarios covering your entire input functional coverage model defined and debugged. The design schedule could be shortened and the verification would no longer be squarely on the critical path. Sound too good to be true? I'm offering exactly such a model: It is called a *behavioral* model.

Behavioral versus Synthesizable Models

Behavioral models are not synthesizable.

Many books, companies and individuals used the term *behavioral* to describe a synthesizable model. This book uses the term differently. A model that can be translated automatically into a gate-level implementation by a synthesis tool, such as Synopsys' Design Compiler, is called a *Register-Transfer-Level* or *RTL* model. It also may be called a *synthesizable* model. This book uses the term behavioral model to identify models that describe the black-box functionality of a design. The Virtual Socket Interface Alliance uses the term *functional* model.

Behavioral code is not just for testbenches.

In "Behavioral versus RTL Thinking" on page 121, I described the characteristics of behavioral code compared with synthesizeable code. Using behavioral descriptions for testbenches is acceptable easily by most design engineers. After all, the testbench will never be implemented in hardware so they never give any thought as to how they would go about it. Their mind hasn't been influenced by an implementation architecture or by a synthesizeable description of the testbench's functionality. They are still open to describing this functionality using behavioral code.

Writing a behavioral model requires a different mindset than writing an RTL model.

Writing a truly behavioral model of a design requires a greater mental leap. You may have already started to think of a design's functionality in terms of state machines, datapaths, operators, memory interfaces and other implementation details. This mindset can be created simply because the functional specification document was written with these implementation details in mind. To write a proper behavioral model, you have to focus on the *functionality*, not the implementation. If the implementation starts to color your thinking, you'll simply write what I call an "RTL++" model.

Example of Behavioral Modeling

RTL++ models may be synthesizeable using behavioral synthesis.

For example, consider the specification in Sample 7-1. How would you write a behavioral description of this functionality? Most would write something similar to the description shown in Sample 7-2. This description clearly is not synthesizeable using logic synthesis tools. However, it happens to be synthesizeable using behavioral synthesis tools such as Synopsys' Behavioral Compiler. The design is synthesizeable behaviorally because the description was tainted by the specification: There is an implicit state machine and everything happens at the active edge of the clock.

Sample 7-1.
Specification of a debounce circuit

The debounce circuit samples the input at every clock cycle. The debounced version of the input changes state only when eight consecutive samples of the input have the same polarity.

Sample 7-2.
RTL++ description of debounce circuit

```
reg debounced;
always @ (posedge clk)
begin: debounce
    if (bouncing != debounced) begin
        repeat (7) begin
            @ (posedge clk);
            if (bouncing == debounced)
                disable debounce;
        end
        debounced <= bouncing;
    end
end
```

A behavioral model cannot be refined into a synthesizeable model.

The objective of a behavioral model is to represent the functionality of a design faithfully, in a way that is easy to write and simulate. The behavioral model is designed to help verification, and indirectly, the implementation. When written properly, a behavioral model cannot be refined into a model synthesizeable by today's logic synthesis tools.

For example, what is the *functionality* of the debounce circuitry specified in Sample 7-1? It prevents pulses on the primary input, narrower than 8 clock periods, from making it to the debounced output. The functionality is similar to a buffer with a significant inertial delay. This behavior can be modeled using a single statement in both Verilog and VHDL, as shown in Sample 7-3 and Sample 7-4, respectively. These statements use the inertial delay model built in each language. If required, please refer to a suitable Verilog or VHDL reference book[1] for a detailed description of inertial delays.

Sample 7-3. Behavioral description of debounce circuitry in Verilog

```
assign #(8*cycle) debounced = bouncing;
```

Sample 7-4. Behavioral description of debounce circuitry in VHDL

```
debounce: debounced <= bouncing after 8 * cycle;
```

Delays cannot be synthesized.

The descriptions in Sample 7-3 and Sample 7-4 are far from being synthesizeable. It is not possible to synthesize a specific inertial delay. The other limitation of these descriptions is the need to know the clock period. It could be specified using a *constant*, a *generic*, or a *parameter*, but the behavioral model would not adjust to different clock periods as the real implementation would. If this is an important requirement, the clock period could be determined at

1. Titles have been suggested in the Preface on page xxii.

runtime by sampling two consecutive edges. Sample 7-5 shows how this sampling could be performed. Notice how the clock cycle is measured only once to improve simulation performance. It is unlikely that the clock period will change significantly during a simulation. Computing the clock period at every clock cycle simply would consume simulation resources without accomplishing additional work.

Sample 7-5.
Measuring the clock period in the debounce circuitry

```
architecture beh of debounce is
   signal cycle: time := 8 * 10 ns;
begin
   process
      variable stamp: time;
   begin
      wait until clk = '1';
      stamp := now;
      wait until clk = '1';
      cycle <= now - stamp;
      wait;
   end process;

   debounced <= bouncing after 8 * cycle;
end beh;
```

Characteristics of a Behavioral Model

They are partitioned for maintenance.

A behavioral model is partitioned differently from a synthesizeable model. The latter is partitioned to help the synthesis process. Partitioning is decided along implementation lines, producing a design with several instances arranged in a wide and shallow structure.

Behavioral models are partitioned according to generally accepted software engineering practices. Behavioral models tend to be partitioned according to main functional boundaries to avoid maintaining one large file, or to allow more than one author to write it. Duplication of function in a model, such as many interfaces of the same type, is also implemented using multiple instances of a single description. Behavioral models tend to have very few instances creating a narrow and shallow structure of large blocks.

They do not use a clock.

A clock signal is an implementation artifice for synchronous design methodologies. These methodologies are functionally irrelevant. A behavioral model does not change state synchronously with a clock. Instead, a behavioral model uses many different synchronization mechanisms—one of which could be a clock edge. While an RTL model continuously recomputes and updates the value of inferred registers, a behavioral model performs computations only when necessary.

Consider the RTL model in Sample 4-3 on page 124: The process labeled *SEQ* is executed every time the clock changes. The signal named *STATE* is assigned at every rising edge of the clock signal, regardless of the value of *NEXT_STATE*.

The equivalent behavioral model in Sample 4-4 on page 124, on the other hand, does not even use a clock. Instead, it acts on the only functionally significant event: the change in *ACK*. This behavioral model changes the only functionally significant state, the state of the *REQ* output.

A clock would be used only when data needs to be sampled or produced synchronously with a clock signal. Examples of synchronous interfaces include PCI or Utopia Level 1. The clock signals for synchronous interfaces are usually externally generated and are not used any further by the behavioral model.

Behavioral models do not use FSMs.

Synthesizeable models are littered with finite state machines. They are the primary synchronous design mechanism for implementing control algorithms. When writing software using a language like C++, you would not usually implement it as a series of cooperating finite state machines. The language does not lend itself very well to that.

Instead, the control algorithm and the data transformations would be part of the control flow of the program. The model's state would depend on the current values of the variables and the location of the statement under execution in the program sequence.

Behavioral models follow a similar strategy. Consider the example in Sample 4-3 on page 124. The state of the RTL model is determined by the value of the state register and the current input values. The same code is executed over and over. On the other hand, the state of the behavioral model shown in Sample 4-4 on page 124 depends only on which *wait* statement is being executed currently.

Data can remain at a high-level of abstraction.

The skills of the hardware engineer reside in mapping a complex functionality into the very primitive resources available in hardware. Everything must be implemented using a binary value, with a small number of bits, and reduced to integer arithmetic. A behavioral model can make use of the high-level data types provided by the language, such as enumerals, floating-point numbers, records and multi-dimensional arrays. The section titled "Data Abstraction" on page 145 illustrates many examples of using high-level data abstraction instead of using representations suitable for implementation.

Data structures are designed for ease-of-use, not implementation.

In a synthesizeable model, the format of the data structures are organized to make implementation possible. For example, imagine that a routing table in a packet router is composed logically of 256-bit records with various fields. The router is specified to support 1,024 possible routes and the table is maintained by an external processor through a 16-bit wide interface.

The physical implementation of the routing table is likely to use a 16-bit RAM with 16K locations. Whenever the routing engine performs a table lookup, it has to read a block of 16 words to build the entire 256-bit routing record.

If the table maintenance via the CPU interface has a much lower frequency than packet routing, a behavioral model would instead optimize the data structure for the table look-up and routing operation. The routing table would be implemented using an array of records with 1,024 locations. It would also probably use a sparse array implementation to minimize memory usage as well. The table would look the same from the CPU's perspective, with each 16-bit access being performed at the right offset within the record identified by the upper 10 bits of addresses. Sample 7-6 shows a Verilog implementation of the CPU access into the routing table of the behavioral model.

Sample 7-6.
Mapping a narrow access in a wide data structure

```
reg [255:0] table [0:1023];

always
begin: cpu_access
   reg [255:0] entry;
   ...
   entry = table[addr[13:4]];
   if (read) data = entry >> addr[3:0];
   else begin
      for (i = 0; i < 16; i = i + 1) begin
         entry[addr[3:0]*16+i] = data[i];
      end
   end
   ...
end
```

Their interfaces are implemented using bus-functional models.

The testbench is a behavioral model of the environment. To make implementation more efficient, Chapters 5 and 6 explained how bus-functional models are used and located in a testcase-independent test harness. The bus-functional models abstract data from the physical level to a functional level where they are simpler to process using behavioral code.

The same strategy can be used when writing a behavioral model. Bus-functional models are used for each interface around the periphery of the model. Data is transformed behaviorally and moved from bus-functional model to bus-functional model according to the function of the device. And as Figure 7-1 shows, you will likely be able to reuse the bus-functional models written for the testbench in your behavioral model.

Figure 7-1.
Structure of a UART test harness and behavioral model

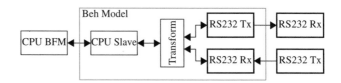

Behavioral models can be written using HVLs.

The use of HVLs is always depicted as a testbench interfacing with the design under verification through the top-level interface. *e* or OpenVera models can access HDL signals anywhere in the design hierarchy. This feature is usually used to perform white-box verification checks. But if HDL signals are not driven by HDL code, they are free to be driven by HVL code. It is possible to write behavioral models of arbitrary HDL blocks using HVLs, as illustrated in Figure 7-2. Thus, behavioral models can be written using the high-level constructs available in HVLs.

Figure 7-2.
Using HVLs
for testbench
and
behavioral
model

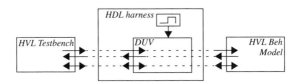

Use an empty
module or *architecture*.

To be able to interface the behavioral model transparently with other HDL models, encapsulate the HVL model in an empty Verilog *module* or VHDL *architecture*. Use the same physical interface as the RTL model. The HVL behavioral model would interface with the signals inside the empty *module* or *architecture*. From the outside, the HDL simulator would have no knowledge that the internals of a particular instance are actually being driven by an external simulation process.

Modeling Reset

Reset is part of the
RTL coding style.

Modeling exceptions can take of lot of time and introduce a lot of intricacies in an otherwise simple algorithm description. When writing a synthesizeable description, modeling the effect of reset on the state elements is defined in the supported coding style. For example, Sample 7-7 shows how an asynchronous reset is modeled to reset a finite state machine. Resetting an entire RTL model is accomplished by including the logic to handle the reset exception in each process that infers a register.

Sample 7-7.
Modeling an asynchronous reset in RTL

```
process (clk, rst)
begin
    if rst = '1' then
        state <= IDLE;
    elsif clk'event and clk = '1' then
        case state is
            . . .
        end case;
    end if;
end process;
```

Behavioral models must reset variables and execution points.

As described in the previous section, the state of a behavioral model is not just composed of the values of the variables. It also includes the location of the statement currently being executed in the sequence of statements describing each concurrent execution thread. To reset a behavioral model, you need not just reset the content of the variables. You must also reset the execution to a specific statement, usually at the top of the process. For example, resetting the process shown in Sample 7-8 would require resetting the variables and signal drivers to their initial values, as well as restarting the execution of the process at the top.

Sample 7-8.
Behavioral process to be reset

```
process
    variable count: integer := 0;
begin
    strobe <= '0';
    wait until go;
    while (go) loop
        count := count + 1;
        wait on sync;
    end loop;
    strobe <= '1';
    wait for 10 ns;
    strobe <= '0';
    wait until ack = '1';
    count := 0;
end process;
```

In VHDL, check for exceptions in all *wait* statements.

Processes in VHDL can be affected by other processes only through signals. For a process to be reset, it has to monitor a reset signal, then take the appropriate action once reset is detected. A single reset signal of type *boolean* is sufficient. It would be set to *true* by a reset control *process* whenever a valid reset condition from any number of sources—such has a hardware, power-up or software reset—is detected.

Using a *boolean* type avoids any misunderstanding about the active level of reset. The activity level of a signal, either high or low, is an implementation detail that we need not concern ourselves with internally. Sample 7-9 shows the process from Sample 7-8 with reset detection and handling. Pretty ugly and unmaintainable if you ask me. An otherwise straightforward sequential description is turned into a complex network of nested *if* statements.

Sample 7-9.
Behavioral
process with
poor reset
detection and
handling

```
process
    variable count: integer := 0;
begin
    strobe <= '0';
    wait until go or reset;
    if not reset then
        while (go and not reset) loop
            count := count + 1;
            wait until sync'event or reset;
        end loop;
        if not reset then
            strobe <= '1';
            wait until reset for 10 ns;
            if not reset then
                strobe <= '0';
                wait until ack = '1' or reset;
            end if;
        end if;
    end if;
    count := 0;
end process;
```

In VHDL, embed
the process body
in a *loop* state-
ment.

In VHDL, the best way to reduce the clutter of nested control flow statements is to embed the body of the process into an infinite loop, as shown in Sample 7-10. The loop iterates during normal operations but is exited whenever a reset condition is detected. The implicit loop around the *process* statement takes the execution of the process back to the top where the initialization code is located. Sample 7-11 shows the resettable process shown in Sample 7-9 with this new control structure. Each *wait* statement must still detect the reset condition, but the sequential description of the algorithm remains almost untouched.

Sample 7-10. Recommended structure of a behavioral process with reset detection and handling

```
process
begin
    -- Initialization
    main: loop
        -- Body of process
        . . .
        exit main when reset;
        . . .
    end loop main;
end process;
```

Sample 7-11. A structured behavioral process with reset detection and handling

```
process
    variable count: integer;
begin
    main: loop
        count := 0;
        strobe <= '0';
        wait until go or reset;
        exit main when reset;
        while (go) loop
            count := count + 1;
            wait until sync'event or reset;
            exit main when reset;
        end loop;
        strobe <= '1';
        wait until reset for 10 ns;
        exit main when reset;
        strobe <= '0';
        wait until ack = '1' or reset;
        exit main when reset;
    end loop main;
end process;
```

In Verilog, disable all the blocks.

Resetting a behavioral model in Verilog is easier and much more elegant. When an exception is detected, all you need to do is disable all the blocks in the model using the *disable* statement. The *always* blocks restart their execution from the top. Note that, as described in "Disabled Scheduled Values" on page 220, pending values assigned using a nonblocking assignment may remain in the event queue and clobber the reset state of a variable in some Verilog simulator.

<table>
<tr><td>Replace *initial* blocks with *always* blocks.</td><td>Only the *initial* blocks present a difficulty. Since they only run once in a simulation, they cannot be disabled since they are no longer active. If they are still active, disabling them would simply make them inactive immediately. To include *initial* blocks in the reset handler, simply replace them with *always* blocks with an infinite *wait* statement at the bottom. Sample 7-12 shows an original Verilog behavioral model. Sample 7-13 shows the same model, this time with the proper handling of reset exceptions using the *disable* statement.</td></tr>
</table>

Sample 7-12. Behavioral model in Verilog

```
initial count = 0;
always
begin
    strobe <= 1'b0;
    wait (go);
    while (go) begin
        count = count + 1;
        @ sync;
    end
    strobe <= 1'b1;
    #10;
    strobe <= 1'b0;
    wait (ack);
end
```

Encapsulate the *disable* statements in a task.

It is good practice to encapsulate all *disable* statements into a single task to perform a reset of a Verilog behavioral model. Multiple reset sources and exception detection can call this task to perform the reset operation. This technique also reduces maintenance to a single location when *always* blocks are added or removed. The *reset* task also can be called using a hierarchical name when a higher-level module in a complex behavioral model needs to reset all its lower-level components. This approach is more efficient than having to assert a reset signal broadcast through the pins of all interfaces in the model. Sample 7-14 shows the reset handler of Sample 7-13 modified to use a task to disable all of the blocks.

Terminate subthreads in OpenVera.

In OpenVera. define a reset event in each *class*. The reset event can be triggered internally when the reset condition is detected, or externally by having the reset event manually triggered. When the reset event is triggered, abort all subthreads started in the class using the *terminate* statement, as illustrated in Sample 7-15.

Sample 7-13.
Behavioral
model with
reset detec-
tion and han-
dling

```
always
begin: init
    count = 0;
    wait (0);
end

always
begin: main
    strobe <= 1'b0;
    wait (go);
    while (go) begin
       count = count + 1;
       @ sync;
    end
    strobe <= 1'b1;
    #10;
    strobe <= 1'b0;
    wait (ack);
end

always
begin
    // Detect reset exception
    . . .
    disable init;
    disable main;
end
```

Sample 7-14.
Encapsulating
the *disable*
statements in a
task

```
task reset;
begin
    disable init;
    disable main;
end
endtask

always
begin
    // Detect reset exception
    . . .
    reset;
end
```

In *e*, use the
rerun() method.

In *e*, define a reset event in each *unit*. The reset event can be triggered from an external HDL signal using a temporal expression or
manually triggered using the *emit* action. In an *on* block, call the
predefined *rerun*() method. If the unit instantiates sub-units, extend

Writing Testbenches: Functional Verification of HDL Models

```
class beh_model {
    event reset;

    slave_if cpu;

    task new() {
        this.cpu = new(...);
        fork
            while (1) {  // Reset detector
                while (this.sigs.$reset != 1'b1) {
                    ...
                }
                trigger(this.reset);
                trigger(this.cpu.reset);
            }
            while (1) {  // Subthread starter
                fork
                    ...
                join none
                sync(ANY, this.reset);
                terminate;
            }
        join none
    }
}
```

the *rerun()* method to trigger the reset event in each sub-unit instance or invoke its *rerun()* method. Sample 7-16 gives an example.

```
unit beh_model is {
    ...
    event clk   is rise('(clk)') @sim;
    event reset is true('(rst)' == 1'b1) @clk;

    on reset {
        me.rerun();
    };

    cpu: slave_if  is instance;
    ram: memory_if is instance;

    rerun() is also {
        emit cpu.reset;
        ram.rerun();
    };
};
```

Writing Good Behavioral Models

Many attempts to write behavioral models fail.

I have seen and heard of many projects where the use of behavioral models was attempted, but without producing much benefit over RTL models. Often, the behavioral model was abandoned in favor of the RTL model as soon as the latter became available. The behavioral model failed to exhibit any the benefits outlined in "The Benefits of Behavioral Models" on page 395.

Writing a good behavioral model requires specialized skills.

Further investigation into those failed attempts usually reveals that the behavioral model was written by experienced hardware designers. Unfortunately, their valuable skills were not appropriate to writing good behavioral models. Their level of thinking was still too close to the implementation and they had difficulty thinking in terms of higher levels of abstraction. Very often, there was the implicit intent of refining the behavioral model into a synthesize-able model. This is a fatal mistake as it is conducive to low-level thinking that yields not a behavioral model, but an RTL++ model.

Focus on the relevant functional details.

All the techniques illustrated in this chapter, as well as in Chapter 4, can be used and still yield a poor behavioral model. A good behavioral model captures the details that are functionally relevant and does not on implementation artifices. For example, the *latency* of a design—the number of clock cycles necessary for an input to be transformed into an output—is usually not functionally relevant[2]. If you insist on writing a model that is clock-cycle accurate with the actual implementation, you may be spending a lot of maintenance effort and adding a lot of complexity for a characteristic that may not be important functionally.

At first glance, latency seems a significant characteristic.

To many, saying that latency may not be a relevant functional detail and should not be modeled sounds like a recipe for disaster. But if you take a step back from your design, ignoring its implementation details, does it *really* matter whether a particular output comes exactly N cycles after the corresponding input was sampled? As long as the *order* of these outputs is the same, is the time at which they come out significant?

Consider the speech synthesizer design illustrated in Figure 3-4 on page 103. To produce audible speech, coefficients must be modified

---

2. But if it *is* relevant, then it should be modeled.

at regular intervals to produce the different sequences of sounds that compose normal speech.

For example, to say "cat", the coefficients would be modified to create the sequence of sounds "k", "a", "a", "a", "t", "t". From these coefficients, a digitized sound waveform should come out at a 8 KHz sample rate. The delay between the time the coefficients are set and the corresponding sound is synthesized is irrelevant, as long as it is under the limit of perception by the user. A similar argument can be made for packet routers: It does not really matter how long it takes for packets to transit through a routing node; what matters is that they eventually come out in the same order.

In some cases, latency *is* significant.

The only time where a "detail" like latency is significant is when the design under verification does not have complete visibility over a system-level "unit of work". A unit of work is the smallest amount of data than can be processed by the system: an atomic operation. For example, a packet router's unit of work is an entire packet. In a speech synthesizer, it is a phoneme. In a hardware tester, it is a complete vector with input and expected output values. If the design under verification only processes a portion of the unit of work, it is important that the latencies in the reconvergent paths are identical so the unit of work is reassembled properly.

For example, the input formatter in a hardware tester, as illustrated in Figure 7-3, only processes the input value. For the corresponding expected output value to be checked at the proper time, it must have the exact same latency as the Expect Delay design.[3] In a packet router, as illustrated in Figure 7-4, if the packet is dismembered to be routed by a different switching node, each node must have an identical latency for the packet to be put back together properly. If you mix behavioral and RTL models in a system-level verification,

---

3. Actually, since the latter is easier to design, its latency is made to match that of the input formatter, whatever it may be.

and each has a different latency, the system-level simulation would become a very effective packet scrambler!

Figure 7-3.
Reconvergent paths in a hardware tester

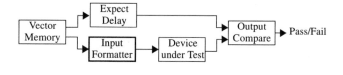

Figure 7-4.
Reconvergent paths in a packet router

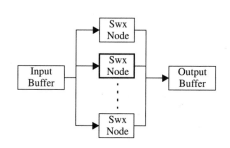

Do not let the testbench dictate what is functionally relevant.

The reason most often cited for making a behavioral model clockcycle accurate with the implementation is to be able to pass the same cycle-oriented testbenches. If the testbenches enforce a specific latency, they are verifying a specific implementation, not a specification.[4] I hope I have explained successfully how to write testbenches that are independent of the latency of the design under verification in Chapters 5 and 6. If your testbenches do not expect a specific latency, then you need not model it.

Details relevant at the system-level can be back-annotated.

An implementation "detail", such as latency, may not be relevant to the functionality of the *stand-alone* design under verification. However, it may be critical for the proper operation of the system-level design. If that is the case, such as the example designs shown in Figure 7-3 and Figure 7-4, the behavioral model still may be modeled as if the latency was not important and perform its transformation in zero-time. At appropriate points in the input or output paths, programmable delay pipelines can be introduced so the exact latency of the implementation can be *back-annotated* into the

---

4. Unless of course a specific latency is required, in which case it should be specified in the specification document. And if something is specified, it should be modeled and verified.

behavioral model. The behavioral model would then model the functionality of the synthesizeable model at a clock-accurate level. Sample 7-17 shows a configurable delay pipeline to adjust the latency of a behavioral model.

Sample 7-17.
Configurable
delay pipeline

```
process (clk)
   constant delay: natural := 1;
   type pipeline_typ is array (integer range <>)
      of data_typ;
   variable pipeline: pipeline_typ(1 to delay);
begin
   if clk = '1' then
      actual_ouput <= pipeline(delay);
      pipeline := output &
                      pipeline(1 to delay - 1);
   end if;
end process;
```

Specify the func-
tionality, not the
implementation.

Another big obstacle to writing good and efficient behavioral models is the level of the specification for the design. If it is written at a very low level, it becomes difficult to abstract significant functionality and discard irrelevant implementation details. I once had to write a behavioral model for a customer whose functional specification was done using technology-independent schematics using a general-purpose drawing tool. Each block was specified independently with no description of the overall functionality. Not only did it make the job of writing RTL code that met timing requirements difficult, it made writing a high-level behavioral model impossible. After 10 weeks, I had a model that was barely faster than the RTL model. But after those 10 weeks, I was able to piece the entire design together in my mind and understand the intended functionality. I scrapped the first model and rewrote it *entirely* in under two weeks. That newer model outperformed the RTL model. Had the specification been written at an appropriate level in the first place, a more effective behavioral model could have been written from the start.

Behavioral Models Are Faster

They are faster to write.

As shown in "Behavioral versus RTL Thinking" on page 121, a behavioral model is much faster to write simply because the functionality is described using significantly fewer statements than an RTL model. Furthermore, behavioral models do not need to meet physical timing or other implementation constraints. They are written with the sole purpose to describe the functionality of a design.

They are faster to debug.

The fewer statements, the fewer bugs. Bugs are easier to identify because of the simpler descriptions. The code is written based on a functional description. The code is not cluttered by directives aimed at a synthesis tool or twisted to be synthesized into specific hardware structures. Behavioral models also tend to use fewer parallel constructs, instead preferring large sequential descriptions in a few processes. Sequential code is much easier to debug than parallel code, since it does not involve synchronization or data exchange intricacies.

They are faster to simulate.

Less code used to describe a function should naturally simulate faster. But the greatest contributor to the increase in simulation speed of a behavioral model over a synthesizeable model is avoiding the use of the synthesizeable subset itself. Look at all the *processes* and *always* blocks used to infer registers. Each and every one of them is sensitive to the clock. If you remember the discussion on event-driven simulation in "The Parallel Simulation Engine" on page 189, you know that this sensitivity causes all of these processes to be scheduled for execution after each event on the clock signal, whether their state changes or not.

In a typical ASIC activity levels are below 40%. This means that over 60% of the processes are evaluated for no reason. A behavioral model only executes when there is useful work to be done. The load it puts on the simulator is much lower. In the small example illustrated in "Contrasting the Approaches" on page 123, the activity in the behavioral model is estimated to be 20 times lower than in the equivalent RTL model.

They are faster to bring to "market".

Being faster to write and debug, a behavioral model takes significantly less time to develop to a level where it can be used in a system-level model. With behavioral models, you are able to start system-level simulations sooner. Because they also simulate faster, you are able to run more of them, on less expensive hardware.

The Cost of Behavioral Models

Behavioral models require additional authoring effort.

Someone has to write these behavioral models. If you use your existing resources, it means that the coding of the RTL model will be delayed. If you do not want to affect the schedule of the synthesizeable model, you will have to hire additional resources to write the behavioral model. But being a completely separate model, it is a task that is easy to parallelize with the implementation effort. And writing a behavioral model is not as costly as writing an RTL model. A behavioral model that is sufficient to start simulating and debugging the testbenches should not take more than two person-weeks to produce. A complete model with all of the functionality of the design under verification should not take more than 5% of the effort required to write an equivalent RTL model.

The maintenance requires additional efforts.

When was the last time you were involved in a design project where the functional specification did not change? Whenever a functional or architectural change is made, the behavioral model needs to be modified. Often, these modifications are dictated by the RTL model because the technology cannot implement the original design and still meet timing requirements. Some of these implementation-driven changes can be planned for and made easy to modify, such as the latency. More significant changes may require rewriting a significant portion of the behavioral model. Toward the end of a project, when schedule pressure is at its greatest, it often leads to the decision of abandoning the behavioral model in favor of focusing on the RTL.[5] However, most of the modifications to an RTL model are made to meet timing goals and do not affect the functionality of the design and thus should not require modification of the behavioral model.

The Benefits of Behavioral Models

Audit the specification.

Most specification reviews I have attended focus on high-level functions and on the spelling and grammatical errors in the document. The missing functional details were often left to be discovered during RTL coding. Decisions regarding these functional details were usually then made according to the ease of implemen-

---

5. An error in my opinion. See the next section titled "The Benefits of Behavioral Models".

tation. There is nothing like writing a model to make you thoroughly read a specification document.

For example, after you've coded a particular function that occurs under some condition, you've come to the *else* part of the *if* statement. What should be done when the condition does *not* occur? Flip, flip, flip through the specification document. Not a word. You've just found a case of incomplete specification! Since you are writing the behavioral model faster than the RTL model, you'll reach that section of the specification earlier than the RTL designers. By the time the RTL model incorporates this functionality, it will have been specified. A similar process occurs with inconsistencies in the specification. When the RTL model is written, there are fewer problems in the specification, and thus it takes less time to write.

Develop and debug the testbenches in parallel with the RTL coding.

Testbenches are implemented using code, just as RTL models are. If the RTL model requires debugging, so do the testbenches. Since a behavioral model is available much earlier than the RTL code, you are able to debug the testbenches earlier as well. You are debugging the behavioral model and the testbenches effectively while the RTL is being written. And because the behavioral model simulates faster than the RTL model, the testbench debug cycles are much shorter.

Once the RTL is available, you will have a whole series of debugged testbenches. Whenever an error is detected, it likely will be due to an error in the RTL model. If you decide to abandon the maintenance of the behavioral model after the RTL is available, debugging the testbenches (which will also need to be modified whenever the RTL is modified significantly) will take much longer. It is important to maintain the behavioral model to keep reaping its benefits for the entire duration of the project.

System verification can start earlier.

Figure 7-5 shows a design process that uses behavioral models for developing the testbenches and the functional verification of the system. Figure 7-6 shows a comparative timeline for a design and verification process with and without behavioral models. The design process is somewhat shortened by using a behavioral model because the testbenches are already debugged. But the greatest saving comes from system verification. The behavioral model is available sooner than the RTL model, so functional verification can start much earlier. Because a behavioral model is much smaller and simulates more efficiently than the equivalent RTL model, you are able to create models of larger systems, execute longer testcases and run on ordinary hardware platform configurations. If the behavioral model is demonstrated to be equivalent to the RTL model, the latter never needs to be brought into the system-level verification. For systems incorporating very large ASICs, a behavioral model may be that which makes system verification even possible.

Figure 7-5.
Design
process
including
behavioral
model

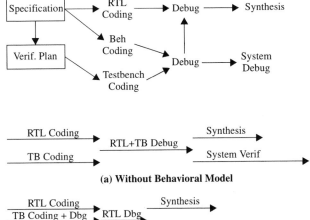

Figure 7-6.
The effect of
behavioral
models on a
project
timeline

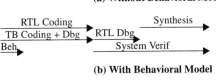

(a) **Without Behavioral Model**

(b) **With Behavioral Model**

It can be used as an evaluation and integration tool by your customers.

If your design is to be available as reusable intellectual property or a chipset, a behavioral model can be a powerful marketing tool. Since it only describes functionality, not implementation, and it is far from being synthesizeable, the behavioral model should not convey intellectual property information.[6] A customer could start using the behavioral model while the legal issues with licensing the RTL model are being resolved. The system-level models could be used as application notes. The behavioral model could be used to start the integration of your design into your customer's design. Since reusing intellectual property is about time-to-market, a behavioral model can be an effective tool to help your customers improve the odds that they will meet their market window.

6. Unless the intellectual property is in the function itself, such as a DSP algorithm.

Demonstrating Equivalence

The RTL and behavioral models must be equivalent.

The greatest benefit from creating a behavioral model comes from system verification. To use it instead of the RTL model in a simulation or as a marketing tool, you have to demonstrate that both are an equivalent representation of the design. I use the term *demonstrate* because I do not think it will ever be possible to *prove* mathematically that they are equivalent.

Equivalence checking can prove that an RTL model is equivalent to a gate-level model or to another RTL model because they are structurally very similar. A properly written behavioral model would use a completely different modeling approach that would be very difficult to correlate mathematically with the equivalent RTL model.

Demonstrate equivalence by using the same test suite.

The only way to demonstrate that the behavioral and the RTL models are equivalent is to verify both of them using the same verification environment. If both models pass the same testcases, from a system-level perspective, it should not matter which one you are using. For a testcase to be executable on both models, it must not depend on a specific implementation. Based on the testcase taxonomy described in "Functional Verification Approaches" on page 12, only black- and grey-box testcases can be used to demonstrate equivalence. Both are executed through the same physical interface. Both do not depend on a particular implementation of the design under verification. The grey-box testcases may not be very relevant to the behavioral model as they are designed to test a particular implementation-specific feature in the RTL model, but should nonetheless execute successfully.

PASS OR FAIL?

This section describes how the ultimate failure or success of a self-checking testbench is determined.

The absence of error is not a sufficient condition.

The goal of a testbench is to determine if the design under verification passes or fails a simulation. But how do you determine if the design passed the simulation? Is it by the absence of error messages? What if the simulation never ran at all? It could be caused by a lack of licenses, or a runtime error such as running out of memory or experiencing a power failure, or a simple syntax error in your source code. You need positive proof that the simulation ran to completion successfully.

Produce and look for a termination message.

Do not rely on a time bomb to terminate your simulation. Nor should you attempt to have the simulation terminate by itself through event starvation. Each simulation should be terminated intentionally. Upon termination, it should produce a message that the simulation was terminated normally. If that message is not present, you must assume that the simulation did not run to completion and failed. To terminate a simulation from within the testbench, use the *$finish* statement in Verilog, or the *assert* statement in VHDL. In OpenVera, use the *exit()* task. In *e*, use the *stop_run()* routine[7]. Sample 7-18 illustrates the use of an explicit termination statement.

Sample 7-18.
Terminating a VHDL simulation

```
test_procedure: process
begin
    ...
    assert false
        report "Simulation terminated normally"
        severity failure;
end process test_procedure;
```

An error in the testbench could prevent error detection.

What if there is a functional problem in your testbench? That error could prevent the testbench from detecting any errors at all. This would clearly be a *false-positive* situation. You should always ensure that your testbench is functionally correct as part of your testcases. Error detection can be verified by injecting errors deliberately in the design under verification. These errors can be intro-

---

7. *e* will go on to execute the clean-up code, but once *stop_run()* is invoked, the simulation will terminate eventually.

duced by simply misconfiguring the design for the expected output. For example, a UART could be configured with the wrong parity setting to verify that the output monitor detects the bad parity.

Provide consistent error message formats.

The final pass or fail judgment could be made by a script parsing the simulation output log file, counting all error messages from all sources. To facilitate the implementation of such a script, use a consistent error format. This style is best accomplished by using a message log package that produces consistent headers, as shown in Sample 7-19.

Sample 7-19.
Simulation log package

```
module log;
integer n_errs;

task warning;
    $write("WARNING at %t: ");
endtask

task error;
    n_errs = n_errs + 1;
    $write("*ERROR* at %t: ");
endtask

endmodule
```

Keep track of success or failure in the log package.

By using a single message log package as shown in Sample 7-20, it is possible for the simulation to keep track of its own success or failure by checking that no error messages were issued. By including a simulation termination function, the final pass or fail indication can be determined by the simulation, without using a script to parse the output log.

_e_ and OpenVera have predefined error reporting procedures.

HVLs come with predefined error message utilities that will declare a failure if any error message is issued. In OpenVera, the predefined task *error*() is used. In *e*, the predefined procedures *warning*(), *error*(), *dut_error*() and *fatal*() are used. The predefined procedures have additional advantages over manually written ones in that they may be able to terminate the simulation automatically, exit out of the simulator, break into the command-line debugger or produce a call stack dump. The effect of an error message, such as whether to continue or break the simulation, is usually user configurable. If any error message is issued through one of the predefined subprograms, the HVL simulator will produce a final message indicating failure.

Sample 7-20.
Determining
pass or fail in
the simulation
log package

```
module log;

integer n_errs;

task warning;
    $write("WARNING at %t: ");
endtask

task error;
    n_errs = n_errs + 1;
    $write("*ERROR* at %t: ");
endtask

task terminate;
begin
    $write("Simulation %0sED\n",
           (n_errs) ? "FAIL" : "PASS");
    $finish;
end
endtask

endmodule
```

Using a script to
parse the simula-
tion output log is
still a good idea.

Using a message log package is not sufficient to determine if a testcase is successful. Other errors could have been generated before the simulation started, when the log package is not available. You still need to confirm the presence of the termination message to verify that the testcase was executed properly in its entirety. Errors could also have been produced by some verification IP using its own log package or from within the HDL model that cannot use the HVL error messaging routines. The output log parsing script can also detect the presence of errors or warnings issued by the simulation management tools, linting tools, syntax errors, elaboration warnings and other possible error conditions not visible to the testbench log package. You can find a link to a configurable output log parser script in the *resources* section of:

http://janick.bergeron.com/wtb.

MANAGING SIMULATIONS

Are you simulating the right model?

You've defined your verification task through a verification plan. You have a test harness with many bus-functional models and utility packages. Several testcases using that test harness have been written and you can choose between the RTL and behavioral model to simulate them. How do you bring all of these components together in a single simulation? How can you reproduce a simulation? And more importantly, how do you make sure that what you simulate is what will be built?

Configuration Management

A configuration is the set of models used in a simulation.

Configuration management is different from source management. Source management, as described in "Revision Control" on page 68, deals with changes to source files and the set of source files making up a particular release. Configuration management deals with the particular set of models you decide to use in a particular simulation. For a specific design, a single configuration would be composed of a specific test function, the test harness used by that test function and the model of the design to be exercised by the testbench. In a system-level simulation, the configuration would also include that particular mix of models used to populate the system model.

It must be easy to specify a particular configuration.

The only information required to define a particular configuration is the identity of the test function, the test harness and the model of the design under verification. The problem is that each configuration component is composed potentially of several source files and design units. Many individuals contribute to the creation of these source files and design units. Their number and names may change throughout the project. It is not realistic to expect every engineer who needs to run a simulation to know exactly what makes up a particular component of the desired simulation. Just as bus-functional models abstract the data from the physical implementation level, configuration management abstracts the details of the structure of a model and the files that describe it.

Use a script to create a configuration.

The most efficient way to abstract the configuration details from the user is to provide a script that expands a test name and an abstraction level for the design under verification into their respective simulation components. Different scripts have to be used for different designs, file system structures and simulators. To simplify the user interface and minimize the amount of repeated information, scripts infer pathnames and expect particular set-up files.

For example, Sample 7-21 shows the command line of a hypothetical script named *sim_design* used to simulate a configuration composed of the test named "*basic_tx*" on the behavioral model. It is followed by a configuration composed of the testcase named "*overflow_rx*" on the RTL model.

Sample 7-21.
Configuration
script com-
mand line

```
% sim_design -b basic_tx
% sim_design -r overflow_rx
```

Verilog Configuration Management

There are many
ways of specify-
ing files.

There are six different ways to include a source file into a Verilog simulation:

1. Specify the filename on the command line.

2. Specify the name of a file containing a list of filenames, using the *-f* option.

3. Specify a directory to search for files likely to contain the definition of a missing module, using the *-y* option. The files used in the simulation depend on the *+libext* command-line option.

4. Specify the name of a file that may contain the definition of missing modules, using the *-v* option.

5. Include a source file inside another using the `include directive. The actual file included in the simulation depends on the *+incdir* command-line option.

6. Locate files in virtual libraries specified in a library search order in a *configuration block*.

Use the *configura-
tion block*.

Of all the mechanisms for specifying input source files in Verilog, only the *configuration block* can be source-controlled and reproduced reliably. The *configuration block* is also the only mechanism that is defined formally as part of the language and not left to the tool implementation.

Use the *-f* option if *configuration blocks* are not available.	*Configuration blocks* are a new (and welcome) addition to Verilog-2001. If your simulator vendor does not support them, complain loudly and use the *-f* option until they do. Since the *-f* option uses a file to specify the actual command-line options used, it can (and must) be source controlled as well.
Use a *configuration block* for each model of the design.	Each available model of the design should be specified using its own *configuration block* (or *-f* file). For example, if you have a behavioral model, two versions of the RTL model (one for FPGAs, the other for the final ASIC) and two gate-level models (one mapped to FPGA gates, the other mapped to ASIC gates), there should be five different *configuration blocks* available.
Use a *configuration block* for the test harness	The test harness should come with its own *configuration block*. The configuration of the test harness would include the configuration of all necessary Verilog bus-functional models instantiated in the harness. If Verilog testbenches are used, the test harness configuration would include the configuration of the self-checking structure.
Compiled simulations are disconnected from the source files.	Using compiled simulation creates a disconnect between the source file and the simulation of the compiled files. What if the source file has changed? How do you make sure that what you are simulating is the proper version of the source files? Configuration management techniques for compiled simulations are outlined in the next section. As all VHDL simulators are compiled simulators, the management of VHDL models must deal with this problem. To eliminate this difficulty and provide a familiar interface to experienced *interpreted* Verilog users, all compiled Verilog simulators provide an interpreted-like mode where all the source files are recompiled every time and simulated immediately.

VHDL Configuration Management

You may not be simulating what you thought was compiled.	All VHDL simulators are *compiled* simulators. During compilation, the individual source files are compiled into libraries and translated to object code. During *elaboration*, a top-level unit is selected and, using the configuration information, a hierarchical model is built by connecting entities and architectures recursively into component instances. The elaborated model is then simulated. A separate command (sometimes two) is used to trigger the compilation and the elaboration. This approach creates a potential disconnect between the source files and what is simulated ultimately. How do you know

that the source files located in your directory are the ones you are simulating?

Use make files.

The most effective way to assure a compiled simulation is up-to-date is to use make files. Make files, and the associated *make* program, were created in the mid-1970s to maintain software programs and make sure that the compiled code was always up to date.

Make files contain dependency rules describing relationships between files. If a file is found to be older than a file it depends on, it is brought up to date using a user-defined command. Nowadays, it is not necessary to know the frustrating syntax[8] of make files. All VHDL and compiled Verilog toolsets can generate a make file from a compiled model. Third-party utilities, such as VMK, can also generate make files automatically from VHDL or Verilog source code.

Use *make* to invoke the simulation.

To ensure that a simulation is always up to date, do not invoke the simulator manually. Some source files might have changed and you would be simulating an out-of-date model. Instead, use the *make* command to invoke the simulator. The program ensures that any source files that have changed since the last compilation are recompiled, in the correct order. There should be a *target* for each testcase available for simulation. The name of the target depends on the tool used to generate the make file. Sample 7-22 shows how to invoke *make* using a specific target.

Sample 7-22.
Invoking *make*

```
% make basic_tx
% make overflow_rx
```

The model should report the name and version of files.

For additional confidence and a positive confirmation of the files and version of the files simulated in a compiled model, you should have each architecture report the name and revision number of the file that contained it during compilation. Sample 7-23 shows how to use a concurrent *assert* statement and RCS keywords to perform this task. All of the *assert* statements are executed at the beginning of the simulation, displaying the filename and revision information. Because they are not sensitive to any signals, they do not execute

8. I personally would like to have a little chat with whomever picked the TAB character as a significant control character!

again. Unfortunately, it is not possible to have *packages* report their source file name and version numbers. A compiled Verilog model could use a *$write* statement and an *initial* block to accomplish the same thing, as shown in Sample 7-24.

Sample 7-23.
VHDL model reporting its filename and revision

```
architecture beh of design is
    ...
begin
    assert false
        report "Configuration: $Header$"
        severity note;
    ...
end beh;
```

Sample 7-24.
Verilog model reporting its filename and revision

```
module design(...);
    ...
initial $write("Configuration: $Header$\n");
    ...
endmodule
```

Use configuration units to define the possible configurations of the design.

VHDL supports the concept of model configuration using configuration units. Configuration declarations are like assembly instructions for a simulation model. They specify which entity and which architecture of that entity should be used for each component instantiation.[9] The author of a model should provide a configuration unit that specifies how to assemble the model in question. The configuration unit becomes an integral part of the source of the model.

For example, as shown in Sample 7-25, there should be a configuration for building the behavioral model of a design, as well as a configuration for building the RTL model of a design, as shown in

---

9. For more information on VHDL configuration, see page 273 of *VHDL Coding Styles and Methodologies*, 2nd edition, by Ben Cohen (Kluwer Academic Publisher)

Sample 7-26. There should also be a configuration for building a model of a board.

Sample 7-25. Configuration for a behavioral model

```
configuration beh_design of design is
for beh  -- architecture of "design"
    . . .
end for;
end beh_design;
```

Sample 7-26. Configuration for an RTL model

```
configuration rtl_design of design is
for rtl  -- architecture of "design"
    . . .
end for;
end rtl_design;
```

Use configuration units for each testcase and test harness.

There should also be a configuration unit to specify the configuration of the test harness. An example is shown in Sample 7-27. If testbenches are implemented using VHDL, each testcase would also be configured using a configuration unit. The configuration of the testcase should include the configuration of the test harness. Sample 7-28 shows an example of a testcase configuration unit.

Sample 7-27. Configuration for a test harness

```
configuration main of harness is
for main  -- architecture of "harness"
    for cpu:cpu_server use ...
    end for;
end for;  -- architecture "main"
end main;
```

Sample 7-28. Configuration for a testcase

```
library harness;
configuration testcase of bench is
for testcase  -- architecture of "bench"
    for th:harness use configuration harness.main;
    end for;    -- configuration "main"
    . . .
end for;      -- architecture "testcase"
end testcase;
```

Generate the final configuration to select the design.

If you examine the configuration for the test harness and the testcase, you will notice it does not include a configuration specification for the design under verification. A default configuration cannot be used because you want the ability to change which model you want to simulate using a specific testcase. Once a model is compiled, the default configuration is selected already.

You could write a different configuration unit for each possible combination of testcase and design under verification. This would duplicate a lot of information and would be cumbersome to maintain should the structure of the testcase or harness be modified. It is easier to generate the final testcase configuration to include the configuration specification to the desired design.

Provide a configuration unit template. The simulation script could parse the VHDL source files to generate a configuration unit automatically. However, it is best to provide a testcase configuration template to the script, with a clearly identifiable placeholder for the configuration specification for the design under verification. Sample 7-29 shows the testcase configuration unit modified to include a placeholder to be expanded by the simulation script. Using the template, the simulation script only needs to replace the placeholder with a reference to the desired design configuration, as shown in Sample 7-30. Notice how the final configuration unit is able to configure a component instantiation in an architecture that was configured previously in its own configuration unit. The generated configuration unit is compiled every time before each simulation.

Sample 7-29. Configuration template for a test harness

```
library harness;
configuration testcase of bench is
for testcase  -- architecture of "bench"
    for th:harness use configuration harness.main;
    for main   -- architecture inside cfg "main"
        for duv:design
            use <design>;
        end for;
    end for;   -- architecture "main"
    end for;   -- configuration "main"
    . . .
end for;       -- architecture "testcase"
end testcase;
```

<table>
<tr>
<td>

Sample 7-30.
Expanded configuration template for a test harness

</td>
<td>

```
library beh_lib;
library harness;
configuration testcase of bench is
for testcase   -- architecture of "bench"
   for th:harness use configuration harness.main;
   for main    -- architecture inside cfg "main"
      for duv:design
         use configuration beh_lib.beh_design;
      end for;
   end for;    -- architecture "main"
   end for;    -- configuration "main"
   ...
end for;       -- architecture "testcase"
end testcase;
```

</td>
</tr>
</table>

OpenVera Configuration Management

Use *make*.

Vera compiles OpenVera source files into object files. The most effective way to assure the object files are up to date is to use make files. Make files, and the associated *make* program, were created in the mid-1970s to maintain software programs and make sure that the compiled code was always up to date.

Make files contain dependency rules describing relationships between files. If a file is found to be older than a file it depends on, it is brought up to date using a user-defined command. An application note in the Vera distribution describes how to generate make files automatically from your OpenVera source code. Note that, for a complete make file to be generated, the code must be compiled manually first.

Report the name and version of files.

For additional confidence and a positive confirmation of the files and version of the files used in a specific simulation run, you should have each *class* and *program* report the name and revision number of the file that contained it during compilation. Sample 7-31 shows how to use a *static* data member and RCS keywords to perform this task in a *class* constructor. Sample 7-32 shows how to perform this task for a *program*.

Use a *vrl* file for the test harness and self-checking structure.

A *vrl* file is a list of object files. There should be a *vrl* file for the transaction-layer harness and a *vrl* file for the self-checking structure. Because *vrl* files cannot include a reference to another *vrl* file and because you cannot specify more than one *vrl* file on the command line using the *+vera_mload* option, the *vrl* file for the self-checking structure must duplicate the content of the transaction-

Sample 7-31.
OpenVera
class report-
ing its file-
name and revi-
sion

```
class some_bfm {
    static local bit config_printed = 0;
    . . .
    task new(...) {
        . . .
        if (!config_printed) {
            printf("Configuration: $Header$\n");
            once = 1;
        }
    }
}
```

Sample 7-32.
OpenVera pro-
gram report-
ing its file-
name and
revision

```
program my_testcase {
    . . .
    printf("Configuration: $Header$\n");
    . . .
}
```

layer test harness *vrl* file. If a *proj* file is going to be used, you can specify multiple *vrl* files.

Use *+vera_vros* to specify stimulus.

The stimulus to be used in a simulation usually involves more than one object file. The *+vera_load* command-line option can be used only once and can specify only a single object file. To specify all of the object files required to run a simulation, in addition to the ones specified in any *vrl* file, use the *+vera_vros* command-line option with a comma-separated list of object files.

e Configuration Management

Use a top-level import file for each major component.

Each major component of the verification environment (e.g., bus-functional models, test harness, scoreboard, self-checking structure, functional coverage collector, random generator, scenario definitions and constraint aspects) should have a top-level file that imports all source files used to implement that component. Whenever that component will be needed by another component in the verification environment, the top-level file of the higher-level component only needs to import the top-level file of the lower-level component.

Specify all required files in a top-level file.

Importing a top-level file should not result in a compilation error because some data types or instances are not found. A top-level file should import all required source files to compile a verification component correctly. Do not worry about having a particular source file imported multiple times from different top-level files: Files with the same name are imported only once.

Avoid relying on SPECMAN_PATH environment variable.

You should not rely on the SPECMAN_PATH environment variable to locate the individual files to import. The value of an environment variable cannot be source-controlled and cannot be reproduced from the source code without including an environment set-up script. Instead, use relative pathnames to locate the imported files that are not in the same directory as the top-level file. Note that relative pathnames are relative to the location of the file that specifies them, not the current working directory. Reserve the SPECMAN_PATH variable for locating the top-level file from third parties that are installed as part of your tool or library distributions, such as the *e*RM package or *e*VCs.

Report the name and version of files.

For additional confidence and a positive confirmation of the files and version of the files used in a specific simulation run, each imported file should report its name and revision number at the beginning of the simulation. Sample 7-31 shows how to use RCS keywords to perform this task in the *sys* struct.

Sample 7-33. *e* file reporting its filename and revision

```
...
<'
package ...;
import ...;

extend sys {
   init() is also {
     outf("Configuration: $Header$\n");
   };
};
...
'>
...
```

Compile the test harness.

By default, the Specman Elite simulator loads then simulates *e* code in a single invocation in interpreted mode. However, any portion of the verification environment can be pre-compiled using the *sn_compile.sh* script and embedded into the Specman Elite base code. Compiled *e* code can be mixed arbitrarily with interpreted *e* code. To get the best possible performance, you should compile the portions of your verification environment that are stable and are no longer changing on a daily basis or from simulation to simulation. The transaction-layer test harness is a prime candidate for compilation. During regression simulations, it is a good idea to compile the self-checking structure and the functional coverage components as well.

SDF Back-Annotation

This section may not be applicable. Back-annotation is a process used only with gate-level models. Due to their large size, they are excruciatingly painful to simulate in terms of performance and resource requirements. The purpose of gate-level simulation is to verify that the synthesis tool has synthesized the RTL description correctly without modifying the functional behavior. The purpose of gate-level simulation is also to verify that there are no timing violations. In most circumstances, these checks are better performed using an equivalence checking static timing analysis tool (see "Equivalence Checking" on page 9).

SDF files are used to model accurate delays. In a gate-level model, each gate is modeled using delays estimated from average output load conditions. However, in a real gate-level netlist, each gate is subject to different output loads: The gates drive different numbers of inputs, and the length of the wires connecting the output of the gate to the driven inputs are different. Each contributes to the load on the output of the gate, producing different loads for different instances of the same gate.

To be more accurate, gate-level simulations are back-annotated with delay values calculated from the physical netlist or the layout geometries. These more accurate delay values are stored in a *Standard Delay File*. The *SDF* file is read by the simulator and each delay value replaces the average delay estimate for each instance. Thus, each gate instance can have a different delay value. The delay between an output pin and each of its driven input pins also can be different.

SDF annotation can take a long time. Gate-level netlists can contain a few million gates and several million pin-to-pin nets or connections. Each must be annotated with a new delay value. This process can be very time consuming and should be minimized whenever possible. If you have to recompile your model for each testcase, you have to perform the back-annotation each time as well.

Use compiled back-annotation whenever possible.	Compiled simulators usually offer compile-time back-annotation of a gate-level model. In that mode, the back-annotation of the delay values is performed once at compile time. Different testcases can be configured to run on the design in separate simulations without requiring that the back-annotation process be repeated.
Concatenate testcases to minimize back-annotation.	If you must use simulation-time back-annotation, you should minimize the time spent back-annotating the gate-level model of the design under verification. This reduction can be accomplished only by minimizing the number of times the simulation is invoked. To invoke the simulation only once for multiple testcases, you need to concatenate each testcase into a single thread that executes each in sequence. This concatenation is not necessary when using Specman Elite as the entire *e* portion of the simulation can be reloaded from scratch without quitting the HDL simulator. If the design is appropriately reset at the start of every *e* simulation, the new testbench can proceed as if it had been simulated independently.
Use a single procedure or task to control each testcase.	You still want each testcase to be written separately, and you want the ability to simulate them independently during development or when SDF back-annotation is not required. It is simple to concatenate each testcase into a sequence, if each testcase is encapsulated in a single *task* or *procedure*. To simulate a particular testcase, you simply have to call the *task* or *procedure* that encapsulates it. The testcases are concatenated by creating a sequence of *task* or *procedure* calls in a sequencing process.
In Verilog, add an *initial* block to execute the testcase standalone.	Sample 7-34 shows the Verilog testcase, originally shown in Sample 6-7 on page 325, encapsulated in a *task*. To execute this testcase in stand-alone mode, an *initial* block calls the task, if the simulation is invoked with the *+standalone* user-defined simulation-time option. To execute this testcase in stand-alone mode, presumably to

debug a problem, you would compile only this testcase with the harness and the design under verification, as shown in Sample 7-35.

Sample 7-34.
Verilog
testcase encap-
sulated in a
task

```
module testcase;

task testcase;
    reg 'ATM_CELL_TYP cell;
begin
    th.write(16'h0001, 16'h0010);
    . . .
    th.atm_port_0.send(cell);
    . . .
end
endtask

initial
begin
    if ($test$plusargs("standalone")) begin
        testcase;
        syslog.terminate;
    end
end
endmodule
```

Sample 7-35.
Command-line
for stand-alone
simulation

```
% verilog -f tests/testcase.f \
         -f beh/design.f
         +standalone
```

Control the
sequence of
testcases using
user-defined
options.

The sequence of testcases is created in a separate module. It contains an *initial* block that can invoke all known testcases. To control which testcases are run and which ones are not, each *task* call is embedded in an *if* statement that tests for a user-defined command-line option. That way, you might run only a subset of testcases instead of all of them. Sample 7-36 shows the structure of the testcase sequencer module. To run a set of testcases, you simply have to specify the name of the desired testcases as a user-defined command-line option. To run all testcases, simply use the *+all_testcases* user-defined option. Sample 7-37 shows an example of each usage with a VCS-compiled simulation. Notice how it was unnecessary to recompile the model to execute different testcases.

Sample 7-36.
Sequencer
module for
simulating
multiple
testcases

```
module sequencer;
initial
begin: sequence
   reg all;

   all = $test$plusargs("all_testcases");

   if (all || $test$plusargs("testcase1") begin
      testcase1.testcase1;
   end
   if (all || $test$plusargs("testcase2") begin
      testcase2.testcase2;
   end
   ...
   syslog.terminate;
end
endmodule
```

Sample 7-37.
Sequencing
different
testcases

```
% vcs -F tests/all.mft -F gate/design.mft \
     -F phy/sdf.mft
% ./simv +testcase3 +testcase7
% ./simv +all_testcases
```

In VHDL, pass all client/server control signals to the encapsulating procedure.

Encapsulating the testcase in a VHDL procedure requires that all client/server control signals be passed through the procedure interface as *signal*-class arguments. Each testcase can be located in individual packages. Sample 7-38 shows the testcase originally shown in Sample 6-15 on page 332 encapsulated in a procedure in a package.

Control the sequence of testcases using top-level generics.

The sequence of testcases is implemented in a process in the architecture instantiating the test harness. The process can invoke all known testcases. To control which testcases are run and which ones are not, each procedure call is embedded in an *if* statement checking if a top-level generic has been defined. That way, you might run only a subset of testcases instead of all of them.

Sample 7-39 shows the structure of the testcase sequencer architecture. To run a set of testcases, you have to override the default value of the top-level generic corresponding to the desired testcase. The type of the top-level generics is *integer* because some VHDL simu-

Sample 7-38.
Encapsulated
client testcase

```
use work.i386sx;
package body testcase is

procedure do_testcase(
    signal to_srv : out i386sx.to_srv_typ;
    signal frm_srv: in  i386sx.frm_srv_typ)
is
    variable data: data_typ;
begin
    ...
    -- Perform a read
    i386sx.read(..., data, to_srv, frm_srv);
    ...
    -- Perform a write
    i386sx.write(..., ..., to_srv, frm_srv);
    ...
end do_testcase;
end testcase;
```

lators may not support user-defined or enumerated types when setting a generic from the command line. To run all testcases, simply override the value of the *all_testcases* top-level generic. Sample 7-40 shows how to select different testcase sequences by setting the top-level generics on the simulation command line under Model-Sim.

Sample 7-39.
Sequencer
architecture
for simulating
multiple
testcases

```
entity sequencer;
    generic (all_testcases: natural := 0;
             testcase1    : natural := 0;
             testcase2    : natural := 0; ...);
end sequencer;

use work.testcase1;
use work.testcase2;
...
use work.i386sx;
...
architecture test of sequencer is
begin

    duv: design;

    run:process
    begin
        if all_testcases + testcase1 > 0 then
            testcase1.do_testcase1(i386sx.to_srv,
                                   i386sx.frm_srv);
        end if;
        if all_testcases + testcase2 > 0 then
            testcase2.do_testcase2(i386sx.to_srv,
                                   i386sx.frm_srv);
        end if;
        ...
        assert false
            report "Simulation completed"
            severity failure;
    end process run;
end test;
```

Sample 7-40.
Sequencing
different
testcases

```
% vsim -Gtestcase3=1 -Gtestcase7=1 sequencer
% vsim -Gall_testcases=1 sequencer
```

Output File Management

Simulations pro-
duce output files.

A simulation usually creates at least one output file. For example, Verilog-XL simulations generate a copy of the output messages in a file named *verilog.log* by default. Another frequently produced output file is the file containing the signal trace information for a waveform viewer. These output files are valuable. They are used to determine if the simulation was successful. They should be saved after each simulation run and parsed or post-processed to determine success or failure.

Multiple simula-
tions can clobber
each other's files.

When you run only one simulation at a time, you can save them by renaming them after the completion of the simulation. That way, you can keep a history of testcases that were run on the design under verification. However, if you run multiple simulations in parallel, usually on different machines, the output files from one simulation can clobber those of another. If you rely on default or hardcoded filenames, you will not be able to run simulations in parallel. You must be able to name files differently for different testcases.

Specify output
filenames on the
command line in
your simulation
run script.

A few default output filenames can be changed from the command line. For example, the *-l* option in Verilog can be used to change the name of the output *log* file. In "Configuration Management" on page 402, I recommended that you use a script to help manage the configuration of a simulation. That same script can also manage the naming of the output files according to the name of the testcase. Sample 7-41 shows how a perl script can use the name of the testcase specified on the command line to rename the output log file in a Verilog simulation.

Sample 7-41.
Simulation run
script

```
require "getopts.pl";
&usage if &getopts("hr") || $opt_h || !@ARGV;

sub usage {
    print <<USAGE;
Usage: $0 [-r] {testcase}
  -r   Use the RTL model instead of behavioral
USAGE
    exit (1);
}

$design = ($opt_r) ? "rtl" : "beh";
$prefix = "verilog -f $design/design.f ";

foreach $test (@ARGV) {
    $command = "$prefix -f tests/$test.f";
    $command .= " -l logs/$test.log";
    system($command);
}
```

In VHDL, use a *string* type constant and generic.

Not all filenames can be renamed from the command line. In VHDL, all file objects are named from within the VHDL model. To make the filenames unique for each testcase, provide the name of the current testcase to the test harness to create unique filenames. In the testcase itself, you must be careful to use the testcase name to generate filenames as well. The simplest way is to have each testcase contain a constant defined as the (unique) testcase name. The value of that constant would be used to generate filenames and passed to the test harness via a generic. Sample 7-42 shows an example. Notice how the string concatenation operator is used to generate a unique filename in the file declarations.

In Verilog, use a parameter.

A similar technique can be used in Verilog. You simply use a *parameter* instead of the constant and pass the testcase name to the test harness as a parameter association. Strings in Verilog are simply bit vectors with eight bits per character. They can be concatenated using the usual concatenation operator to create unique filenames. Sample 7-43 shows an example.

Use a *string* expression in HVLs.

In OpenVera and *e*, filenames are specified as string expressions. These expressions can include a testcase name specified as a string

---

Writing Testbenches: Functional Verification of HDL Models 419

Sample 7-42.
Generating
unique file
names in
VHDL

```
architecture test of bench is
    constant name: string := "testcase";
    component harness
        generic (name: string);
    end component;
begin

    th: harness generic map (name);

    process
        file results: text is out name & ".out";
    begin
        . . .
    end process;
end test;
```

Sample 7-43.
Generating
unique file
names in Ver-
ilog

```
module testbench is
parameter name = "testcase";

harness #(name) th ();

initial
begin
    integer results;
    results = $fopen({name, ".out"});
    . . .
end
endmodule
```

variable. The final filename can be built using the *sprintf()* and
appendf() predefined routines respectively, as shown in Sample 7-
44.

Sample 7-44.
Generating
unique file
names in *e*

```
extend sys {
    testname: string;
    keep testname == "corner_case";
};
. . .
var fname: string = appendf("%s.out",
                            sys.testname);
var fp    : file = files.open(fname, "w",
                            "Scratch File");
. . .
```

Seed Management

Seeds contribute random stability.

The main concern with random stimulus is reproducing a simulation that detected a functional error. Random stability allows the same input sequence to be generated if the same initial seed is used, even in the presence of changes in the source code. Any instance that is not affected by randomization-related source code changes—such as additional constraints or random objects—will always produce the same pseudo-random sequence in two different simulations if the same seed is used, even if other instances are affected by such changes. Random stability involves more than using the same seed. A C++ model using the *random()* function is not randomly stable as any change in randomization-related code will affect all subsequent calls to the function in the entire model. With random stability, the effects are localized in the modified instance.

Don't always use the default seed.

Vera and Specman Elite use a default seed unless a different seed is specified. Most people keep using the default seed over and over until the simulation runs error free, then they consider their job done. Using the same seed will generate the same input sequence. You will not be verifying or debugging your environment under different conditions. Before declaring your environment or bus-functional model "done", verify it with different seeds.

Pick random seeds.

To pick random seeds using Specman Elite, use the *-seed=random* option to the *test* or *gen* command. To pick a random seed using Vera, use the *+vera_random_seed=...* command-line option, where "..." is a random value generated by a suitable function in your simulation run script or the output of the simple C program shown in Sample 7-45. Do not generate a random seed based on the current system time from within Vera because, by the time the random seed is set, some constructors may already have been invoked, initializing the local random source for those instances from the default seed.

Sample 7-45.
Generating a
random seed

```
#include <stdlib.h>
#include <time.h>
main() {
    srandom(time(NULL));
    printf("%d\n", random());
    exit(0);
}
```

Whatever seed is being used in a simulation, its value must be known so it can be reused. Vera and Specman Elite display the value of the seed used at the beginning of the simulation, as shown in Sample 7-46 and Sample 7-47. This display makes it possible to associate the results in an output log file with a particular seed.

Sample 7-46.
Random seed
display from
Specman Elite

```
Specman> test -seed=random
Doing setup ...
Generating the test using seed 7005084...
Starting the test ...
...
```

Sample 7-47.
Random seed
display from
Vera

```
% vera_cs +vera_random_seed='random' ...
...
WARNING: override default seed 1 with command
line seed 715573055
...
```

Associate output
files with seeds.

Simulation results are the product of a simulation run with a specific seed. Performing another simulation run, using the exact same code, but with a different seed will yield different results. It is therefore important to associate results with a specific seed. Once this association exists, results can be reproduced. They also can be graded to identify which seeds contribute most toward the final verification objective.

Include the seed in
all output file
names.

If the same output filename is used by two simulations of the same code but using different seeds, the results from the first simulation will be lost. You should include the seed value in all output pathnames. This technique can be done by putting all output files in a seed-specific directory or by including the seed value in the filename itself. Specman Elite already includes the seed value in the functional coverage database output file name. But any file created by the user, the default *specman.elog* file or any Vera output file is not, by default, associated with the seed.

REGRESSION

A regression
ensures backward
compatibility.

A regression suite ensures that modifications to the design remain backward compatible with previously verified functionality. Many times, a change in the design made to fix a problem detected by a testcase, will break functionality that was verified previously. Once a testbench is written and debugged to simulate successfully, you must make sure that it continues to be successful throughout the duration of the design project.

Running Regressions

Regressions are
run at regular
intervals.

As individual self-checking testbenches are completed, they are added to a master list of testbenches included in the regression simulation. This regression simulation is run at regular intervals, usually nightly. For directed testcases, simulations are run one after another. For random-based testbenches, simulations are run repeatedly using different seeds. As the number of testbenches or the size of the functional coverage space grows, it may not be possible to complete a full regression simulation overnight.

Divide directed
testcases into two
groups.

Directed testcases can then be classified into two groups: One group is run every night, while the second group is included only in regressions run over a weekend. Testcases selected for the first group should be the ones that verify the basic functionality of the design.

Rank seeds.

With random-based testbenches, rank[10] seeds based on their incremental contribution to the overall functional coverage goal. Select the seeds that provide the greatest contribution and start the regressions simulation with those. If there is still time left in the regression period, continue with additional, randomly selected seeds.

Testbenches may
have a fast mode
to speed up regres-
sions.

Another approach is to provide a *fast mode* to testcases where only a subset of the functionality is verified during overnight regression simulations, or simulations are run for shorter period of time with the same seed. The full-length simulation would be performed only during individual simulations or regression simulations over a weekend. In Verilog or OpenVera, the fast mode could be turned on using a user-defined command-line option, as shown in Sample 7-

10. In Specman Elite, see the *rank cover* command.

48. In VHDL, it could be turned on using a top-level generic, as shown in Sample 7-49. In *e* it could be turned on by loading an additional aspect composed of additional constraints and method extensions.

Sample 7-48.
Implementing a fast mode in a Verilog testbench

```
% verilog ... +fastmode

module testcase;
...
initial
begin
    // Repeat only 4 times in fast mode
    repeat (($test$plusarg("+fastmode"))?4:256)
        begin
            ...
        end
    syslog.terminate;
end
endmodule
```

Sample 7-49.
Implementing a fast mode in a VHDL testbench

```
entity bench is
    generic (fast_mode: integer := 0);
end bench;

architecture test of bench is
begin
    process
        variable repeat: integer := 256;
    begin
        -- Repeat only 4 times in fast mode
        if fast_mode /= 0 then
            repeat := 4;
        end if;
        for I in 1 to repeat loop
            ...
        end loop;
        assert false
            report "Simulation completed"
            severity failure;
    end process;
end test;
```

Use a script to run regressions.

A regression script could invoke each testbench in the regression test suite, for a specific number of seeds, using the simulation configuration script used to invoke individual simulations, as discussed in "Configuration Management" on page 402. If the number and duration of testbenches in the regression suite make it impossible to run a regression simulation in the allotted time, you will want to consider parallel simulations. If you do, it is necessary that testbenches be designed to produce results independently from each other, as discussed in "Output File Management" on page 418. Parallel simulations can be managed using readily available utilities, such as *pmake* or *LSF*.

Regression Management

Check out a fresh view with local copies.

Not all source files are suitable for regression runs. If you are using your revision control system properly, you should be checking in files at times convenient for you, not convenient for the regression run. The latest version of a file might contain code that was not verified at all or that might even have syntax errors. You do not want to waste a regression simulation on files that were not debugged properly. Before running a regression, you should checkout a complete view of the source control database, populated with local copies whose revisions are tagged as being suitable for regression testing. This tag is applied by verification and design engineers once they have confidence in the basic functionality of the code and are ready to submit that particular revision of the testbench or the design to regression. Sample 7-50 shows an example of tagging a particular view of a file system, then checkout the particular files associated with a tag at a later time using CVS.

Sample 7-50. Tagging and retrieving a particular revision of a view

```
% cvs tag -R regress
. . .
% cvs update -dA -rregress
```

Put a time bomb in all simulations.

One of the greatest killers of regression simulations, second only to the infinite loop, is the simulation that never terminates. A simulation will run forever if a condition you are waiting for never occurs. The clock generator keeps the simulation alive by generating events continuously. Time advances until the maximum value is reached, which, in modern simulators using 64-bit time values, will take a long time! To prevent a testcase from hanging a regression simulation, include a time bomb in all simulations. This time bomb should go off after a delay long enough to allow the normal operations of the testcase to complete without interruption. Sample 7-51 shows a time bomb, used with a concurrent procedure call in Sample 7-52.

The procedure could be modified to include a signal argument that, when triggered with an event, would reset the fuse.

Sample 7-51.
Time bomb
procedure

```
package body bomb_pkg is

procedure timebomb(constant fuse: in time) is
begin
    wait for fuse;
    assert false
        report "Boom!"
        severity failure;
end timebomb;

end bomb_pkg;
```

Sample 7-52.
Using the time
bomb proce-
dure

```
use work.bomb_pkg.all;
architecture test of bench is
begin
    bomb: timebomb(fuse => 12 ms);

    test_procedure: process
    begin
        ...
    end process test_procedure;
end test;
```

Do not rely on a time bomb for normal termination.

The time bomb should be used only to prevent runaway simulations from running forever. It should not be used to terminate a testcase under normal conditions. It would be impossible to distinguish between a successful completion of the testcase and a deadlock condition. Furthermore, the time bomb would require fine tuning every time the testbench or design is modified to avoid the testcase from being interrupted prematurely or wasting simulation cycles by running for too long.

Automatically generate a report after each regression run.

Once the regression simulation is completed, the success or failure of each testcase in the regression suite should be checked using the output log scan script (see "Pass or Fail?" on page 399.) The results are then summarized into a single regression report outlining which particular testbench and seed was successful or unsuccessful. It is a good idea to have the regression script mail the report to all the engineers in the design team to ensure that the design remains backward compatible at all times. Reviewing this report also should be the first item on the agenda in any design team meeting.

SUMMARY

Write a behavioral model to help debug your verification environment sooner and faster.

Behavioral models are not the same as RTL models but must pass the same verification suite.

Behavioral models enable system-level verification.

Carefully model exceptions in behavioral models.

Use a common error reporting mechanism.

Use a script to look for the absence of error messages and the presence of the termination message to declare if a simulation completed successfully.

Manage your models and the components of the verification environment using configuration techniques.

Have compiled code report the filename and version number in the output log file.

Have a mechanism for reporting—and later specifying—a seed used for a particular simulation run.

Separate output files for each testbench and each seed used to simulate each testbench.

Run regression simulations at regular intervals, using a tagged version of the design and verification environment.

Include a time bomb in all simulations to prevent a single testbench from blocking an entire regression run.

APPENDIX A   CODING GUIDELINES

There have been many sets of coding guidelines published for hardware description languages, but historically they have focused on the synthesizeable subset and the target hardware structures. Writing testbenches involves writing a lot of code and also requires coding guidelines. These guidelines are designed to enhance code maintainability and readability, as well as to prevent common or obscure mistakes.

Verisity has *e* coding guidelines.

Verisity has published coding guidelines it recommends be used when coding *e*. These guidelines can be found in the shareware section of the Verification Vault (http://www.verificationvault.com).

Guidelines are structured from the generic to the specific.

The guidelines presented here are reproduced with permission from the *Reuse Methodology Field Guide* from Qualis Design Corporation (http://www.qualis.com). They are organized from the general to the specific. They start with general coding guidelines that should be used in any language. They are followed by guidelines specific to hardware description languages. Verilog- and VHDL-specific guidelines follow after that. Note: A guideline applicable to a more specific context can contradict and supersede a more general guideline.

Define guidelines as a group, then follow them.

Coding guidelines have no functional benefits. Their primary contribution is toward creating a readable and maintainable design. Having common design guidelines makes code familiar to anyone familiar with the implied style, regardless of who wrote it. The primary obstacle to coding guidelines are personal preferences. It is important that the obstacle be recognized for what it is: personal taste. There is no intrinsic value to a particular set of guidelines. The value is in the fact that these guidelines are shared by the entire group. If even one individual does not follow them, the entire group is diminished.

DIRECTORY STRUCTURE

Use an identical directory structure for every project.

Using a common directory structure makes it easier to locate design components and to write scripts that are portable from one engineer's environment to another. Reusable components that were designed using a similar structure will be more easily inserted into a new project.

Example project-level structure:

```
.../bin/      Project-wide scripts/commands
   doc/       System-level specification documents
   SoCs/      Data for SoCs/ASICs/FPGA designs
   boards/    Data for board designs
   systems/   Data for system designs
   mech/      Data for mechanical designs
   shared/    Data for shared components
```

At the project level, there are directories that contain data for all design components for the project. Components shared, unmodified, among SoC/ASIC/FPGA, board and system designs are located in a separate directory to indicate that they impact more than a single design. At the project level, shared components are usually verification and interface models.

Some "system" designs may not have a physical correspondence and may be a collection of other designs (SoCs, ASICs, FPGAs and boards) artificially connected together to verify a subset of the system-level functionality.

Each design in the project has a similar structure. Example of a design structure for an SoC:

```
SoCs/name/          Data for ASIC named "name"
        doc/        Specification documents
        bin/        Scripts specific to this design
        beh/        Behavioral model
        rtl/        Synthesizeable model
        syn/        Synthesis scripts & logs
        phy/        Physical model and SDF data
        verif/      Verif env and simulation logs
SoCs/shared/        Data for shared ASIC components
```

Components shared, unmodified, between SoC designs are located in a separate directory to indicate that they impact more than a single design. At the SoC level, shared components include processor cores, soft and hard IP and internally reused blocks.

Use relative pathnames.

Using absolute pathnames requires that future use of a component or a design be installed at the same location. Absolute pathnames also require that all workstations involved in the design have the design structure mounted at the same point in their file systems. The name used may no longer be meaningful, and the proper mount point may not be available.

If full pathnames are required, use preprocessing or environment variables.

Put a Makefile with a default 'all' target in every source directory.

Makefiles facilitate the compilation, maintenance, and installation of a design or model. With a Makefile the user need not know how to build or compile a design to invoke "make all". Top-level makefiles should invoke *make* on lower level directories.

Example "all" makefile rule:

```
all: subdirs ...

SUBDIRS = ...
subdirs:
      for subdir in $(SUBDIRS); do \
          (cd $$subdir; make); \
      done
```

VHDL Specific

Locate the directories containing the libraries as subdirectories of the source directories.

It is a good idea to name a library according to the VHDL toolset it corresponds to. This naming convention makes it possible to use more than one VHDL toolset on the same source installation.

Example:

```
.../SoC/name/beh/
              work.nc/     NC library
              work.msim/   ModelSim library
              work.vss/    VSS library
          rtl/
              work.nc/     NC library
              work.msim/   ModelSim library
              work.vss/    VSS library
```

Create a file that lists all required libraries (other than WORK) and lists the full relative path name to the directory containing the source files for that library.

Having this file makes it easier to locate all source files required by a design or a portion of a design. This file also can be processed by a script that automatically generates the VHDL toolset library map file, which associates logical library names and container directories.

Verilog Specific

Create a file that lists all required source files and command-line options for simulating the design in every directory that contains a Verilog description of a design (or sub-design).

> The file, called a manifest file, is used with the *-f* option when invoking the Verilog simulator.

To include a sub-design in a higher-level design, include the manifest file for the sub-design using the *-f* option in the higher-level manifest file.

> This structure effectively creates hierarchical manifest files.

Specify files loaded using the `include directive using a complete relative pathname.

> Requiring the use of a *+incdir* option on the Verilog command line makes it impossible to determine, from the source code only, which files are required to describe a model completely. The exact command line used is also required to reproduce any problems. If a complete relative pathname is specified, it becomes easy to locate all files required to make up a complete model.

GENERAL CODING GUIDELINES

These guidelines are intended to be used for any programming or scripting language. Additional guidelines for HDL and language-specific descriptions can be found in the section titled "HDL Coding Guidelines" on page 452.

Comments

Put the following information into a header comment for each source file: copyright notice, brief description, revision number, original author name and contact data, and current author name and contact data.

Example (PERL script under RCS):

```
#! /usr/local/bin/perl
#
# (c) Copyright 1999, Qualis Design Corporation
# All rights reserved.
#
# This file contains proprietary and confidential
# information. The content or any derivative work
# can be used only by or distributed to licensed
# users or owners.
#
# Description:
#    This script parses the output of a set of
#    simulation log files and produces a
#    regression report.
#
# Original author: John Q. Doe <jdoe@qualis.com>
# Current author : Jane D. Hall <jhall@qualis.com>
# Revision       : $Revision$
#
```

Use a trailer comment describing revision history for each source file.

The revision history should be maintained automatically by the source management software. Because these can become quite lengthy, put revision history at the bottom of the file. This location eliminates the need to wade through dozens of lines before seeing the actual code.

Example (shell script under RCS):

```
#
# History:
#
# $Log$
#
```

Use comments to describe the intent or functionality of the code, not its behavior.

Comments should be written with the target audience in mind: A junior engineer who knows the language, but is not familiar with the design, and must maintain and modify the design without the benefit of the original designer's help.

Example of a useless comment (C):

```
/* Increment i */
i++;
```

Example of a useful comment (C):

```
/
* Move the read pointer to the next input element
*/
i++;
```

Preface each major section with a comment describing what it does, why it exists, how it works and assumptions on the environment.

A major section could be a *process* in VHDL, an *always* block in Verilog, a *function* in C, a *unit* in *e*, a *class* in OpenVera or a long sequence of code in any language.

It should be possible to understand how a piece of code works by looking at the comments only and by stripping out the source code itself. Ideally, it should be possible to understand the purpose and structure of an implementation with the source code stripped from the file, leaving only the comments.

Describe each input and output of subprograms in individual comments.

> Describe the purpose, expected valid range, and effects of each input and output of all subprograms or other coding unit supported by the language. Whenever possible, show a typical usage.
>
> Example (PERL):

```
#
# Subroutine to locate all files matching a given
# pattern under a given directory path
#
sub scandir {      # Returns array of filenames
   local($dir,     # Dir to recursively scan (str)
         $pattern) # Filename pattern (regexp str)
         = @_;
   . . .
}
```

Delete bad code; do not comment-out bad code.

> Commented-out code begs to be reintroduced. Use a proper revision control system to maintain a track record of changes.

Layout

Lay out code for maximum readability and maintainability.

> Saving a few lines or characters only saves the time it takes to type it. Any cost incurred by saving lines and characters will be paid every time the code has to be understood or modified.

Use a minimum of three spaces for each indentation level.

> An indentation that is too narrow (such as 2), does not allow for easily identifying nested scopes. An indentation level that is too wide (such as 8), quickly causes the source code to reach the right margin.

Write only one statement per line.

> The human eye is trained to look for sequences in a top-down fashion, not down-and-sideways. This layout also gives a better opportunity for comments.

Example of poor code layout (PERL):

```
$| = 1; print "Continue (y/n) ? [y] ";
$ans = <STDIN>; last if $ans =~ m/^\s*[nN]/;|
```

Example of good code layout (PERL):

```
# Prompt the user...
$| = 1;
print "Continue (y/n) ? [y] ";

# Read the answer
$ans = <STDIN>;

# Get out if answer started with a "n" or "N"
last if $ans =~ m/^\s*[nN]/;
```

Limit line length to 72 characters. If you must break a line, break it at a convenient location with the continuation statement and align the line properly within the context of the first token.

Printing devices are still limited to 80 characters in width. If a fixed-width font is used, most text windows are configured to display up to 80 characters on each line. Relying on the automatic line wrapping of the display device may yield unpredictable results and unreadable code.

Example of poor code layout (Verilog):

```
#10
expect = $realtobits((coefficient * datum)
         + 0.5);
```

Example of good code layout (Verilog):

```
#10 expect = $realtobits((coefficient * datum)
                         + 0.5);
```

Use a tabular layout for lexical elements in consecutive declarations, with a single declaration per line.

> A tabular layout makes it easier to scan the list of declarations quickly, identifying their types, classes, initial values, etc. If you use a single declaration per line, it is easier to locate a particular declaration. A tabular layout also facilitates adding and removing a declaration.

> Example of poor declaration layout (VHDL):

```
signal counta, countb: integer;
signal c: real := 0.0;
signal sum: signed(0 to 31);
signal z: unsigned(6 downto 0);
```

> Example of good declaration layout (VHDL):

```
signal counta: integer;
signal countb: integer;
signal c     : real                     := 0.0;
signal sum   : signed   (0 to     31);
signal z     : unsigned (6 downto 0 );
```

If supported by the language, use named associations when calling subprograms or instantiating subunits. Use a tabular layout for lexical elements in consecutive associations, with a single association per line.

> Using named associations is more robust than using port order. Named associations do not require any change when the argument list of a subprogram or subunit is modified or reordered. Furthermore, named associations provide for self-documenting code as it is unnecessary to refer to another section of the program to identify what value is being passed to which argument. A tabular layout makes it easier to scan the list of arguments being passed to a subprogram quickly. If you use one named association per line, it is easier to locate a particular association. A tabular layout also facilitates adding and removing arguments.

Example of poor association layout (Verilog):

```
fifo in_buffer(voice_sample_retimed,
               valid_voice_sample, overflow, ,
               voice_sample, 1'b1, clk_8kHz,
               clk_20MHz);
```

Example of good association layout (Verilog):

```
fifo in_buffer(.data_in    (voice_sample),
               .valid_in   (1'b1),
               .clk_in     (clk_8kHz),
               .data_out   (voice_sample_retimed),
               .valid_out  (valid_voice_sample),
               .clk_out    (clk_20MHz),
               .full       (overflow),
               .empty      ());
```

Syntax

Do not use abbreviations.

Some languages, particularly scripting languages, allow using an abbreviated syntax, usually as long as the identifiers are unique prefixes. Long form and command names are self-documenting and provide a more consistent syntax than various abbreviations. If additional commands are later added to the language, abbreviations that used to be unique may now conflict with the new commands and require modification to remain compatible with the newer versions.

Example of poor command syntax (DC-shell):

```
re -f verilog design.v
```

Example of good command syntax (DC-shell):

```
read -format verilog design.v
```

Encapsulate repeatedly used operations or statements in subprograms.

By using subprograms, maintenance is reduced significantly. Code only needs to be commented once and bugs only need to be fixed once. Using subprograms also reduces code volume.

Example of poor expression usage (Verilog):

```
// sign-extend both operands from 8 to 16 bits
operand1 = {{8 {ls_byte[7]}}, ls_byte};
operand2 = {{8 {ms_byte[7]}}, ms_byte};
```

Example of proper use of subprogram (Verilog):

```
// sign-extend an 8-bit value to a 16-bit value
function [15:0] sign_extend;
    input [7:0] value;
    sign_extend = {{8 {value[7]}}, value};
endfunction

// sign-extend both operands from 8 to 16 bits
operand1 = sign_extend(ls_byte);
operand2 = sign_extend(ms_byte);
```

Use a maximum of 50 consecutive sequential statements in any statement block.

Too many statements in a block create many different possible paths. This layout makes it difficult to grasp all of the possible implications. It may be difficult to use a code coverage tool with a large statement block. A large block may be broken down using subprograms.

Use no more than three nesting levels of flow-control statements.

Understanding the possible paths through several levels of flow control becomes difficult exponentially. Too many levels of decision making may be an indication of a poor choice in processing sequence or algorithm. Break up complex decision structures into separate subprograms.

Example of poor flow-control structure (C):

```c
if (a == 1 && b == 0) {
    switch (val) {
    4:
    5: while (!done) {
            if (val % 2) {
                odd = 1;
                if (choice == val) {
                    for (j = 0; j < val; j++) {
                        select[j] = 1;
                    }
                    done = 1;
                }
            } else {
                odd = 0;
            }
        }
        break;
    0: for (i = 0; i < 7; i++) {
            select[j] = 0;
        }
        break;
    default:
        z = 0;
    }
}
```

Example of good flow-control structure (C):

```
void
process_selection(int val)
{
    odd = 0;
    while (!done) {
        if (val % 2) {
            odd = 1;
        }
        if (odd && choice == val) {
            for (j = 0; j < val; j++) {
                select[j] = 1;
            }
            done = 1;
        }
    }
}

if (a == 1 && b == 0) {
    switch (val) {
    0: for (i = 0; i < 7; i++) {
            select[j] = 0;
        }
        break;
    4:
    5: process_selection(val);
        break;
    default:
        z = 0;
    }
}
```

Debugging

Include a mechanism to exclude all debug statements automatically.

Debug information should be excluded by default and should be enabled automatically via a control file or command-line options. Do not comment out debug statements and then uncomment them when debugging. This approach requires significant editing. When available, use a preprocessor to achieve better runtime performance.

Example of poor debug statement exclusion (Verilog):

```
// $write("Address = %h, Data = %d\n",
//         address, data);
```

Example of proper debug statement exclusion (Verilog):

```
`ifdef DEBUG
   $write("Address = %h, Data = %d\n",
          address, data);
`endif
```

NAMING GUIDELINES

These guidelines suggest how to select names for various user-defined objects and declarations. Additional restrictions on naming can be introduced by more specific requirements.

Capitalization

Use lowercase letters for all user-defined identifiers.

Using lowercase letters reduces fatigue and stress from constantly holding down the Shift key. Reserve uppercase letters for identifiers representing special functions.

Do not rely on case mix for uniqueness of user-defined identifiers.

The source code may be processed eventually by a case-insensitive tool. The identifiers would then lose their uniqueness. Use naming to differentiate identifiers.

Example of bad choice for identifier (C):

```
typedef struct RGB {
    char red;
    char green;
    char blue;
} RGB;

main () {
    RGB rgb;
    ...
}
```

Example of better choice for identifier (C):

```
typedef struct rgb_struct {
    char red;
    char green;
    char blue;
} rgb_typ;

main () {
    rgb_typ rgb;
    ...
}
```

In a case-insensitive language, do not rely on case mix for adding semantic to identifiers.

Instead of using the case of identifiers to document variable types, use naming (prefix, suffix) to self-document identifiers. Consistent case usage for a given identifier cannot be enforced by the compiler and therefore may end up being used incorrectly.

Example of poor choice of identifier (VHDL):

```
package Pci is
    type command is (MEM, IO, CONFIG);
    procedure readCycle(ADDRESS: in ...;
                             data: out ...);
end Pci;
```

Example of proper choice of identifier (VHDL):

```
package pci_pkg is
   type command_typ is (MEM, IO, CONFIG);
   proceedure read_cycle(address_in: in ...;
                            data_out: out ...);
end pci_pkg;
```

Use uppercase letters for constant identifiers (runtime or compile-time).

The case differentiates between a symbolic literal value and a variable.

Example (Verilog):

```
module block(...);
...
`define DEBUG
parameter WIDTH = 4;
...
endmodule
```

Separate words using an underscore; do not separate words by mixing upper-case and lowercase letters

It can be difficult to read variables that use case to separate word boundaries. Using spacing between words is more natural. In a case-insensitive language or if the code is processed through a case-insensitive tool, the case convention cannot be enforced by the compiler.

Example of poor word separation (C):

```
readIndexInTable = 0;
```

Example of proper word separation (C):

```
read_index_in_table = 0;
```

Identifiers

Do not use reserved words of popular languages or languages used in the design process as user-defined identifiers.

Not using reserved words as identifiers avoids having to rename an object to a synthetic, often meaningless, name when translating or generating a design into another language. Popular languages to consider are C, C++, Verilog, VHDL, PERL, OpenVera and *e*.

Use meaningful names for user-defined identifiers, and use a minimum of five characters.

Avoid acronyms or meaningless names. Using at least five characters increases the likelihood of using full words.

Example of poor identifier naming (VHDL):

```
if e = '1' then
    c := c + 1;
end if;
```

Example of good identifier naming (VHDL):

```
if enable = '1' then
    count := count + 1;
end if;
```

Name objects according to function or purpose; avoid naming objects according to type or implementation.

This naming convention produces more meaningful names and automatically differentiates between similar objects with different purposes.

Example of poor identifier naming (Verilog):

```
count8 <= count8 + 8'h01;
```

Example of good identifier naming (Verilog):

```
addr_count <= addr_count + 8'h01;
```

Do not use identifiers that are overloaded or hidden by identical declarations in a different scope.

If the same identifier is reused in different scopes, it may become difficult to understand which object is being referred to.

Example of identifier overloading (Verilog):

```
reg [7:0] address;

begin: decode
   integer address;

   address = 0;
   . . .
end
```

Example of good identifier naming (Verilog):

```
reg [7:0] address;

begin: decode
   integer decoded_address;

   decoded_address = 0;
   . . .
end
```

Use suffixes to differentiate related identifiers semantically.

The suffix could indicate object kind such as: type, constant, signal, variable, flip-flop etc., or the suffix could indicate pipeline processing stage or clock domains.

Minimize identifiers in shared name spaces.

A shared name space is shared among all of the components implemented using the same language. When components define the same identifier in a shared name space, a collision will occur when they are integrated in the same simulation. Minimize your consumption of shared name spaces.

Each language has a different type and number of shared name spaces:

Table 1-1.
Shared name spaces

Language	Name Space
Verilog	Module, primitive name 'define symbol
VHDL	Library name
C/C++	Non-static function name #define symbol, macro
OpenVera	Program name Program variable, task, func. Class, enum. type name #define symbol, macro
e	Filename Struct, unit, enum. type name define macro name *sys* unit members

Use prefixes to differentiate identifiers in shared space.

When declaring an identifier in a shared name space, prefix it with a unique prefix that will ensure it will not collide with a similar identifier declared in another component. The suffix used has to be unique to the author or the authoring group or organization.

Example of poor shared identifier naming (*e*):

```
extend sys {
    debug: bool;
};
```

Example of good shared identifier naming (*e*):

```
extend sys {
   jb_debug: bool;
};
```

Constants

Use symbolic constants instead of "magic" hard-coded numeric values.

Numeric values have no meaning in and of themselves. Symbolic constants add meaning and are easier to change globally. This result is especially true if several constants have an identical value but a different meaning. If the language does not support symbolic constants, use a preprocessor or a variable named appropriately.

Example of poor constant usage (C):

```
int table[256];

for (i = 0; i <= 255; i++) ...
```

Example of good constant usage (C):

```
#define TABLE_LENGTH 256

int table[TABLE_LENGTH];

for (i = 0; i < TABLE_LENGTH; i++) ...
```

HDL & HVL Specific

Number multi-bit objects using the range N:0.

Using this numbering range avoids accidental truncation of the top bits when assigned to a smaller object (Verilog). This convention also provides for a consistent way of accessing bits from a given direction. If the object carries an integer value, the bit number represents the power-of-2 for which this bit corresponds.

Example (Verilog):

```
parameter width = 16;

reg [      7:0] byte;
reg [     31:0] dword;
reg [width-1:0] data;
```

Example (VHDL):

```
generic(width: integer := 16);
...
variable byte : unsigned (      7 downto 0);
variable dword: unsigned (     31 downto 0);
variable data : unsigned (width-1 downto 0);
```

Do not specify a bit range when referring to a complete vector.

If the range of a vector is modified, all references would need to be changed to reflect the new size of the vector. Using bit ranges implicitly means that you are referring to a subset of a vector. If you want to refer to the entire vector, do not specify a bit range.

Example of poor vector reference (VHDL):

```
signal count: unsigned(15 downto 0);
...
count(15 downto 0) <= count(15 downto 0) + 1;
carry <= count(15);
```

Example of proper vector reference (VHDL):

```
signal count: unsigned(15 downto 0);
...
count <= count + 1;
carry <= count(count'left);
```

Preserve names across hierarchical boundaries.

Preserving names across hierarchical boundaries facilitates tracing signals up and down a complex design hierarchy. This naming convention is not applicable when a unit is instantiated more than once or when the unit was not designed originally within the context of the current design.

Filenames

Use filename extensions that indicate the content of the file.

Tools often switch to the proper language-sensitive function based on the filename extension. Use a postfix on the filename itself if different (but related) contents in the same language are provided. Using postfixes with identical root names causes all related files to show up next to each other when looking up the content of a directory.

Example of poor file naming (Verilog):

```
design.vt        Testbench
design.vb        Behavioral model
design.vr        RTL model
design.vg        Gate-level model
```

Example of good file naming (Verilog):

```
design_tb.v      Testbench
design_beh.v     Behavioral model
design_rtl.v     RTL model
design_gate.v    Gate-level model
```

HDL CODING GUIDELINES

The following guidelines are specific to HDL descriptions. These guidelines are presented in addition to the guidelines outlined for general coding and naming. Additional guidelines will be needed when describing a design to be synthesized.

Structure

Use a single compilation unit in a file.

A file should contain a single module (Verilog), or a single entity, architecture, package, package body or configuration (VHDL). This structure facilitates locating various model components. For VHDL, it further reduces the amount of recompilation that may be required.

Layout

Declare ports and arguments in logical order according to purpose or functionality; do not declare ports and arguments according to direction.

Group port declarations that belong to the same interface. Grouping port declarations facilitates locating and changing them to a new interface. Do not order declarations output first, data input second, and control signals last because it scatters related declarations.

VHDL Specific

Label all processes, concurrent statements and loops.

Labeling makes referring to a particular construct much easier. If a particular construct is not named, some features in debugging tools may not be available. Labeling also provides for an additional opportunity to document the code.

Example of a named loop:

```
scan_bits_lp: for i in data'range loop
   ...
   exit scan_bits_lp when data(i) = 'X';
end loop scan_bits_lp;
```

Example of a named process:

```
clock_generator: process
begin
    wait for 50 ns;
    CLK <= not CLK;
end process clock_generator;
```

Label closing "end" keywords.

The "begin" and "end" keywords may be separated by hundreds of lines. Labeling matching end keywords facilitates recognizing the end of a particular construct. If the VHDL-87 syntax does not support a labeled end keyword, add the label using a comment.

Example:

```
component FIFO
    generic (...);
    port (...);
end component; -- FIFO
```

Use inline range constraints instead of subtypes.

Because type and subtype names are not syntactically different, using too many subtypes makes it hard to remember what type remains compatible with what other type.

Example of subtype constraints:

```
subtype address_styp is
    std_logic_vector (15 downto 0);
subtype count_styp   is
    integer range 15 downto 0;

signal address: address_styp;
signal count   : count_styp;
```

Example of inline range constraints:

```
signal address: std_logic_vector (15 downto 0);
signal count   : integer          range 0 to 15;
```

Do not use ports of mode *buffer* and *linkage*.

Buffer ports have special requirements when instantiated in a higher level unit. Use an "out" port instead. If internal feedback is required, use an internal feedback signal. I am still not sure what linkage ports were designed for. Coolant fluid?

Example using an internal feedback signal:

```
port(was_buffer_mode: out std_logic);
...
signal was_buffer_mode_int: std_logic;
...
was_buffer_mode <= was_buffer_mode_int;
```

Do not use blocks with ports and generics.

Ports and generics on blocks can be used to rename signals and constants already visible, thus creating a second name for an object. Using ports and generics on blocks reduces maintainability. Use blocks only when a local declarative region is required (e.g., to configure instantiations in a generate statement or declare an intermediate signal).

Do not use guarded expressions, signals and assignments, driver disconnect and signal kinds *bus* and *register*.

These features are scheduled to be removed from the language. Furthermore, they are used so little that tools may be unreliable when using them.

Use the logical library name WORK when referring to units in the same library.

Using this logical name makes it possible for a design to be moved or copied into another library with a different name without requiring any modifications. It also eliminates the need for a particular library name to hold the design.

Using WORK is similar to using the relative directory name ".", whereas using the actual library name is similar to using a full path-name. The former is portable to a different environment. The latter is not.

Verilog Specific

Start every module with a `resetall directive.

Compiler directives remain active across file boundaries. A module inherits the directives defined in earlier files. This inheritance may create compilation-order dependencies in your model. Using the `resetall directive ensures that your module is not affected by previously defined compiler directives and will be self-contained properly.

Avoid using `define symbols.

`define symbols are global to the compilation and may interfere with other symbols defined in another source file. For constant values, use parameters. If `define symbols must be used, undefine them by using `undef when they are no longer needed.

Example of poor style using `define symbols:

```
`define CYCLE 100
`define ns    * 1
always
begin
    #(`CYCLE/2 `ns);
    clk = ~clk;
end
```

Example of good style avoiding `define` symbols:

```
parameter CYCLE = 100;
`define ns * 1
always
begin
   #(CYCLE/2 `ns);
   clk = ~clk;
end
`undef ns
```

Use a nonblocking assignment for registers used outside the *always* or *initial* block where the register was assigned.

Using nonblocking assignments prevents race conditions between blocks that read the current value of the reg and the block that updates the reg value. This assignment guarantees that simulation results will be the same across Verilog simulators or with different command-line options.

Example of coding creating race conditions:

```
always @ (s)
begin
   if (s) q = q + 1;
end

always @ (s)
begin
   $write("Q = %b\n", q);
end
```

Example of good portable code:

```
always @ (s)
begin
    if (s) q <= q + 1;
end

always @ (s)
begin
    $write("Q = %b\n", q);
end
```

Assign regs from a single *always* or *initial* block.

Assigning regs from a single block prevents race conditions between blocks that may be setting a reg to different values at the same time. This assignment convention guarantees that simulation results will be the same across Verilog simulators or with different command-line options.

Example of coding that creates race conditions:

```
always @ (s)
begin
    if (s) q <= 1;
end

always @ (r)
begin
    if (r) q <= 0;
end
```

Example of good portable code:

```
always @ (s or r)
begin
    if (s)      q <= 1;
    else if (r) q <= 0;
end
```

Do not disable tasks with output or inout arguments.

The return value of output or inout arguments of a task that is disabled is not specified in the Verilog standard. Disable the inner begin/end block instead of disabling tasks with output or inout arguments. This technique guarantees that simulation results will be the same across Verilog simulators or with different command-line options.

Example of coding with unspecified behavior:

```
task cpu_read;
    output [15:0] rdat;
begin
    . . .
    if (data_rdy) begin
        rdat = data;
        disable cpu_read;
    end
    . . .
end
endtask
```

Example of good portable code:

```
task cpu_read;
    output [15:0] rdat;
begin: do_read
    . . .
    if (data_rdy) begin
        rdat = data;
        disable do_read;
    end
    . . .
end
endtask
```

Do not disable blocks containing nonblocking assignments with delays.

What happens to pending nonblocking assignments performed in a disabled block is not specified in the Verilog standard. Not disabling this type of block guarantees that simulation results will be the same across Verilog simulators or with different command-line options.

Example of coding with unspecified behavior:

```
begin: drive
    addr <= #10 16'hZZZZ;
    ...
end

always @ (rst)
begin
    if (rst) disable drive;
end
```

Do not read a wire after updating a register in the right-hand side of a continuous assignment, after a delay equal to the delay of the continuous assignment.

The Verilog standard does not specify the order of execution when the right-hand side of a continuous assignment is updated. The continuous assignment may be evaluated at the same time as the assignment or in the next delta cycle.

If you read the driven wire after a delay equal to the delay of the continuous assignment, a race condition will occur. The wire may or may not have been updated.

Example creating a race condition:

```
assign qb = ~q;
assign #5 qq = q;
initial
begin
    q = 1'b0;
    $write("Qb = %b\n", qb);
    #5;
    $write("QQ = %b\n", qq);
end
```

Do not use the bitwise operators in a Boolean context.

Bitwise operators are for operating on bits. Boolean operators are for operating on Boolean values. They are not always interchangeable and may yield different results. Use the bitwise operators to indicate that you are operating on bits, not for making a decision based on the value of particular bits.

Some code coverage tools cannot interpret a bitwise operator as a logical operator and will not provide coverage on the various components of the conditions that caused the execution to take a particular branch.

Example of poor use of bitwise operator:

```
reg [3:0] BYTE;
reg VALID
if (BYTE & VALID) begin
   ...
end
```

Example of good use of Boolean operator:

```
reg [3:0] BYTE;
reg VALID
if (BYTE != 4'b0000 && VALID) begin
   ...
end
```

APPENDIX B GLOSSARY

AO	Aspect-oriented.
AOP	Aspect-oriented programing.
ASIC	Application-specific integrated circuit.
ATM	Asynchronous Transfer Mode.
ATPG	Automatic test pattern generation.
BFM	Bus-functional model.
CAD	Computed aided design.
CPU	Central processing unit.
CRC	Cyclic redundancy check.
CTS	Clear to send.
DFT	Design for test.
DFV	Design for verification.
DSP	Digital signal processing.

DTR	Data terminal ready.
EDA	Electronic design automation.
EPROM	Erasable programmable read-only memory.
*e***RM**	*e* reuse methodology developer manual.
*e***VC**	*e* language verification component.
FAA	Federal aviation agency (US government).
FCS	Frame check sequence (ethernet).
FIFO	First in, first out.
FPGA	Field-programmable gate array.
FSM	Finite state machine.
GB	Gigabytes.
Gb	Gigabits.
ID	Identification.
HDL	Hardware description language
HEC	Header error check (ATM).
HVL	Hardware verification language.
IEEE	Institute of electrical and electronic engineers.
IP	Internet protocol, intellectual property
LAN	Local area network.
LFSR	Linear feedback shift register.
LLC	Link layer control (ethernet).

MAC	Media access control (ethernet).
MII	Media independent interface (ethernet).
MPEG	Moving picture expert group.
NASA	National aeronautic and space agency (US government).
NNI	Network-network interface (ATM).
OO	Object-oriented.
OOP	Object-oriented programming.
OVL	Open verification library.
PC	Personal computer.
PCI	PC component interface.
PLL	Phase-locked loop.
RAM	Random access memory.
RGB	Red, green and blue (video).
ROM	Read-only memory.
RTL	Register transfer level.
SDF	Standard delay file.
SDH	Synchronous digitial hierarchy (european SONET).
SNAP	(ethernet).
SOC	System on a chip.
SOP	Subject-oriented programming.
SONET	Synchronous optical network (north-american SDH).

TCM	Time-consuming method (*e*).
UART	Universal asynchronous receiver transmitter.
UFO	Unidentified flying object.
UNI	User-network interface (ATM).
VHDL	VHSIC hardware description language.
VHSIC	Very high-speed integrated circuit
VLAN	Virtual local area network (ethernet).
VPI	Virtual path identifier (ATM).

AFTERWORDS

This book should have given you the necessary skills to plan, implement and manage a best-in-class verification process. The methodologies and techniques will need to be tuned to your specific requirements. Think of this book as providing you with a set of *Lego* blocks. It is now up to you to put them together to build the infrastructure you envision.

Training classes are available.

If you would like to attend a formal training class covering the techniques presented in this book, I recommend the language and methodology classes[1] offered by *Qualis Design Corporation* (www.qualis.com). Just like this book, they focus on the methodology and how to implement it efficiently, not the tools. These classes are taught by professional engineering consultants (sometimes by myself) who spend most of their time applying these techniques on leading-edge designs. The instructors draw on their extensive industry experience to answer any question you may have on verification, adapting the techniques to your circumstances, often going beyond the content of the class material.

Join the on-line *verification guild.*

Send me email at janick@bergeron.com and ask to be added to the *verification guild* mailing list. It is a moderated on-line forum to discuss verification-related issues. Verification languages, behavioral modeling, testbench structures, detailed syntax of a waveform

---

1. Of course I am going to recommend them: I wrote the bulk of these classes myself!

data trace command, scripts, PERL, makefiles, hardware emulation are some of the topics discussed on the list. It is also a forum for debating the content of this book as well as future books, papers and articles on verification. This list is to verification what John Cooley's *esnug*[2] is to synthesis.

Tell me where I erred.

Despite the best effort of several reviewers and many read-throughs, there are errors in this book. From simple grammatical errors in the text, to syntax errors in the code samples, to functional bugs in the algorithms. I maintain a list of errors that were found in this edition of the book in the *errata* section at:

```
http://janick.bergeron.com/wtb
```

If you find an error that is not listed, please let me know via email. Errors will be corrected in the next edition.

---

2. John can be reached at `jcooley@world.std.com`.

INDEX

Bus-functional models 269–290
  and HVLs 274
  asynchronous in *e* 287
  asynchronous in OpenVera 281
  callback procedures 300, 314
  calling HDL procedures from
      HVL 274
  client/server processes
          abstracting   procotol   in
              VHDL 329
  client/server processes in VHDL 327
  concurrent invocation 272
  configuration 289
  CPU 269
  encapsulating in VHDL 327
  error injection 315
  multiple server instances in
      VHDL 334
  nonblocking 311
  OpenVera interface model 274
  packaging in *e* 142
  packaging in OpenVera 139
  packaging in Verilog 139
  packaging in VHDL 137
  reusing 20
  split transactions 310
  static binding in OpenVera 276
  synchronous in *e* 282
  synchronous in OpenVera 276
  system-level control 313
  transaction-level interface 307
  transactions 272
  using qualified names in VHDL 333
  vs. response monitor 294

C

Callback procedures 300
Capitalization
  naming guidelines 443
Class 166
  virtual 177
Clock multipler 235
Clock signals
  aligning 233
  asynchronous 236

  multiplier 235
  parameters, random generation 239
Clocks signals
  time resolution 231
Code coverage 46–54
  code-related metrics 79
  expression coverage 52
  FSM coverage 52
  path coverage 51
  statement coverage 48
Code reviews 32
Coding guidelines 429–460
Comments
  guidelines 434
  quality of 129
Component-level features 99
Concurrency
  definition of 189
  misuse of 197
  problems with 191
  with fork/join statement 199
Configuration
  randomly generating 117
Configuring the design 338
Connectivity
  definition of 189
Constants
  naming guidelines 449
Constrainable generator 116
Co-simulators 39
Costs for optimizing 126
Coverage
  code 46
          expression 52
          FSM 52
          path 51
          statement 48
    functional 55
          cross 59
          item 57
          transition 60
Coverage-driven verification 109–119
CPU bus-functional models 269
Cross-coverage 59
Cycle-accurate behavioral models 392

model checking 10
FPGA verification 92
FSM coverage 52
Functional coverage 55–62
  cross 59
  definition 56
  from features 111
  item 57
  model 111
  transition 60
Functional verification
  black-box 12
  grey-box 15
  purpose of 11
  white-box 13

G

Generating clocks 230–241
  aligning 233
  asynchronous domains 236
  multiplier 235
  parameters, random generation 239
  timescale 231
Generating reset signals 241–245
Generators
  constraints 116
  design 115
  random 354
  slaves 299
  specification 115
Global control signals 332
Grey-box verification 15
Guidelines
  capitalization 443
  code layout 436
  code syntax 439
  comments 434
  constants 449
  debugging 442
  directory structure 430–433
  filenames 451
  general coding 434–443
  HDL code layout 452
  HDL code structure 452
  HDL coding 452–460
  HDL specific naming 449

  identifiers 446
  naming 443–451
  Verilog specific 455
  VHDL specific 452

H

Hardware modelers 43
HDL vs HVL xxiv
High-level modeling 121–226, ??–226
Hook procedures 300
HVL vs. HDL xxiv

I

Identifiers
  naming guidelines 446
Implementation assertions 65
Inheritance 173
  vs. instance 176
Intellectual property 42–44
  behavioral models 397
  hardware modelers 43
  make vs. buy 42
  protection 43
Interface model
  asynchronous in *e* 287
  asynchronous in OpenVera 281
  clock domains in OpenVera 280
  driving event in *e* 286
  in OpenVera 274
  sampling event in *e* 286
  static binding in OpenVera 276
  synchronous in *e* 282
  synchronous in OpenVera 275
Issue tracking 74–78
  computerized system 77
  grapevine system 75
  Post-it system 76
  procedural system 76

L

Language, choosing xxiii–xxvi
Linked lists 160
Linting tools 26–33
  code reviews 32
  limitations of 27
  with *e* 32

feedback between stimulus and
design 263
reference signals 230
simple 246–250
aligning waveforms 233
synchronous data 246
Stimulus, directed 352–354
Stimulus, random 354–373
Structure
coding guidelines for 452
System configuration 341
System-level features 99
System-level transaction interface 313
System-level verification 92

T

Test harness 320–322
encapsulation 320
VHDL 325–336
global control signals 332
implementation 332
multiple bus-functional server
instances 334
Testbenches
and formal verification 8
random 113
stopping 114
verifying 107
Testbenches, architecting 319–373
autonomous generation and
monitoring 341–351
multiple server instances in
VHDL 334
test harness in VHDL 332
Testbenches, self-checking 256–261,
341–351
data tagging 343
failure modes 341
golden vectors 258
hard coded 342
input and output vectors 257
reference model 345
scoreboarding 348, 351
simple operations 260
test harness, integration with 350

transaction-level 350
transfer function 347
Testing
and verification 16–18
scan-based 17
Time
definition of 189
Time resolution 231
Timescale 231
Timestamping transactions 298
Transaction descriptor 305
Transaction-level interface 307–316
Transactions
completion status 312
error injection 315
multiple possible 304
retries 312
split 310
system-level control 313
variable length 308
Transfer function 347
vs. reference model 347
Transition coverage 60
Type I error 21
Type II error 21

U

Unit-level verification 90

V

Verification
and design reuse 19–20
and testing 16–18
ASIC and FPGA 92
black-box verification 12
board-level 93
checking result of transformation 5
cost 21
definition of 1
designing for 18
formal verification 8
functional verification 11, 12–15
grey-box verification 15
importance of 2–4
improving accuracy of 7, 107
need for specifying 86

REVIEWS

"Brilliant. Janick Bergeron has built on his ground-breaking first version of *Writing Testbenches* in this second edition, deeply embedding the key additional topics of Hardware Verification Languages, and coverage-driven random-based verification. I look forward to the third edition, which will no doubt add a discussion of the SystemC verification library capabilities and extend the treatment to full C++-based verification methods."

> \- Grant Martin, Fellow
> Cadence Berkeley Labs

"In the latest edition, Mr. Bergeron continues to keeps pace with the industry while providing world-class solution to the verification problem. His latest edition embraces the verification paradigm shift to HVLs and explains how to use them to achieve higher confidence in a design in less time.

Mr. Bergeron not only explains how to verify today's complex designs, but also walks through the entire design process—from selecting the proper tools and creating a verification plan, to knowing when the verification effort is (finally) finished."

> \- Chris Macionski, Senior Engineer
> Qualis Design Corp.

"When I first heard about this book, I had a healthy amount of skepticism based on the superficial nature in which most books tend to treat verification—lots of details about file IO, PLI etc. and a great deal of hand waving when it comes to the important issues. But this book is different in that it was the first book that truly is about verification and nothing else. It does not avoid the issues, but tackles them head on. If there can ever be such a thing as a 'classic book' in the EDA field, then this is most certainly a candidate for that honor. Many companies out there now owe their current verification methodologies to this book. From it they have learned the secrets of efficiency, effectiveness and re-use as they apply to verification, and have made their own efforts that much more valuable to their companies."

                    - Brian Bailey, Chief Technologist
                      Mentor Graphics Corp.

"A must have bible for understanding verification issues and techniques with HDLs and HVLs, and for writing effective, readable and reusable testbenches within a best-in-class verification process."

                    - Ben Cohen, VhdlCohen Training